Water Resources Engineering

Handbook of Essential Methods and Design

Anand Prakash

Cataloging-in-Publication Data on file with the Library of Congress

Published by American Society of Civil Engineers
1801 Alexander Bell Drive
Reston, Virginia 20191
www.pubs.asce.org

Any statements expressed in these materials are those of the individual authors and do not necessarily represent the views of ASCE, which takes no responsibility for any statement made herein. No reference made in this publication to any specific method, product, process or service constitutes or implies an endorsement, recommendation, or warranty thereof by ASCE. The materials are for general information only and do not represent a standard of ASCE, nor are they intended as a reference in purchase specifications, contracts, regulations, statutes, or any other legal document.

ASCE makes no representation or warranty of any kind, whether express or implied, concerning the accuracy, completeness, suitability, or utility of any information, apparatus, product, or process discussed in this publication, and assumes no liability therefore. This information should not be used without first securing competent advice with respect to its suitability for any general or specific application. Anyone utilizing this information assumes all liability arising from such use, including but not limited to infringement of any patent or patents.

ASCE and American Society of Civil Engineers—Registered in U.S. Patent and Trademark Office.

Photocopies: Authorization to photocopy material for internal or personal use under circumstances not falling within the fair use provisions of the Copyright Act is granted by ASCE to libraries and other users registered with the Copyright Clearance Center (CCC) Transactional Reporting Service, provided that the base fee of $18.00 per article is paid directly to CCC, 222 Rosewood Drive, Danvers, MA 01923. The identification for ASCE Books is 0-7844-0683-9/03/$18.00. Requests for special permission or bulk copying should be addressed to Permissions & Copyright Dept., ASCE.

Copyright © 2004 by the American Society of Civil Engineers.
All Rights Reserved.
ISBN 0-7844-0674-X
Manufactured in the United States of America.

CONTENTS

List of Figures vii
List of Tables viii
Preface xi
Acknowledgments xii
Author's Disclaimer xii

Chapter 1. Introduction 1
    Water Resources Engineering 1
    Planning of Water Resources Engineering Projects 2
    Documentation of Water Resources Engineering Studies 2

Chapter 2. Hydrologic Analyses 5
    Estimation of Peak Flows 6
        Rational Method 6
        Regression Equations 19
        Statistical Analysis of Available Data 22
    Surface Runoff Hydrographs 25
        Unit Hydrograph 29
        Kinematic Wave Method 32
        Design Storm Duration and Depth 32
        Soil Losses 33
        Snowmelt and Snow Loads 37
        Baseflow 39
        Combining Hydrographs and Routing through Channels 39
        Probable Maximum Flood Hydrograph 40
        Risk Analysis and Estimation of Failure Probabilities 46
        Benefit-Cost Analysis 49
    Reservoir Operation Studies 50
        Generation of Streamflow Sequences 50
        Flow Duration Analysis 53
        Mass Curve (Rippl) Analysis 54
        Estimation of 7-day Average 10-yr Low Flow 56
    Hydrologic Models 62
        Models for Development and Routing of Storm Runoff Hydrographs ... 62
        Continuous Flow Simulation Models 64
        Other Useful Models 65

Chapter 3. Hydraulic Analysis 67
    Classification of Flows 67
    Steady Uniform Flow in Open Channels 68
    Design of Nonerodible Channels 69
    Design of Erodible Channels 72

CONTENTS

       Maximum Permissible Velocity 72
       Nonsilting, Nonscouring Velocity 74
       Design of Vegetated Channels 78
   Flow Through Bends 82
   Freeboard 83
   Water Surface Profiles 83
   Critical Flow 86
   Hydraulics of Weirs and Spillways 87
       Broad-Crested and Sharp-Crested Weirs 87
       Ogee Crest 88
       Flow along Steep Slopes 91
       Hydraulic Jump 91
       Unsteady Flow 99
       Overland Flow along Slopes 100
       Soil Erosion on Slopes 103
       Sediment Yield Analysis 105
       Wind Erosion 110
       Sediment Transport Analysis 111
       Open-Channel Dispersion 114
       Far-Field Dispersion 117
       Estimation of Dispersion Coefficients 124
       Dissolved Oxygen Concentrations in Streams 126
   Pipe Flow 130
       Laminar Flow in Pipes 130
       Turbulent Flow in Pipes 131
       Hydraulics of Diffusers 136
       Water Hammer 143
   Hydraulic Models 144

Chapter 4. Groundwater **147**
   Occurrence of Groundwater and Types of Porous Media 147
   Properties of Porous Media 148
       Permeability and Hydraulic Conductivity 148
   One-Dimensional Steady-State Groundwater Flow 151
       Darcian Flow 151
       Non-Darcian Flow 153
   Steady-State Radial Flow 155
       Flow to a Single Well 155
       Groundwater Mound 156
       Capture Zone 157
       Partially Penetrating Well 159
       Radius of Influence 160
       Well Interference 161
       Induced Recharge 162
       Recharge by Precipitation 163
       Seawater Intrusion 164
       Effect of Barometric Pressure Fluctuations 166
       Subsidence 167
       Safe Yield, Specific Capacity, and Efficiency 169

	Transient (Unsteady) Groundwater Flow	170
	Unsteady One-Dimensional Flow	170
	Bank Storage	171
	Flow toward Drains and Drain Spacing	174
	Response of Groundwater Levels to River Stage Fluctuations	177
	Unsteady Radial Flow to a Well Fully Penetrating a Confined Aquifer	178
	Unsteady Radial Flow to a Well Fully Penetrating an Unconfined Aquifer	183
	Estimation of Aquifer Parameters	184
	Slug Tests	188
	Contaminant Transport in Saturated Zone	192
	Flow and Contaminant Transport through Fractured Rock	199
	Contaminant Transport through Unsaturated (Vadose) and Saturated Soil Zones	200
	Gas Phase Transport	203
	Monitoring Well Installation	206
	Groundwater Flow and Transport Models	207

Chapter 5. Hydraulic Designs ... 209
Introduction ... 209
Channel Transitions ... 209
   Flood Control ... 212
   Erosion Protection ... 217
   Dams and Reservoirs ... 236
Stilling Basins and Energy Dissipation Devices ... 263
   Hydroelectric Power ... 270

Chapter 6. Economic Analysis ... 289
Estimates of Costs and Benefits of Water Resources Engineering Projects ... 289
   Benefit-Cost Analysis ... 290
   Evaluation of Water Resources Engineering Project Alternatives ... 292

Chapter 7. Environmental Issues and Monitoring ... 299
Introduction ... 299
Environmental Impact Statement ... 301
   Definition and Format of an EIS ... 301
   Content of an EIS ... 302
Environmental Impacts of Water Resources Engineering Projects ... 303
   Environmental Impacts of Dams ... 303
   Environmental Impacts of Other Water Resources Engineering Projects ... 305
Evaluation and Analysis of Impacts ... 306
Environmental Monitoring ... 308
   Introduction ... 308
   Components of Environmental Monitoring ... 309
Evaluation of Environmental Significance of Projects ... 312
   Significance for Fish Habitat Sustainability and Enhancement ... 312
   Significance for Recreational Activities ... 313
Remedial Investigation and Feasibility Studies (RI/FS) ... 314

Definition and Scope 314
Regulatory Perspective. 318
Relevant Hydrologic Processes and Simulations. 319
Remediation Technologies 324

References **327**
Index **341**

Figures

2-1(a).	IDF curves (arithmetic scale)	16
2-1(b).	IDF curves (log scale)	16
2-2.	Typical surface runoff hydrograph	29
2-3.	Watershed map	40
2-4.	Line diagram of hydrologic network	43
2-5.	Hurricane-prone watershed	47
2-6(a).	Mass curves of flow and demand	58
2-6(b).	Mass curve analysis	59
2-7.	Flow duration curve	61
3-1.	Split flow schematic	85
3-2.	Typical ogee crest shape	89
3-3.	Hydraulic jump in a horizontal channel	92
3-4.	Typical specific energy curve	95
3-5.	Typical Blench curve	95
3-6.	Drop with sloping apron	96
3-7(a).	Hydraulic jump on mild slope	97
3-7(b).	Hydraulic jump on steep slope	97
3-8.	Overland flow along embankment slope	102
3-9.	Schematic of low-level outlet	134
3-10.	Schematic of water supply lines	137
4-1.	Schematic of tile drains	153
4-2.	Schematic of saltwater interface and upconing	165
4-3.	Schematic of slug test well	189
4-4.	Pollutograph reaching water table	202
4-5.	Schematic of gaseous diffusion	204
5-1.	Channel contraction and expansion	211
5-2.	Typical groin	215
5-3(a).	Weighted creep length	233
5-3(b).	Line diagram of drop structure floor	234
5-4.	Morning glory spillway discharge curve	257
5-5.	Siphon spillway schematic	258
5-6.	Plan view of four cycles of labyrinth spillway	262
5-7.	Schematic of ski jump bucket	270
5-8.	Simple surge tank	279
5-9.	Line diagram of draft tube	282

TABLES

2-1.	Commonly used values of runoff coefficients	7
2-2.	Typical values of K_n	10
2-3.	Approximate overland flow velocities (cm/s)	11
2-4.	Typical values of Manning's n for sheet flow	12
2-5.	Typical values of retardance coefficient	13
2-6(a).	Rainfall depth-duration-frequency table	15
2-6(b).	Rainfall intensity-duration-frequency table	15
2-7.	Rainfall data for mine site	17
2-8.	Annual maximum 10-min rainfall (mm)	19
2-9.	10-min annual maximum screened rainfall (mm)	20
2-10.	Regression coefficients and errors of prediction for USGS regression equation for Kansas	21
2-11.	Estimated 100-yr peak flows	21
2-12.	Selected values of frequency factor for normal or log-normal distribution	23
2-13.	Statistical analysis of annual peak flows of Ohio River at Louisville, Kentucky	26
2-14.	Time distribution and sequencing of design storms	34
2-15.	Typical runoff curve numbers (AMC II and $I_a = 0.2\ S_a$)	36
2-16(a).	Hydraulic parameters of subwatersheds	41
2-16(b).	Kinematic wave parameters for Subwatersheds 11 and 12	41
2-16(c).	Sequence of rainfall-runoff computations	42
2-17(a).	PMP depths	46
2-17(b).	Subwatershed parameters	46
2-17(c).	Sequence of rainfall-runoff computations for hurricane-prone watershed	47
2-17(d).	PMP depths for 5- to 30-min durations	47
2-18(a).	Computation of expected damage protection or benefits	49
2-18(b).	Computation of benefit-cost ratio	50
2-19(a).	Monthly streamflow and rainfall data for gauged watershed	52
2-19(b).	Known monthly rainfall and estimated monthly flows	53
2-20(a).	Monthly streamflow data (m³/s)	55
2-20(b).	Mass curve analysis with variable demand	55
2-20(c).	Mass curve analysis with constant demand	57
2-21(a).	Daily streamflow data (1 year) (m³/s)	60
2-21(b).	Flow duration table	61
2-22(a).	7-day average annual low flows	63
2-22(b).	Computation of 7-day average low flow	63
2-22(c).	Statistical parameters of 7-day average annual low flows	64
3-1.	Typical values of Manning's roughness coefficient	70
3-2.	Suitable side slopes for nonerodible channels	71
3-3.	Trial computations for concrete-lined channel	72
3-4.	Typical values of Manning's roughness coefficients and maximum permissible velocities	73

3-5.	Suitable side slopes for erodible channels	73
3-6.	Approximate values of coefficient a in Shields equation	76
3-7.	Retardance classes for different vegetal covers	79
3-8.	Guide for selection of retardance class	80
3-9.	Permissible velocities for channels lined with vegetation	80
3-10(a).	VR and n values for different retardation classes	81
3-10(b).	n and VR values for different retardation classes	81
3-11.	Coefficients and exponents of ogee profile equation	89
3-12(a).	Discharge coefficients for vertical-faced ogee crest with varying values of P	90
3-12(b).	Discharge coefficients for vertical-faced ogee crest for other than design head	90
3-13.	Toe velocities for spillways with downstream face slopes of $0.6H:1V$ to $0.8H:1V$	91
3-14.	Selected values of x, y, m, n, and l	94
3-15.	Length of jump on horizontal floor	96
3-16.	Hydraulic jump calculations for location of horizontal floor below sloping aprons	99
3-17.	Typical values of roughness factor, c	101
3-18.	Overland flow computations using Horton's method	103
3-19.	Typical values of soil erodibility factor, K	107
3-20.	Suspended sediment load and steamflow data	109
3-21.	Coefficients in sediment transport capacity equation	110
3-22.	Soil loss in tons/acre (t/ha) on unridged surface, WEG = 2, C = 30%, in Wyoming	112
3-23.	Particle size distribution of annual sediment load	114
3-24.	Computations of trap efficiency	115
3-25.	Rates of seepage of contaminated water	122
3-26.	Concentration distribution in stream	123
3-27.	First-order biodegradation rate constants and Henry's law constants for selected chemicals	123
3-28.	Typical values of volatilization rate constants	124
3-29.	Longitudinal dispersion coefficients for selected streams	125
3-30.	Transverse dispersion coefficients for selected streams	125
3-31.	Solubility of oxygen in water at atmospheric pressure	129
3-32(a).	Typical values of Manning's n for pipes	132
3-32(b).	Typical values of Hazen-Williams, C_H	133
3-33.	Typical values of pipe wall roughness, e	133
3-34.	Saturated vapor pressure of water	134
3-35.	Computed hydraulic parameters at selected diffuser ports	141
4-1.	Porosity, particle unit weight, and dry bulk unit weight of porous media	149
4-2.	Physical properties of pure water	149
4-3.	Typical values of permeability (k) and hydraulic conductivity (K)	150
4-4.	Typical values of $Wm^{0.50}$	154
4-5(a).	Coordinates of capture zone for a single well in uniform flow	159
4-5(b).	Coordinates of capture zone for two wells in uniform flow	159
4-6.	Maxey-Eakin coefficients	164
4-7(a).	Water table elevations near shoreline	165
4-7(b).	Position of saltwater–freshwater interface	166
4-8.	Typical values of specific storage	171

TABLES

4-9.	Typical values of specific yield	172
4-10.	Computations of water level fluctuations	178
4-11.	Well function $W(u)$ for confined aquifers	180
4-12.	Distances of additional wells from original well	182
4-13.	Time intervals to record time-drawdown data	186
4-14.	Time-drawdown data for observation well ($r = 60$ m)	188
4-15.	Approximate values of dimensionless coefficients A, B, and C	190
4-16.	Slug test data	192
4-17.	Typical values of solubility, K_{oc}, λ, density, and ν	193
4-18.	Values of K_d for selected inorganic substances (metals)	194
4-19.	Typical values of organic carbon content in soils	195
5-1.	Riprap gradation	222
5-2.	Hydraulic parameters of channels	222
5-3.	Estimated riprap sizes	223
5-4.	Parameters for estimation of scour depths	226
5-5.	Estimated scour depths	227
5-6.	Design parameters for launching apron	227
5-7.	Reservoir sedimentation rates	239
5-8.	Reservoir trap efficiencies	242
5-9.	Ratio of over-water to over-land wind speed	243
5-10.	Variation of discharge coefficient of ogee crest with submergence (H_s = crest elevation − tailwater elevation and C_s = modified discharge coefficient)	248
5-11.	Radii and coordinates of centers of arcs forming upstream spillway profile	249
5-12.	Coordinates of points on upstream spillway profile	249
5-13.	Computed coordinates of points on downstream spillway face	250
5-14.	Computed coordinates of points on upstream spillway face	250
5-15.	Hydraulic computations for side channel spillway	253
5-16.	Discharge coefficients for morning glory spillway crest	254
5-17.	Computations of jet profile trough shaft of morning glory spillway	256
5-18.	Rating curve for weir and orifice flows	257
5-19.	Toe velocities for stepped and unstepped spillways	261
5-20.	Preliminary dimensions of stepped spillways	262
5-21.	Length of stilling basins II and III	266
5-22.	Heights of baffle blocks and end sill for basin III	267
5-23.	Conjugate depths and length of jump on sloping aprons	268
5-24.	Flow duration and head-discharge table	273
5-25.	Computations of annual energy generation	274
5-26.	Typical values of contraction coefficients for pipes	277
5-27.	Typical values of bend loss coefficients for pipes (90° bend)	277
5-28.	Computations of surge height	281
5-29.	Thicknesses and hydraulic conductivities of formations above tunnel bottom	287
6-1.	Computation of expected damage for a flood control project	291
6-2.	Computation of incremental costs and benefits	293
6-3.	Description of alternatives and weights of evaluation criteria	295
6-4.	Scoring criteria for tangible factors	296
6-5.	Scoring criteria for intangible factors	296
6-6.	Evaluation matrix	296
6-7.	Weighted average scores, Case 1	297
6-8.	Alternative weighted average scores, Case 2	297

PREFACE

The typical water resources engineer in consulting practice is interested in developing solutions to real-world hydrologic and hydraulic problems in a relatively quick manner using simple methods of analyses and designs. In many practical situations, significant decisions have to be made based on limited amounts of data, and there are limited resources and little time to collect all the data and information required to perform sophisticated studies involving state-of-the-art modeling and analyses. Often, relatively simple calculations are made to develop project plans that may involve substantial financial investments and may have significant social, pecuniary, and environmental implications. Complex and sophisticated modeling and analyses, if necessary, are undertaken as special studies.

This book includes methods and equations that are applicable to situations with various levels of data availability, particularly where available site-specific information is inadequate. The presentation focuses on how to solve a practical problem with minimum literature search and attempts to fit well-known theoretical equations to real-world conditions. Methods are presented to develop preliminary designs, which must be refined or modified by additional numerical or physical modeling, experimentation, and field investigations. Some of the simplifications, approximations, and methods of analysis may appear trivial and even crude to a specialist in ground- and surface-water hydrology, fluid mechanics, water quality, and other related subjects, but these may be reasonable for a practicing water resources engineer. Generally, final designs would involve additional structural analyses, geotechnical analyses and investigations, preparation of detailed drawings, and cost estimates.

The expected readership of this book includes a consulting engineer with a bachelor's degree in engineering or applied sciences with some professional experience, graduate students who plan to study commonly used methods of analysis to enter the consulting industry, and practicing engineers who have to review studies or designs prepared by others. The reader is expected to have knowledge of fundamental hydraulics, fluid mechanics, and hydrology and must have access to standard texts on these subjects.

The material presented in this book includes answers to specific problems with commonly used formulas and equations and references for values of relevant constants and parameters. The computational procedures range from "back of the envelope" to somewhat detailed calculations, allowing for various levels of data availability. References are provided for commonly used models. Detailed discussion of complex and sophisticated models and research subjects is avoided. The size limitation of this book precludes discussions of the fundamentals of basic subjects, principles, and mathematical derivations. The emphasis has been on presenting material that may be directly usable by a practicing water resources engineer.

Acknowledgments

The author is grateful to his wife, Chandra Kanta, for her support, incentive, and encouragement during the completion of this book. Thanks are also due to the author's employer, URS Corporation (formerly Dames & Moore), Rolling Meadows, Illinois, for their support, and to Suzanne Coladonato and Charlotte McNaughton of ASCE for their help and encouragement. Finally, the author is grateful to his daughter, Manju P. Sharma, and son, Dr. Ajay Prakash, for their help.

Author's Disclaimer

The methods presented in this book are intended to be used for feasibility-level and preliminary designs with appropriate factors of safety and modifications based on professional judgment. Conclusions and results derived from precise or imprecise applications of the methods are the sole responsibility of the user. The author, author's employer, and publisher hold no responsibility for the consequences of such conclusions and results. The material presented does not reflect the policies or practices of the author's employer.

CHAPTER 1

INTRODUCTION

Water Resources Engineering

The subject of water resources engineering includes methods of hydrologic, hydraulic, and groundwater analyses related to the planning and design of remediation, water supply, flood control, and navigation facilities, and different types of hydraulic structures; ground- and surface-water flow and quality monitoring; feasibility and environmental impact analyses for various water-related projects; and designs of appurtenant hydraulic structures. The analyses may vary from the use of empirical or analytical equations to simple or sophisticated computer models, depending on the requirements of specific projects. The designs may include preliminary and final sizes of various components of a hydraulic structure. The scope of the discipline is so broad that it engulfs virtually all aspects of water-related studies and designs. From the viewpoint of a practicing water resources engineer, the subject of water resources engineering may include basic elements of a number of water-related disciplines (e.g., surface water hydrology; groundwater hydrology; fluid mechanics; open-channel hydraulics; sediment transport; and design of hydraulic structures, including dams, spillways, channels, navigation and flood control facilities, water supply systems, and shore protection, hydropower, and irrigation structures). Abundant published literature is available that addresses various specialized topics within each of the above-mentioned disciplines. However, because of limitations of data, budgets, or objectives and scope of specific projects, the practicing water resources engineer is often required to address most of these subjects on a somewhat basic level of detail. Specialized studies, where warranted, are referred to specialists in the respective disciplines.

Earlier water resources development projects focused mainly on engineering aspects. Experience with the operation of past water resources engineering projects has highlighted some of their adverse impacts on other natural resources and the environment. With increasing population and diminishing natural resources, concern is growing about holistic impacts of water resources engineering projects. Planning and design of a water resources engineering project today and in years and decades to come must consider their impacts on other resources (e.g., aquatic biota, ecosystem, aesthetics, and recreation). It must include quantitative prediction of the volume and volumetric flow rate of water, methods to control water volume and flow to serve various needs of the society, management of limited water resources in terms of quantity and quality, and, above all, interdisciplinary consequences of the proposed plan.

The details of various topics addressed in this book are commensurate with the requirements of a practicing water resources engineer, and the abridged descriptions of various equations and simplified analytical techniques presented in different chapters are intended to serve as "pocket book" rather than textbook material. Practical examples are included that illustrate specific methods of computations or analyses. The reader must refer to other relevant publications for sophisticated analyses and theoretical details of the methods described in this book. Examples of such references include McCuen (1998), ASCE (1996), Maidment (1993), Bras (1990), and Ponce (1989) for hydrologic analyses (Chapter 2); Martin and McCutcheon (1999), Chapra (1997), Brater et al. (1996), Potter and Wiggert (1991), and Tchobanoglous and Burton (1991) for hydraulic analysis (Chapter 3); Zheng and Bennett (2002), Fetter (2001), Charbeneau (2000), Delleur (1999), Fetter (1999), Batu (1998), Domenico and Schwartz (1998), and Anderson and Woessner (1992) for groundwater (Chapter 4); Mays (1999), Simons and Senturk (1992), Zipparro and Hansen (1993), USBR (1987), Barfield et al. (1981) for hydraulic designs (Chapter 5); and Linsley et al. (1992) for economic analysis (Chapter 6). A number of other relevant publications are included in the list of references.

Planning of Water Resources Engineering Projects

A typical project that a water resources engineer is required to plan, analyze, and design may include hydrologic analysis, hydraulic analysis, groundwater evaluation, design of hydraulic structures, economic analysis of water resources development projects, and evaluation of environmental impacts of water-related activities. Planning for the completion of such projects involves the following:

- **Identification of objectives.** This includes a list of specific goals or products that the project is expected to achieve or provide.
- **Scoping analysis.** This includes identification of sequential technical tasks required to be completed to accomplish the stated objectives (e.g., data collection, field inspection, analyses including computer modeling, preparation of designs and drawings, and report preparation).
- **Requirements for software and other equipment.** This includes identification of computer models and equipment (e.g., AutoCAD, GIS facilities, and equipment for field surveying and data collection) that would be required to complete the required analyses or prepare designs of proposed hydraulic structures.
- **Cost estimate.** This includes estimation of man-hours and other activities that may impact cost of the project (e.g., field surveys and monitoring, site inspections, communications and presentations, analyses, and production of reports, drawings, and construction plans).
- **Schedule.** This includes preparation of a schedule for the completion of each technical task along with relevant documentation. It must be noted that collection of field data and relevant site-specific information from different sources are fairly time-consuming tasks and usually have to be completed before other significant tasks can be undertaken.

Documentation of Water Resources Engineering Studies

Report preparation—or documentation of the methods and findings—constitutes an important part of water resources engineering studies. The organization and details of the contents of a study report depend on the type of study (e.g., surface water, groundwater, water quality, environmental impact, modeling, remediation, and feasibility-level design study), scope of study, and type of readership (e.g., general public, regulatory agency, planners and designers, or construction contractors). A typical water resources engineering study report may include the following basic elements:

- **Title.** The title should be brief and should indicate the primary objective of the study (e.g., "Hydrologic Study of Silver Creek Basin in Kansas"; "Evaluation of Groundwater Supply Potential in Sarasota County, Florida").

- **Table of Contents.** The Table of Contents should include Section and Subsection Titles, Lists of Tables and Figures, References, and Appendices, all with page numbers.

- **Executive Summary.** If the report is voluminous, it is advisable to include an Executive Summary that describes main findings and limitations of the study. For relatively short study reports, main findings and limitations may be included in the Conclusions and Recommendations section.

- **Introduction.** This should include a brief description of the problem being analyzed; objectives, scope, and overall approach of the study; and reference to the intended readership or recipients of the report. In addition, a brief description of the site location and hydrologic environment in the site vicinity may be included. This may include nearby streams and lakes, mean annual precipitation, snowfall, surface runoff, and free surface water evaporation in the region. For a design report, this may include location and purpose of the hydraulic structures.

- **Hydrologic Characteristics of Watersheds or Study Area.** For a study related to surface water hydrology, this should include areal extent, soil types, soil covers, hydraulic length and slopes of subwatersheds, and other information relevant to the estimation of times of concentration and lag times; precipitation depths of desired durations and return periods; and information about snowfall and snowmelt. For a design study, this may include description of salient features of the water body where the hydraulic structure is located (e.g., peak flow; 7-day, 10-year low flow; and drainage area). For a groundwater study, this may include delineation of the study area along with hydraulic boundaries (e.g., streams, lakes, and groundwater divides); information on average precipitation, infiltration, and evapotranspiration; and location and sizes of lakes and wetlands in the study area.

- **Data Collection and Analysis.** This should include site-specific and regional hydrologic data collected from different sources (e.g., precipitation and streamflow data for gauging stations in the vicinity; hydrogeologic data for aquifers; topographic survey information; and ground- and surface-water uses). Voluminous raw data, which may not be available in the cited references, should be included in appendices. Methods to screen and analyze the data and to extract or develop values or data sets to be used in the study should be included in this section. In addition, it should include limitations on the accuracy of the data and data analysis and justification for using the selected data sets, values, or methods of analysis.

- **Analytical, Numerical, or Other Studies.** This should include the methods of analysis or simulations including equations, description of numerical or physical models with implicit or explicit assumptions, and appropriate references. The methods and results of model calibration and validation should be included. Sensitivity analyses to illustrate the sensitivity of the results to variations in data values within plausible limits also should be documented in this section. The details to be included may vary depending on the objective, scope, and recipients of the study report. In design studies, this may include methods and calculations to develop various design dimensions.

- **Results.** This should include results of the study along with limits on their accuracy. Methods to verify and demonstrate the reasonableness of the results should be described (e.g., comparison with similar information for other sites, estimates based on other simpler or empirical methods, and published values for similar conditions).

- **Conclusions and Recommendations.** This should include carefully worded conclusions of the study with clear statement of the limitations of the results. The recommendations should include appropriate caveats and need for refinement by additional studies, if pertinent. The wordings must be clear to avoid misinterpretation by potential readers.

- **Tables and Figures.** The text must be clarified through tabular information, figures, and photographs. In many cases, valuable information may be presented in a concise fashion through tables and figures.

- **References.** Key information used for the study must have a cited reference, which the reader may consult to verify or get additional relevant information. The cited reference should be complete and should include author(s), year of publication, title, and publisher.

- **Appendices.** Information that is used for the study but is not available in the cited references and cannot be included in the main text (without distracting the reader) should be included in appendices.

In practice, it is advisable to prepare a draft of the report for review by peers, editors, or other potential recipients. The draft should be finalized after incorporating responses to the review comments. For clarity of presentation, major sections may be divided into subsections containing information about separate subtopics.

CHAPTER 2

HYDROLOGIC ANALYSES

The subject of hydrology includes study and analysis of the occurrence, circulation, and distribution of water through the hydrologic cycle, which includes the transfer of moisture from the ocean, to the atmosphere and land surface, and back to the ocean. Hydrologic analyses are required for most projects involving planning, design, construction, rehabilitation, remediation, or feasibility evaluation of various types of facilities. Although the types of hydrologic analyses required for different projects may be somewhat different, the basic principles and methodologies are generally the same. Commonly used hydrologic analyses for different types of projects include the following:

1. Community Development Projects
   - Rainfall intensity-duration-frequency or rainfall-depth-duration-frequency analysis.
   - Estimation of pre- and post-development peak flows for design of storm drainage systems, sizing of culverts and bridges, flood insurance studies, and floodplain delineation.
   - Development and routing of storm runoff hydrographs for design of retention/detention basins and wetlands.
   - Water yield analysis for streams, reservoirs, and watersheds.

2. Mining Projects
   - Estimation of peak flows for design of diversion channels.
   - Development and routing of surface runoff hydrographs for design of sedimentation basins and tailings ponds.
   - Evaluation of water supply potential of surface streams.
   - Evaluation of low-flow characteristics of streams receiving mine wastewater discharge.
   - Estimation of surface runoff producing potential of watersheds, including snowmelt runoff.
   - Estimation of pre- and post-mining flooding conditions for environmental reports.

3. Dams, Reservoirs, and Spillways
   - Generation of sequences of streamflows to evaluate watershed yield.
   - Hydrologic analyses to determine dependability of available water supply.

- Hydrologic analyses to determine required reservoir storage.
- Development of design-basis flood hydrographs for spillways contributed by rainfall, snowmelt, or both.
- Reservoir routing for sizing of spillways, reservoirs, and dams.
- Reservoir operation analysis for single and multiple use of available water.
- Low streamflow analysis to determine instream flow requirements.
- Dam-break analysis for safety evaluation and risk analysis.
- Evaluation of pre- and post-project flooding and instream flow conditions for environmental reports.

4. Hydropower Projects
    - Generation of sequences of streamflows to evaluate watershed yield.
    - Hydrologic analyses to determine dependability of available water supply.
    - Reservoir operation studies to determine firm and peak energy generation potential.
    - Evaluation of pre- and post-project daily and monthly streamflow patterns.

5. Nuclear Power Projects
    - Estimation of local and general storm probable maximum precipitation.
    - Development of probable maximum flood and design-basis flood hydrographs.
    - Development of dam-break flood hydrographs.
    - Development of combined event hydrographs (e.g., snowmelt combined with less than the probable maximum precipitation event).
    - Estimation of low streamflows to evaluate impacts of cooling water withdrawal for and wastewater discharges from the power plant.
    - Estimation of probable maximum snow load on safety-related structures.

In general, environmental reports for most development projects require the evaluation of pre- and post-project daily and monthly instream flows, flood hydrographs and peak flows, and low streamflows in the site vicinity.

Hydrologic analyses required for the above-mentioned facilities are described in subsequent sections of this chapter.

Estimation of Peak Flows

Estimation of peak flows is required for hydraulic designs of bridges and culverts, for water surface profile analyses for flood insurance studies, and for evaluation of flooding potential at different sites. It is desirable to estimate peak flows by several different methods and select reasonable values by judgment. Some commonly used methods are described here. Methods to develop surface runoff hydrographs are described in the section of this chapter entitled "Surface Runoff Hydrographs." Surface runoff hydrographs also may be used to estimate peak flows.

Rational Method

The rational formula for estimation of peak flows is

$$Q = 0.2755 \, CIA \tag{2-1}$$

where

Q = peak flow in m³/s

C = runoff coefficient (dimensionless) estimated by judgment in light of typical values given in Table 2-1

I = rainfall intensity in mm/h for the required return period, corresponding to a duration equal to the time of concentration of the watershed

A = watershed area in km²

This method is useful for estimation of peak flows of different return periods for watersheds smaller than 2.5 km², although the basic principles of the method may be applicable to larger drainage areas as well.

Runoff Coefficients

Commonly used values of runoff coefficients are included in Table 2-1 (ASCE 1976). The values given in Table 2-1 are applicable for storms of 5- to 10-year return periods. Higher values may be used for higher return periods and tight clayey soils. For watersheds that

Table 2-1. Commonly used values of runoff coefficients

Description of area	Runoff coefficient
Downtown business	0.70 to 0.95
Neighborhood business	0.50 to 0.70
Single-family residences	0.30 to 0.50
Detached multi-unit residential areas	0.40 to 0.60
Attached multi-unit residential areas	0.60 to 0.75
Residential (suburban)	0.25 to 0.40
Apartments	0.50 to 0.70
Light industrial	0.50 to 0.80
Heavy industrial	0.60 to 0.90
Parks or cemeteries	0.10 to 0.25
Playgrounds	0.20 to 0.35
Railroad yard	0.20 to 0.35
Unimproved	0.10 to 0.30
Asphalt and concrete pavement	0.70 to 0.95
Brick pavement	0.70 to 0.85
Roofs	0.75 to 0.95
Lawns on sandy soils (flat to 2% slope)	0.05 to 0.10
Lawns on sandy soils (2 to 7% slope)	0.10 to 0.15
Lawns on sandy soils (slope, 7%)	0.15 to 0.20
Lawns on heavy soils (flat to 2% slope)	0.13 to 0.17
Lawns on heavy soils (2 to 7% slope)	0.18 to 0.22
Lawns on heavy soils (slope, 7%)	0.25 to 0.35

Source: ASCE (1976).

include different types of areas, a weighted or composite runoff coefficient may be calculated using the relationship

$$C = (C_1A_1 + C_2A_2 + C_3A_3 + \ldots + C_nA_n)/A_t \qquad (2\text{-}2)$$

where

C = composite runoff coefficient

$C_1, C_2, C_3, \ldots, C_n$ = runoff coefficients applicable to areas $A_1, A_2, A_3, \ldots, A_n$, respectively

n = number of different types of areas within the watershed

A_t = total area = $A_1 + A_2 + A_3 + \ldots + A_n$

Time of Concentration

The time of concentration is defined as the time taken by surface runoff to travel from the remotest point in the watershed to the point where peak flow is to be estimated. Various methods have been proposed for estimating the time of concentration (McCuen et al. 1984; USBR 1977, 1987). It is a good practice to use at least three different methods to estimate the time of concentration. Within the range of these estimates, the final value should be selected by judgment. A few relatively simple and useful methods are shown below.

Kirpich Method (USBR 1977)

$$t_c = (0.87\ L^3/H)^{0.385} \qquad (2\text{-}3)$$

where

t_c = time of concentration (h)

L = length of the longest watercourse (km)

H = difference in elevation between the upper end of the watershed and the location at which flow is to be estimated (m)

This method results in relatively low estimates of t_c (Prakash 1987).

Soil Conservation Service (SCS) Curve Number Method (USDA 1972, 1985)

$$t_L = 1.347\ L_f^{0.8}(S_a + 2.54)^{0.7}/\left(1900\sqrt{S_p}\right) \qquad (2\text{-}4)$$

$$t_L = 0.6\ t_c \qquad (2\text{-}5)$$

$$S_a = (2540/CN) - 25.4 \qquad (2\text{-}6)$$

where

t_L = basin lag (h)

L_f = hydraulic length of watershed (m)

CN = curve number of the watershed

S_a = potential maximum retention (cm)

S_p = average land slope of the watershed (percentage)

This method has been found to result in relatively large values of t_L (Prakash 1987).

Snyder's Method (Chow 1964)

$$t_L = 0.7517 \, C_t \, (L \, L_{ca})^{0.3} \tag{2-7}$$

where

L_{ca} = length along the longest watercourse from the location where flow is to be estimated to the centroid of the watershed (km)

C_t = a coefficient

The value of C_t may be taken to be 2.0 for fairly mountainous watersheds similar to the Appalachian Highlands, 0.4 for watersheds similar to those in southern California, 0.7 to 1.0 for those similar to the Sierra Nevada areas, and 8.0 for watersheds bordering the eastern Gulf of Mexico. An analysis of 20 basins in the North and Middle Atlantic states resulted in the empirical relationship

$$C_t = 0.6 / \sqrt{S} \tag{2-8}$$

where S = basin slope (m/m). Where possible, it is advisable to use calibrated values of C_t.

U.S. Bureau of Reclamation Method (USBR 1987)

$$t_L = 4.6169 \, K_n [L \, L_{ca}/S^{0.5}]^{0.33} \tag{2-9}$$

where

S = slope of the longest watercourse (m/m)

K_n = a coefficient (typical values are given in Table 2-2)

Stream Hydraulics Method

The watershed is divided into different segments along the main watercourse based on roughness characteristics and slope. The length of the flow path and the average flow velocity for each segment are estimated. Then,

$$t_c = [L_1/V_1 + L_2/V_2 + L_3/V_3 + \ldots + L_n/V_n]/3600 \tag{2-10}$$

Table 2-2. Typical values of K_n

Region	Watershed size (km²)	K_n^a
Great Plains in Colorado, Kansas, Oklahoma, Nebraska, New Mexico, Wyoming, and North Dakota	5.2 to 10,280	0.070 for basins with considerable overland flow
		0.030 for basins with well-defined drainage network
Rocky Mountain watersheds in Colorado, Wyoming, Utah, Oregon, Montana, Idaho, and New Mexico	3.4 to 6,500	0.260 for 100-yr floods
		0.130 to 0.160 for general storm PMF
		0.050 to 0.073 for thunderstorm events
Southwest desert, Great Basin, and Colorado Plateau in Arizona, California, and parts of Colorado	6.0 to 12,250	0.070 for basins with coniferous forests at higher elevations
		0.042 for desert terrains
Sierra Nevada, California	53.4 to 5,370	0.150 for basins with substantial coniferous growth
		0.064 for basins with well-developed drainage networks
Coast and Cascade Ranges in California, Oregon, and Washington	8.7 to 1,980	0.150 for basins with very heavy coniferous growth extending into the overbank floodplain
		0.080 for low-lying basins with considerably sparser vegetation
Urban basins in California, Texas, Kentucky, Virginia, and Maryland	0.5 to 238	0.033 for basins with low-density or partial development with only minor flood water collection facilities
		0.013 for basins with high-density development with a good collection system

[a] Values between indicated upper and lower limits may be used for basins with intermediate characteristics.
Source: USBR (1987).

where

n = number of segments

$L_1, L_2, L_3, \ldots, L_n$ = lengths (m) of watershed segments with different roughness characteristics and slopes

$V_1, V_2, V_3, \ldots, V_n$ = overland or channel flow velocities (m/s) in the respective watershed segments

Table 2-3. Approximate overland flow velocities (cm/s)

Land slope (%)	Surface type						
	Type 1	Type 2	Type 3	Type 4	Type 5	Type 6	Type 7
0.5	6	11	15	20	22	34	43
1.0	8	15	21	27	31	46	61
2.0	11	21	31	40	43	64	88
3.0	13	27	40	49	55	82	107
4.0	15	31	43	55	61	91	122
5.0	17	37	49	61	70	104	137
10.0	24	49	67	88	98	150	198
20.0	35	69	98	122	137	210	274
30.0	43	82	119	152	171	256	366
40.0	49	98	137	180	198	290	396
50.0	55	107	152	198	213	335	427
60.0	61	122	168	213	244	366	488

Source: USDA (1972, 1985).

The flow velocities may be estimated using Manning's formula for 2-yr peak or bankfull discharge for well-defined channels and the values given in Table 2-3 for overland flow (USDA 1972, 1985). "Type" in Table 2-3 refers to:

- Type 1—Overland flow on forest areas with heavy ground litter and hay meadow
- Type 2—Overland flow on fallow or minimum tillage cultivation areas, contour or strip-cropped lands, and woodland
- Type 3—Overland flow on short grass pasture
- Type 4—Overland flow on straight row cultivated areas
- Type 5—Overland flow on nearly bare and untilled areas and alluvial fans in western mountain regions
- Type 6—Flows in grassed waterways
- Type 7—Sheet flow on paved areas and in small upland gullies

Sheet Flow Equation (USDA 1986)

An empirical equation to estimate travel time for sheet flow of less than about 90 m on plane surfaces or in the headwaters of streams is

$$t_c = [0.0289(nL)^{0.8}]/[P_2^{0.5} S^{0.4}] \quad (2\text{-}11)$$

where

t_c = travel time (h)

L = flow length (m)

P_2 = 2-yr, 24-h rainfall (cm)

S = land slope (m/m)

n = Manning's roughness coefficient

Typical values of Manning's roughness coefficient for sheet flow are included in Table 2-4.
Average flow velocities for shallow concentrated flows on relatively flat slopes less than 0.005 m/m can be estimated by (USDA 1986):

$$\text{Unpaved surfaces: } V = 4.9176\sqrt{S} \qquad (2\text{-}12)$$

$$\text{Paved surfaces: } V = 6.1957\sqrt{S} \qquad (2\text{-}13)$$

where

V = flow velocity (m/s)

S = slope of water course (m/m)

Kerby-Hathaway Method (McCuen et al. 1984)

$$T_c = 0.02407\, L_f^{0.47}\, n^{0.47}\, S_f^{-0.235} \qquad (2\text{-}14)$$

where

L_f = straight line distance (m) from the most distant point in the watershed to the point under consideration measured parallel to the slope

S_f = mean slope of the basin (m/m)

n = retardance coefficient or Manning's roughness coefficient

Typical values of the retardance coefficient, n, are given in Table 2-5 (ASCE 1959).

Table 2-4. Typical values of Manning's n for sheet flow

Surface description	Manning's n
Smooth (concrete, asphalt, gravel, or bare soil)	0.011
Fallow (no residue)	0.05
Cultivated soils, residue cover ≤ 20%	0.06
Cultivated soils, residue cover > 20%	0.17
Grass (short prairie)	0.15
Grass (dense)	0.24
Grass (Bermuda)	0.41
Range (natural)	0.13
Woods (light underbrush)	0.40
Woods (dense underbrush)	0.80

Source: USDA (1986).

Table 2-5. Typical values of retardance coefficient

Type of surface	Retardance coefficient, n
Smooth impervious surface	0.02
Smooth bare packed soil	0.10
Poor grass, cultivated row crops, or moderately rough bare surface	0.20
Pasture or average grass	0.40
Deciduous timberland	0.60
Coniferous timberland, deciduous timberland with deep forest litter or dense grass	0.80

Source: ASCE (1959).

The aforementioned methods were developed for specific sizes and types of watersheds. However, their application to a variety of watersheds is quite common. Appropriate values of times of concentration should be selected with due consideration of watershed conditions.

Example 2-1: Estimate the time of concentration for a 12-km² watershed in the four corners area of the southwestern United States (corner of Colorado, Utah, New Mexico, and Arizona). Relevant watershed parameters are:

Watershed length = 5.18 km

$S = 0.057$ m/m

$CN = 76$

$K_n = 0.045$

$C_t = 0.4$

$L_{ca} = 2.59$ km

The surface soils are nearly bare with some alluvial fans.

Solution: Use several methods to compute t_c and select the appropriate value by judgment.

1. Kirpich Method—(Eq. (2-3))
   $H = 5.18 \times 1000 \times 0.057 = 295.26$ m
   $t_c = [0.87(5.18)^3/295.26]^{0.385} = 0.71$ h

2. SCS Method—(Eqs. (2-4), (2-5), and (2-6))
   $S_a = (2540/76) - 25.4 = 8.021$ cm
   $t_L = 1.347(5180)^{0.8}(8.021 + 2.54)^{0.7}/[1900(0.057 \times 100)] = 1.448$ h
   $t_c = 1.448/0.6 = 2.41$ h

3. Snyder's Method—(Eq. (2-7))
   $t_L = 0.7517 \times 0.4 (5.18 \times 2.59)^{0.3} = 0.655$ h
   $t_c = 0.655/0.6 = 1.09$ h

4. USBR Method—(Eq. (2-9))
   $t_L = 4.6169 \times 0.045[(5.18 \times 2.59)/(0.057)^{0.5}]^{0.33} = 0.785$ h
   $t_c = 0.785/0.6 = 1.31$ h

5. Stream Hydraulics Method—(Eq. (2-10))
   overland flow velocity = 74 cm/s (from Table 2-3)
   $t_c = 5.18 \times 1000/(0.74 \times 3600) = 1.94$ h

The estimates vary from 0.71 to 2.41 h. The median value of 1.31 h appears to be a reasonable approximation. It is also close to the average of the five estimates.

Rainfall Intensity

Rainfall intensities for different return periods and corresponding to different durations (e.g., times of concentration) can be obtained from NOAA Atlas 2 for the eleven western states: Montana, Wyoming, Colorado, New Mexico, Idaho, Utah, Nevada, Arizona, Washington, Oregon, and California (NOAA 1973). Some states and counties have developed precipitation intensity-duration-frequency (IDF), depth-duration-frequency, and depth-area-duration (DAD) curves for areas in their jurisdictions using relevant precipitation data, [e.g., Bulletin 70 for the State of Illinois (Huff and Angel 1989)]. For other areas in the United States, precipitation depths taken from Technical Paper No. 40 (TP-40) may be used with a multiplying factor of about 1.20 to account for uncertainties and extreme storm events that may have occurred during the last three decades since the publication of TP-40 (Hershfield 1961).

An IDF is a plot of precipitation intensity on the y-axis and duration on the x-axis with return period indicated on each intensity-duration curve. As convenient, either arithmetic or logarithmic scales for both axes may be used for these plots. A depth-duration-frequency curve is similar to the IDF curve except that precipitation intensity is replaced by precipitation depth. The DAD curve is a plot of watershed area on a log scale on the y-axis and precipitation depth on an arithmetic scale on the x-axis with duration indicated on each depth-area curve. As convenient, the x and y axes may be interchanged.

For areas where limited precipitation data are available, preliminary values of precipitation intensity or depth may be estimated by (Ponce 1989; Rouse 1950):

$$i = a/(t+b)^m \quad \text{or} \quad d = at/(t+b)^m \qquad (2\text{-}15)$$

where

i = precipitation intensity (mm/h)

d = precipitation depth (mm)

t = time duration (h)

a, b, and m are empirical coefficients

For the sake of simplicity, the exponent $m = 1$. Values of the other empirical coefficients may be determined by substituting available data for rainfall depths or intensities and averaging the computed values for different sets of data points.

Example 2-2: Prepare IDF curves for use in the design of the drainage system for an industrial site near Joliet, Illinois.

Solution: The rainfall depths of various durations and return periods for Joliet, Illinois, obtained from Huff and Angel (1989) are shown in Table 2-6(a). The corresponding intensities are shown in Table 2-6(b). The IDF curves for durations of 5 min to 2 h, plotted on an arithmetic scale, are shown in Figure 2-1(a); those for durations of 10 min to 24 h, plotted on a log-log scale, are shown in Figure 2-1(b).

Table 2-6(a). Rainfall depth-duration-frequency table

Duration (h)	Rainfall depth for different return periods (mm)					
	2-yr	5-yr	10-yr	25-yr	50-yr	100-yr
24	81.28	103.63	123.19	153.42	182.12	215.14
18	74.68	95.25	113.28	141.22	167.64	197.87
12	70.61	90.17	107.19	133.35	158.50	187.20
6	60.96	77.72	92.46	115.06	136.65	161.29
3	52.07	66.29	78.74	98.30	116.59	137.67
2	48.01	61.21	72.64	90.42	107.44	127.00
1	38.10	48.77	57.91	72.14	85.60	101.09
0.500 (30 min)	29.97	38.35	45.47	56.64	67.31	79.50
0.250 (15 min)	21.84	27.94	33.27	41.40	49.28	58.17
0.167 (10 min)	17.78	22.86	27.18	33.78	40.13	47.24
0.083 (5 min)	9.65	12.45	14.73	18.29	21.84	25.91

Source: Huff and Angel (1989).

Table 2-6(b). Rainfall intensity-duration-frequency table

Duration (h)	Rainfall intensity for different return periods (mm/h)					
	2-yr	5-yr	10-yr	25-yr	50-yr	100-yr
24	3.39	4.32	5.13	6.39	7.59	8.96
18	4.15	5.29	6.29	7.85	9.31	10.99
12	5.88	7.51	8.93	11.11	13.21	15.60
6	10.16	12.95	15.41	19.18	22.78	26.88
3	17.36	22.10	26.25	32.77	38.86	45.89
2	24.00	30.61	36.32	45.21	53.72	63.50
1	38.10	48.77	57.91	72.14	85.60	101.09
0.500 (30 min)	59.94	76.71	90.93	113.28	134.62	159.00
0.250 (15 min)	87.38	111.76	133.10	165.61	197.10	232.66
0.167 (10 min)	106.68	137.16	163.06	202.69	240.79	283.46
0.083 (5 min)	115.82	149.35	176.78	219.46	262.13	310.90

Source: Huff and Angel (1989).

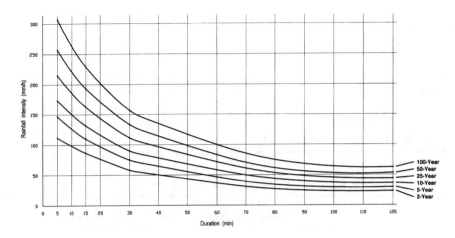

Figure 2-1(a). IDF curves (arithmetic scale)

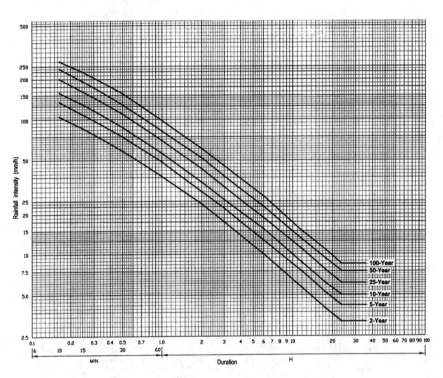

Figure 2-1(b). IDF curves (log scale)

Table 2-7. Rainfall data for mine site

	Duration (min)		
Return period	10	30	60
2-yr high rainfall depth (mm)	19.2	36.9	54.3
5-yr high rainfall depth (mm)	22.1	41.7	62.7

Example 2-3: Five years of rainfall data for a remote mine site are included in Table 2-7. Develop an approximate IDF table usable for planning purposes and extrapolate it up to a 10-yr return period.

Solution: Using Eq. (2-15) with $m = 1$, $i = d/t$, and available 2-yr rainfall depths for 10-, 30-, and 60-min durations,

$i = 19.2 \times 60/10 = 115.2 = a/(0.167 + b)$

$i = 36.9 \times 60/30 = 73.8 = a/(0.5 + b)$

$i = 54.3 = a/(1.0 + b)$

Solving the first two simultaneous equations results in $a = 68.38$ and $b = 0.4266$; solving the first and third equations results in $a = 85.56$ and $b = 0.5757$; and solving the second and third equations results in $a = 102.75$ and $b = 0.8923$. The averages of the three sets of values are $a = 85.56$ and $b = 0.6315$. Thus, a preliminary IDF table may be developed from the equation

$i(\text{2-yr}) = 85.56/(t + 0.6315)$.

Similarly, the three equations for the 5-yr rainfall depths are

$i = (22.1 \times 60/10) = 132.6 = a/(0.167 + b)$

$i = (41.7 \times 60/30) = 83.4 = a/(0.5 + b)$

$i = 62.7 = a/(1.0 + b)$

The solution of these equations in the same order as before results in $a = 74.85$, $b = 0.3975$; $a = 99.08$, $b = 0.5802$; and $a = 126.31$, $b = 1.0145$. Using the averages of these values results in the equation

$i(\text{5-yr}) = 100.08/(t + 0.6641)$

In the absence of any other data, the estimated precipitation intensities for 2- and 5-year return periods may be converted to the corresponding depth, using the relationship $d = it$. The depths for each duration may be plotted on a Gumbel probability paper, and the 10-year precipitation depths for the corresponding durations may be obtained by extrapolation (see the section of this chapter entitled "Statistical Analysis of Available Data"). Alternatively,

18 WATER RESOURCES ENGINEERING

assume the 2-year rainfall depths to be approximately equal to the respective means; for example, for a duration of 60 min, mean = $\overline{X}$ = 54.3 mm. Then, using Eq. (2-23),

K (5-yr) = $-0.7797 \{0.5772 + \ln [\ln (5/4)]\}$ = 0.71946.

Using Eq. (2-21),

$62.7 = 54.3 + 0.71946 \, s$.

This gives s = 11.6754. Reusing Eqs. (2-23) and (2-21),

K (10-yr) = 1.30457

d (10-yr) = 54.3 + 1.30457 × 11.6754 = 69.53 mm

i (10-yr) = 69.53 × 60/60 = 69.53 mm/h

The 10-yr values for other durations may be similarly estimated.

Note that these results are preliminary and may be useful for planning purposes only. They must be modified as soon as additional data are available.

For situations where rainfall data for a few years (e.g., 5 years) for several gauging stations (e.g., 10 stations) within the same climatic region are available, the data may be combined and assumed to be equivalent to 50 years of data at one station. This approximation is known as the Station Year Method (Chow 1964). Where appropriate, stochastic methods to extend available data also may be used. Using the extended data, statistical methods described in the section entitled "Statistical Analysis of Available Data" may be used to estimate rainfall depths of different return periods.

Example 2-4: Data processing of rainfall records from six continuous recording rain gauge stations in a remote mining area (approx. 8 km × 8 km) indicated the annual maximum 10-min values shown in Table 2-8. Estimate the 10-min 2-, 10-, 50-, and 100-yr rainfall depths for the design of the drainage system for the facility.

Solution: The rain gauge stations are located within a relatively small mining area, which is within the same climatic region. Therefore, the Station Year approximation is used. The values at different stations pertaining to the same date may represent one and the same storm and may not necessarily provide additional independent data points. Note that there are three values for November 28, 1993; two for November 29, 1993; three for November 15, 1995; two for February 14, 1996; three for April 14, 1997; and two for January 12, 1997. To develop an approximate equivalent record for more than 6 years at one station within the site area, only the highest value observed on each of these dates is considered. This results in 25 independent data points as shown in Table 2-9, approximating 25 yr of 10-min annual maximum rainfall values at the site.

Normally, the best-fitting probability distribution should be used for frequency analysis using these data points. However, the Fisher-Tippett Type I (i.e., Gumbel) distribution has been found to fit rainfall data (Hershfield 1961; NOAA 1973). Therefore, this distribution is used to estimate the 2-, 10-, 50-, and 100-yr 10-min rainfall depths for the site area. Using functions available in standard software packages (e.g., Excel), the mean, $\overline{X}$, and standard deviation, s, of the 25 values in Table 2-9 are found to be 19.532 and 3.803, respectively. Using Eqs. (2-24), (2-25), and (2-26),

$\alpha = 1.2826/3.803 = 0.337$

$u = 19.532 - 0.5772/0.337 = 17.819$

Therefore,

$P\,(10\text{-yr}) = 17.819 - \ln\,[-\ln\,(1 - 0.10)]/0.337 = 24.50\text{ mm},$

$P\,(50\text{-yr}) = 17.819 - \ln\,[-\ln\,(1 - 0.02)]/0.337 = 29.40\text{ mm, and}$

$P\,(100\text{-yr}) = 17.819 - \ln\,[-\ln\,(1 - 0.01)]/0.337 = 31.47\text{ mm}.$

It must be noted that these estimates are preliminary and may be useful for design of the site drainage system using reasonable factors of safety.

Regression Equations

The U.S. Geological Survey has developed different sets of regression equations to estimate peak flows of different return periods for ungauged sites in different states of the United States and Puerto Rico (USGS 1994). The reported standard errors of these equations are

Table 2-8. Annual maximum 10-min rainfall (mm)

	Station 1		Station 2		Station 3	
Year	Date	Rainfall	Date	Rainfall	Date	Rainfall
1993	November 28	18.0	November 28	19.0	November 29	19.8
1994	December 10	23.4	January 25	16.6	November 16	18.0
1995	January 4	19.8	November 15	20.6	November 15	24.8
1996	January 24	16.0	February 14	17.6	February 14	19.2
1997	April 14	25.5	April 14	19.4	January 12	17.2
1998	November 2	29.6	January 15	17.0	October 24	22.8

	Station 4		Station 5		Station 6	
Year	Date	Rainfall	Date	Rainfall	Date	Rainfall
1993	November 29	21.2	November 28	22.0	—	—
1994	March 18	14.4	February 13	19.4	November 24	16.8
1995	November 23	23.4	November 15	21.0	November 27	18.6
1996	November 15	14.2	January 12	18.8	April 18	18.6
1997	January 12	18.0	April 14	21.8	November 16	15.0
1998	—	—	November 17	19.6	May 10	15.6

Table 2-9. 10-min annual maximum screened rainfall (mm)

Date	10-min maximum rainfall
11/28/93	22.0
11/29/93	21.2
1/25/94	16.6
2/13/94	19.4
3/18/94	14.4
11/16/94	18.0
11/24/94	16.8
12/10/94	23.4
1/4/95	19.8
11/15/95	24.8
11/23/95	23.4
11/27/95	18.6
1/12/96	18.8
1/24/96	16.0
2/14/96	19.2
4/18/96	18.6
11/15/96	14.2
1/12/97	18.0
4/14/97	25.5
11/16/97	15.0
1/15/98	17.0
5/10/98	15.6
10/24/98	22.8
11/2/98	29.6
11/17/98	19.6

relatively high. However, they are useful for verification of the reasonableness of peak flows estimated by other methods. These equations are reported in the foot-pound-second (FPS) system of units. It is generally convenient to use them in the FPS system and convert the results to the SI system.

Example 2-5: Using the HEC-1 model, the 100-yr peak flow of a creek in Kansas is estimated to be 367 m^3/s. Use USGS regression equations to verify the reasonableness of the estimated 100-yr peak flow. Relevant parameters required for the use of the regression equations are:

$DA = 61.4$ km^2 (23.70 mi^2)

mean annual precipitation for the basin = 71 cm (28 in.)

Solution: The generalized form of the regression equations for drainage areas of 0.17 to less than 30 mi^2 (0.44 to 77.7 km^2) in Kansas is (USGS 2000a):

$$Q = a\, DA^b\, P^{b1} \tag{2-16}$$

where

Q = peak flow in cubic feet per second (cfs)

DA = drainage area (mi^2)

P = mean annual precipitation (in.)

$a, b, b1$ = regression coefficients obtained from USGS (2000a)

The values of the regression coefficients for the 100-yr return period, along with the estimated errors of predictions, are shown in Table 2-10. Using Eq. (2-16) with the coefficients given in Table 2-10,

$$Q_{100} = 19.80 \times (23.70)^{0.634} \times (28)^{1.288} = 10{,}770 \text{ cfs} = 305 \text{ m}^3/\text{s}.$$

The estimated 100-yr peak flows and standard errors of the estimate based on the regression equation are shown in Table 2-11.

In view of the values given by the USGS regression equation, 367 m^3/s appears to be a reasonably conservative estimate.

Table 2-10. Regression coefficients and errors of prediction of USGS regression equation for Kansas

Return period	Regression coefficients[a]			Error (%)
	a	b	b1	
100-yr		0.634		+71
	19.80		1.288	−44

[a] For drainage areas ranging from 0.17 to less than 30 mi^2.
Source: USGS (1994).

Table 2-11. Estimated 100-yr peak flows

Subwatershed	Peak flow estimated by HEC-1[a]	Regression equation		
		Estimated peak flow	Prediction error (+)	Prediction error (−)
61.4 km^2	367 m^3/s	305 m^3/s	217 m^3/s	134 m^3/s

[a] See Example 2-8.

Statistical Analysis of Available Data

The following four common statistical parameters are required for statistical analysis of data:

$$\text{Mean} = \overline{X} = \Sigma\, X/n \tag{2-17}$$

$$\text{Standard deviation} = s = [\Sigma\, (X - \overline{X})^2/(n-1)]^{0.5} \tag{2-18}$$

$$\text{Skew coefficient} = G = n\, \Sigma\, (X - \overline{X})^3/[(n-1)(n-2)\, s^3] \tag{2-19}$$

$$\text{Kurtosis} = k = [n(n+1)/\{(n-1)(n-2)(n-3)\}] \times$$
$$[\Sigma\, \{(X - \overline{X})/s\}^4] - [3(n-1)^2/\{(n-2)(n-3)\}] \tag{2-20}$$

where

X = value of the variable

n = number of data points

Σ = from 1 to n

For normal distribution, $G = 0$ and $k = 3$. For the Gumbel (Extreme Value Type I) distribution, $G = 1.1396$ and $k = 5.4$.

If annual peak flow data are available for the site, peak flows of higher return periods may be estimated by statistical analysis. In practice, peak flows of the desired return period should be estimated using several probability distributions and the adopted value should be selected by judgment, giving more weight to the better-fitting and log-Pearson Type III distributions. The normal or log-normal distribution is acceptable if the skew coefficient is small. The Gumbel distribution has a constant skew coefficient of 1.1396. The log-Pearson Type III distribution is applicable for any known skew coefficient (USWRC 1981). A simple method to determine the goodness-of-fit is to plot the annual peak flows on several probability papers (e.g., normal, log-normal, and Gumbel probability papers) and identify the best-fitting distribution by observation.

Computational steps for statistical analysis of annual peak flows are as follows:

- Arrange the annual peak flow values in descending order of magnitude. Assign a rank, m, to each value, with the highest value being 1. Assign different (successive) ranks even if two or more values are equal. This can be done on a spreadsheet.

- Compute logarithms (to the base 10) of all the values.

- Determine mean, $\overline{X}$, standard deviation, s, and skew coefficient, G, of all values and of their logarithms. This can be done using available functions in standard software packages (e.g., Excel).

- Compute plotting position for each value using the Weibull formula, $m/(n+1)$, where n is the total number of data points. Use these plotting positions for preparing straight line graphs on different probability papers. The plotting position is taken to be the probability scaled on the abscissa of the probability paper.

- Estimate peak flow of the desired return period by visual observation or extrapolation from the probability distribution plot, or, alternatively, by using the equation

$$Q_T = \overline{X} + Ks \tag{2-21}$$

where

Q_T = peak flow of a return period of T yr

K = frequency factor obtained from tables (USWRC 1981; Chow 1964)

For normal distribution, the values, Q_T, $\bar{X}$, and s pertain to untransformed annual peak flows, and K is the standard normal deviate, which may be obtained from statistical tables corresponding to $P = 1/T$, where P = probability that the indicated value will be equaled or exceeded (e.g., $P = 0.01$ for the 100-yr peak flow). If tables of frequency factors for the log-Pearson Type III distribution are used, then tabulated values of K for $G = 0$ may be used for the normal distribution (USWRC 1981). Commonly used values of K for normal or lognormal distribution are given in Table 2-12.

For the log-normal and log-Pearson Type III distributions, the values of $\bar{X}$ and s used in Eq. (2-21) are computed from the logarithms of the annual peak flows, and the estimated value of Q_T is the logarithm of the desired peak flow. The frequency factor for the log-normal distribution is the same as for the normal distribution, and for the log-Pearson Type III distribution is that corresponding to the previously computed skew coefficient, G, of the logarithms of the annual peak flows. These values are tabulated in statistical tables (e.g., USWRC 1981). Alternatively, approximate values of K for the log-Pearson Type III distribution may be estimated by the equation

$$K \text{ (LP Type III)} = 2/G \{[(K_n - G/6) G/6 + 1]^3 - 1\} \qquad (2\text{-}22)$$

where K_n is the value of K for the normal distribution.

Table 2-12. Selected values of frequency factor for normal or log-normal distribution

Return period (yr)	P (probability of exceedance)	Frequency factor (K)
2	0.50	0
2.5	0.40	0.25335
3.33	0.30	0.52440
5	0.20	0.84162
10	0.10	1.28155
20	0.05	1.64485
25	0.04	1.75069
40	0.025	1.95996
50	0.02	2.05375
100	0.01	2.32635
200	0.005	2.57583
500	0.002	2.87816
1,000	0.001	3.09023
2,000	0.0005	3.29053
10,000	0.0001	3.71902

Source: USWRC (1981).

For Gumbel distribution, Q_T, $\overline{X}$, and s pertain to untransformed annual peak flows and K may be estimated from the equation:

$$K(\text{Gumbel}) = -0.7797 \{0.5772 + \ln[\ln T - \ln (T-1)]\} \quad (2\text{-}23)$$

Alternatively, Q_T for the Gumbel distribution may be estimated by the following equations:

$$\alpha = 1.2826/s \quad (2\text{-}24)$$

$$u = \overline{X} - 0.5772/\alpha \quad (2\text{-}25)$$

$$1 - P = 1 - 1/T = \exp[-\exp\{-\alpha(Q_T - u)\}] \quad (2\text{-}26a)$$

or,

$$Q_T = u - \{\ln[-\ln(1 - P)]\}/\alpha \quad (2\text{-}26b)$$

The Gumbel distribution is used to define the mean annual flood. Setting $Q_T = \overline{X}$, Eq. (2-26) gives

$$1 - P = 1 - 1/T = \exp[-\exp\{-\alpha(\overline{X} - u)\}]$$

Also, from Eq. (2-25),

$$-\alpha(\overline{X} - u) = -0.5772.$$

Thus, $T = 2.33$ yr = return period of the mean annual flood.

For more refined statistical analyses of annual peak flows, refer to standard texts on application of statistical methods in hydrology (e.g., Haan 1977; Yevjevich 1972a, 1997).

Example 2-6: Estimate the 500-yr peak flow of the Ohio River at Louisville, Kentucky, using the annual peak flow data given in Table 2-13.

Solution:

1. Normal Distribution

    Using Eq. (2-21) with $K = 2.87816$ from Table 2-12, $\overline{X} = 14329.4$, and $s = 3677.0$ from Table 2-13,

    $$Q_{500} = 14329.4 + 2.87816 \times 3677.0 = 24{,}912 \text{ m}^3/\text{s}$$

2. Log-Normal Distribution

    Using Eq. (2-21) with $K = 2.87816$ from Table 2-12, $\overline{X} = 4.142405$, and $s = 0.110875$ from Table 2-13,

    $$\text{Log}(Q_{500}) = 4.142405 + 2.87816 \times 0.110875 = 4.46152$$

    Therefore, $Q_{500} = 28{,}941 \text{ m}^3/\text{s}$.

3. Log-Pearson Type III (LP Type III) Distribution

Using $\overline{X} = 4.142405$, $s = 0.110875$, and $G = -0.24$ from Table 2-13, and K (LP Type III) = 2.588996 (by interpolation from tables for the frequency factor for log-Pearson Type III distribution for $G = -0.24$), Eq. (2-21) gives,

$$\text{Log}\,(Q_{500}) = 4.142405 + 2.588996 \times 0.110875 = 4.42946$$

Therefore, $Q_{500} = 26{,}882\ \text{m}^3/\text{s}$. Alternatively, using Eq. (2-22) for log-Pearson Type III distribution,

$$K\,(\text{LP III}) = 2/(-0.24)[\{(2.87816 + 0.24/6)\\ (-0.24/6) + 1\}^3 - 1] = 2.59$$

which is nearly the same as the value interpolated from the tables.

4. Gumbel Distribution

Using Eq. (2-23) to estimate the frequency factor for Gumbel distribution,

$$K\,(\text{Gumbel}) = -0.7797[0.5772 + \ln\{\ln(500/499)\}] = 4.3947$$

Therefore, using Eq. (2-21) with $\overline{X} = 14329.4$ and $s = 3677.0$,

$$Q_{500} = 14329.4 + 4.3947 \times 3677.0 = 30{,}489\ \text{m}^3/\text{s}$$

Alternatively, using Eqs. (2-24), (2-25), and (2-26b) with $P = 0.002$ for $T = 500$ yr,

$\alpha = 1.2826/3677.0 = 0.0003488$
$u = 14329.4 - 0.5772/0.0003488 = 12675$
$Q_{500} = 12675 - \ln[-\ln(1 - 0.002)]/0.0003488 = 30{,}489\ \text{m}^3/\text{s}$

The estimated 500-yr peak flow varies from 24,912 to 30,489 m^3/s. The estimated value using LP Type III distribution is near the middle of this range. So, a value of 27,000 m^3/s appears to be a reasonable estimate.

Surface Runoff Hydrographs

This involves development of a surface runoff hydrograph for the watershed at the point of interest. A hydrograph is a graphical plot (or tabular presentation) of flows against time. Surface runoff hydrographs resulting from storm events of specified durations are described here. The peak of T-yr storm runoff hydrograph is assumed to represent the T-yr peak flow, although this may not always match the statistically estimated T-yr peak flow. Methods to develop the probable maximum flood (PMF) hydrograph are described in the section of this chapter entitled "Probable Maximum Flood Hydrograph." A typical surface runoff hydrograph consists of a slowly rising approach limb, a relatively faster rising limb, and a receding limb that connects to the baseflow hydrograph (Figure 2-2).

Usually, surface runoff hydrographs are developed using computer models, such as HEC-HMS (USACE 2002), HEC-1 (USACE 1991a), TR-20 (USDA 1983a), and SEDIMOT-II

Table 2-13. Statistical analysis of annual peak flows of Ohio River at Louisville, Kentucky

Year	Q (m³/s)	Ordered Q	Log (Q)	Rank	Plotting position
1872	11070	31427	4.4973	1	0.0085
1873	9881	23868	4.3778	2	0.0171
1874	11835	23358	4.3684	3	0.0256
1875	14439	22225	4.3468	4	0.0342
1876	15572	21772	4.3379	5	0.0427
1877	14524	21235	4.3270	6	0.0513
1878	6852	20187	4.3051	7	0.0598
1879	10136	19932	4.2995	8	0.0684
1880	14638	19451	4.2889	9	0.0769
1881	11835	18177	4.2595	10	0.0855
1882	17667	17979	4.2548	11	0.0940
1883	21235	17979	4.2548	12	0.1026
1884	23358	17865	4.2520	13	0.1111
1885	11637	17667	4.2472	14	0.1196
1886	15742	17611	4.2458	15	0.1282
1887	15742	17469	4.2423	16	0.1367
1888	9541	17327	4.2387	17	0.1453
1889	8296	17242	4.2366	18	0.1538
1890	16874	17214	4.2359	19	0.1624
1891	15572	17214	4.2359	20	0.1709
1892	11778	16931	4.2287	21	0.1795
1893	14213	16874	4.2272	22	0.1880
1894	7843	16818	4.2258	23	0.1966
1895	11325	16789	4.2250	24	0.2051
1896	11920	16676	4.2221	25	0.2137
1897	16818	16648	4.2214	26	0.2222
1898	17214	16591	4.2199	27	0.2308
1899	15714	15940	4.2025	28	0.2393
1900	8267	15883	4.2009	29	0.2479
1901	15940	15855	4.2002	30	0.2564
1902	12797	15798	4.1986	31	0.2649
1903	14185	15770	4.1978	32	0.2735
1904	12146	15770	4.1978	33	0.2820
1905	11637	15770	4.1978	34	0.2906
1906	13279	15742	4.1971	35	0.2991
1907	20187	15742	4.1971	36	0.3077
1908	15119	15714	4.1963	37	0.3162
1909	15770	15572	4.1923	38	0.3248
1910	14071	15572	4.1923	39	0.3333
1911	12061	15515	4.1908	40	0.3419
1912	14185	15232	4.1828	41	0.3504
1913	21772	15147	4.1803	42	0.3590
1914	11806	15119	4.1795	43	0.3675
1915	14609	15119	4.1795	44	0.3761

Table 2-13. (*Continued*)

Year	Q (m³/s)	Ordered Q	Log (Q)	Rank	Plotting position
1916	15119	15062	4.1779	45	0.3846
1917	14864	14892	4.1729	46	0.3932
1918	13307	14864	4.1721	47	0.4017
1919	13675	14864	4.1721	48	0.4103
1920	15232	14836	4.1713	49	0.4188
1921	10278	14751	4.1688	50	0.427
1922	15515	14638	4.1655	51	0.4359
1923	13477	14609	4.1646	52	0.4444
1924	15770	14609	4.1646	53	0.4530
1925	10193	14609	4.1646	54	0.4615
1926	11976	14581	4.1638	55	0.4701
1927	16676	14581	4.1638	56	0.4786
1928	12231	14524	4.1621	57	0.4872
1929	14581	14439	4.1595	58	0.4957
1930	10815	14439	4.1595	59	0.5043
1931	10617	14439	4.1595	60	0.5128
1932	15147	14326	4.1561	61	0.5214
1933	19932	14213	4.1527	62	0.5299
1934	11608	14185	4.1518	63	0.5385
1935	15798	14185	4.1518	64	0.5470
1936	17611	14071	4.1483	65	0.5556
1937	31427	13958	4.1448	66	0.5641
1938	9796	13958	4.1448	67	0.573
1939	17979	13930	4.1439	68	0.5812
1940	16931	13675	4.1359	69	0.5897
1941	7758	13505	4.1305	70	0.5983
1942	11382	13477	4.1296	71	0.6068
1943	17469	13307	4.1241	72	0.6154
1944	13109	13279	4.1231	73	0.6240
1945	23868	13194	4.1204	74	0.6325
1946	13505	13165	4.1194	75	0.6410
1947	9966	13109	4.1176	76	0.6496
1948	19451	13080	4.1166	77	0.6581
1949	14439	12911	4.1109	78	0.6667
1950	17242	12797	4.1071	79	0.6752
1951	14892	12231	4.0875	80	0.6838
1952	15855	12146	4.0844	81	0.6923
1953	8890	12118	4.0834	82	0.7008
1954	6710	12061	4.0814	83	0.7094
1955	17214	12033	4.0804	84	0.7179
1956	14836	11977	4.0783	85	0.7265
1957	13958	11920	4.0763	86	0.7350
1958	15770	11891	4.0752	87	0.7436
1959	14581	11835	4.0732	88	0.7521

(*continued*)

Table 2-13. (*Continued*)

Year	Q (m³/s)	Ordered Q	Log (Q)	Rank	Plotting position
1960	10249	11835	4.0732	89	0.7607
1961	16789	11806	4.0721	90	0.7692
1962	17865	11778	4.0711	91	0.7778
1963	16648	11636	4.0658	92	0.7863
1964	22225	11637	4.0658	93	0.7949
1965	13194	11608	4.0648	94	0.8034
1966	14609	11382	4.0562	95	0.8120
1967	18177	11325	4.0540	96	0.8205
1968	16591	11070	4.0442	97	0.8291
1969	9966	11014	4.0419	98	0.8376
1970	14326	10815	4.0340	99	0.8461
1971	13165	10617	4.0260	100	0.8547
1972	14751	10278	4.0119	101	0.8632
1973	15062	10249	4.0107	102	0.8718
1974	15883	10193	4.0083	103	0.8803
1975	14609	10136	4.0059	104	0.8889
1976	13930	9966	3.9985	105	0.8974
1977	14439	9966	3.9985	106	0.9060
1978	17327	9881	3.9948	107	0.9145
1979	17979	9796	3.9911	108	0.9231
1980	11014	9541	3.9796	109	0.9316
1981	11891	8890	3.9489	110	0.9402
1982	13080	8296	3.9188	111	0.9487
1983	14864	8267	3.9174	112	0.9573
1984	12911	7843	3.8945	113	0.9658
1985	12033	7758	3.8897	114	0.9744
1986	13958	6852	3.8358	115	0.9829
1987	12118	6710	3.8273	116	0.9914
Sum		1662210	1662210	480.519	
Mean ($\bar{X}$)		14329.4	14329.4	4.142405	
Std. deviation (s)		3677.0	3677.0	0.110875	
Skew (G)		1.01088	1.010888	−0.24026	

(Wilson et al. 1984). The most common method to develop a surface runoff hydrograph is to convolute the unit hydrograph ordinates (UHOs) with rainfall excess increments of unit duration arranged in an appropriate sequence. Convolution is a process of multiplication and summation. Rainfall excess is precipitation minus losses. The following data are required to generate a surface runoff hydrograph:

- Watershed parameters including drainage area and time of concentration or lag time to determine the unit hydrograph for the basin
- Duration and time distribution of precipitation and snowmelt runoff

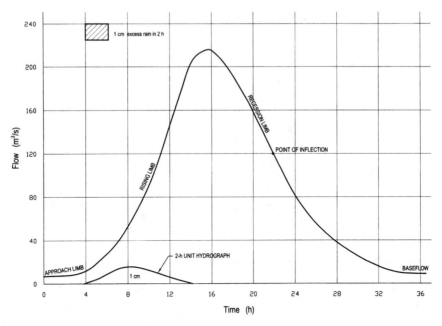

Figure 2-2. Typical surface runoff hydrograph

- Soil losses
- Baseflow hydrograph

The entire watershed is divided into subwatersheds, each representing the drainage area of a significant tributary of the main stream. Identification of significant tributary subwatersheds is based on judgment. The larger the number of subwatersheds, the more complex the hydrologic network. A surface runoff hydrograph has to be developed for each subwatershed, routed along interconnecting channels, and combined with surface runoff hydrographs from other subwatersheds at appropriate locations. A line diagram of the hydrologic network should be prepared that shows the location of each subwatershed and the flow path of surface runoff from all subwatersheds to the point where the composite hydrograph for the entire watershed is to be developed. Sometimes, dividing a watershed into different numbers of subwatersheds may result in appreciably different peak flows at the watershed outlet. If all hydrograph parameters are selected consistently and weighted hydrograph parameters are used to reflect the contribution of minor tributaries within each subwatershed, then the difference in the number of subwatersheds in which the watershed is divided may not result in significant differences (i.e., greater than about 10%) in the estimated peak flows for the entire watershed (Prakash 1987).

Methods to estimate the time of concentration or lag time are described in the section of this chapter entitled "Rational Method."

Unit Hydrograph

The unit hydrograph is a hydrograph of direct runoff from a watershed resulting from a unit depth of effective rainfall (i.e., precipitation minus losses), uniformly distributed over the watershed area occurring during a specified period of time known as the unit duration.

The unit duration should be sufficiently small so that the precipitation rate during this period may be assumed to be constant. Generally, a value smaller than 0.29 times the lag time is considered reasonable (USACE 1991a). The two methods to develop a unit hydrograph for a watershed or subwatershed are the direct and indirect (synthetic) methods. In the direct method, subwatershed area, observed outflow hydrograph at the subwatershed outlet, estimated baseflow parameters or baseflow hydrograph, precipitation associated with the observed outflow hydrograph, precipitation loss parameters, and lag time are used as input to derive the unit hydrograph. For this method to be applicable, the observed outflow hydrograph should be the result of an isolated storm with uniform intensity throughout its duration and uniform distribution over the entire subwatershed. In the HEC-1 model, this method is incorporated in the form of "optimization methodology" (USACE 1991a). (Refer to other texts for details, e.g., ASCE 1996.) The practitioner is often required to develop and use synthetic unit hydrographs computed by indirect methods. Some relatively simple indirect methods to develop unit hydrographs follow:

Dimensionless Unit Hydrograph (USDA 1972, 1985; USBR 1987)

Various agencies (e.g., USDA and USBR) have developed generalized dimensionless unit hydrographs (i.e., plots or tables of dimensionless discharge against dimensionless time) applicable to subwatersheds of different sizes in different regions. The HEC-1 model has incorporated the SCS dimensionless unit hydrograph values (i.e., t/t_p versus q/q_p) within its code, where t = time in hours at which unit hydrograph ordinate (UHO) is q (m³/s), t_p = time from the beginning to the peak of unit hydrograph in hours, and q_p = peak ordinate of the unit hydrograph (m³/s). If subwatershed area, A (km²), unit duration, t_r(h), and lag time, t_L(h), are specified, then

$$t_p = t_r/2 + t_L \tag{2-27}$$

$$q_p = 2.08\, A/t_p \tag{2-28}$$

As stated previously, $t_r \approx 0.29 t_L$. For any time, t, the ratio, t/t_p, is computed from Eq. (2-27) and the corresponding value of q/q_p is obtained from the tables from which q (UHO at time t) is computed. From a practitioner's point of view, this is one of the more convenient methods because it requires only one parameter, t_L, to be estimated, in addition to the area of the subwatershed.

Clark's Unit Hydrograph (ASCE 1996; USACE 1991a, 2002)

To use this method, the subwatershed area is divided into several zones by isochrones, which are loci of points of equal travel times up to the subwatershed outlet. The area between each isochrone and subwatershed outlet is planimetered and expressed as a dimensionless time-area curve or table between $A^* = A(i)/A$ and $t^* = t(i)/t_c$, where $A(i)$ = area and $t(i)$ = travel time from isochrone i to subwatershed outlet. If a site-specific time-area curve is not available, then the equations of the HEC-1 and HEC-HMS models may be used:

$$A^* = 1.414(t^*)^{1.5} \quad 0 \leq t^* < 0.5 \tag{2-29a}$$

$$1 - A^* = 1.414(1 - t^*)^{1.5} \quad 0.5 \leq t^* < 1.0 \tag{2-29b}$$

Although not absolutely necessary, the time step of computations is usually taken to be the same as the selected unit duration, Δt. The UHOs of Clark's unit hydrograph can be computed by the following:

$$B(t) = A\,[A^*(t) - A^*(t - \Delta t)] \tag{2-30}$$

$$q(t) = 2C_0 B(t) + C_1 q(t + \Delta t) \tag{2-31}$$

$$U(t) = 0.5[q(t - \Delta t) + q(t)] \tag{2-32}$$

$$C_0 = \Delta t / (2K + \Delta t) \tag{2-33}$$

$$C_1 = 1 - 2\,C_0 \tag{2-34}$$

where K = storage coefficient in units of time. This method is useful only when a site-specific or regional value of K is available or can be estimated. Methods to estimate K may be found in other references (e.g., USACE 1960; Ponce 1989; Bras 1990).

Snyder's Unit Hydrograph

This method does not produce the complete unit hydrograph. It provides the base width in hours, peak flow (m³/s), and widths at 50% and 75% of the peak flow in hours and is useful only when calibrated values of Snyder's parameters, C_p and t_L, for the region or subwatershed are available. The parameter, t_L, is estimated from Eq. (2-7), and C_p is a coefficient in the equation

$$q_p = 2.78\,C_p A / t_L \tag{2-35}$$

Typical values of C_p are 0.94 for watersheds similar to those in southern California, 0.63 for those similar to fairly mountainous Appalachian Highlands, and 0.31 for those similar to sections of states bordering on the eastern Gulf of Mexico. (Details of the method are provided in other texts, e.g., ASCE 1996; Chow 1964.) Knowing C_p and t_L, the HEC-1 model may be used to develop Snyder's unit hydrograph. This model uses a trial and error method to obtain the corresponding Clark's parameters and then generates the unit hydrograph. An initial value of Clark's parameter, K, is assumed; the time of concentration is estimated from the relationship, $t_c = t_L/0.6$; Clark's UHOs are computed for the desired unit duration, Δt. Then, trial values of Snyder's parameters are computed by:

$$C'_p = q'_p (t'_p - 0.5\,\Delta t) / (2.78\,A) \tag{2-36}$$

$$t'_L = 1.048(t'_p - 0.75\,\Delta t) \tag{2-37}$$

where

t'_p = time when q'_p occurs on the computed Clark's UHOs

q'_p = maximum UHO

The initial assumed values of t_c and K are adjusted to compensate for differences between computed C'_p and t'_L and given values of C_p and t_L. A new set of C'_p and t'_L is recomputed until these values are in close agreement to the given values within an acceptable tolerance limit (e.g., 1.0%). The final set of values is then used to develop Clark's UHOs.

Kinematic Wave Method

This method is incorporated in the HEC-1 and HEC-HMS models (USACE 1991a, 2002) and is useful to develop surface runoff hydrographs for subwatersheds that constitute lateral flows to streams. It also is useful to develop surface runoff hydrographs for flows across linear boundaries of subwatersheds. The basin area contributing lateral flow to the stream may be divided into separate subunits based on land slopes and surface cover types or direction of flow. For instance, overland flow to a stream from the left and right sides of its bank may be computed using two different subunits of the same basin. Surface runoff computations have to be made separately for each subunit. The rainfall excess is assumed to flow laterally to the stream through a wide rectangular channel. The length of the flow path, L, land slope, S, Manning's roughness factor, n, and percentage of basin area that this subunit represents are used as given parameters. The roughness factor for overland flow typically varies from 0.5 for dense vegetation to 0.10 for very shallow depths on surfaces paved with concrete or asphalt. The roughness factor and land slope are used to determine kinematic wave parameters, α and m. For a wide channel where the wetted perimeter is nearly equal to the channel width, the kinematic wave approximation for discharge, Q, is

$$Q = \alpha A^m \tag{2-38}$$

where

$\alpha = (1/n) S^{1/3}$

$m = 5/3$

Rainfall excess, q, per unit width of the flow path constitutes the inflow to the channel. The resulting continuity equation with the kinematic wave approximation is

$$\partial A/\partial t + \alpha m A^{m-1} \partial A/\partial x = q \tag{2-39}$$

A discretized form of Eq. (2-39) is solved numerically to develop a surface runoff hydrograph for overland flow per unit width of the subunit. The average width of the subunit is estimated by dividing its area by the average length, L. The surface runoff hydrograph ordinates per unit width are multiplied by this width to obtain the surface runoff hydrograph for the entire subunit. Surface runoff hydrographs from these subunits are combined with the hydrograph entering the stream reach or subwatershed at the upstream edge of each subunit. Further details of the method are provided in other texts (e.g., USACE 1991a, 2002).

Design Storm Duration and Depth

The duration of design precipitation for the watershed is determined using the estimated time of travel of surface runoff from the upper edge of the watershed to the point of interest through overland flow paths, channels, and impoundments. A study of annual maximum discharge records of selected streams in Maryland indicated that a 24-h storm duration may be appropriate for drainage areas of 5 to 130 km² (Levy and McCuen 1999). Smaller storm durations may be used for smaller watersheds with smaller travel times of surface runoff, and a

24- to 96-h duration may be necessary for larger watersheds. Precipitation depths of the desired return periods and durations can be obtained from NOAA Atlas 2 for the eleven western states (NOAA 1973), TP-40 for other contiguous states (Hershfield 1961), Technical Paper No. 47 for Alaska (Miller 1963), Technical Paper No. 43 for Hawaii (USDOC 1962), and Technical Paper No. 42 for Puerto Rico and the Virgin Islands (USDOC 1961). Methods to estimate areal average precipitation depths from point estimates also are included in these publications.

For convolution with UHOs, precipitation depths have to be divided into time increments equal to the unit duration and sequenced to approximate the time distribution of precipitation in the storm event of interest. Various methods have been proposed for time distribution and sequencing of incremental precipitation. If hyetographs of rainfall events in the site vicinity are available, then the distribution and sequence of incremental precipitation for the design storm should be selected to be as close to the observed ones as possible. A few commonly used methods are presented here.

SCS Method

The SCS has expressed time distribution and sequencing of 24-h rainfall by four curves applicable to different regions of the United States. Approximate distributions of precipitation depths convenient for computer use are included in Table 2-14 (USDA 1986; Ponce 1989).

The Type I distribution is applicable to Hawaii, the coastal side of the Sierra Nevada in southern California, and the interior regions of Alaska. Type IA represents areas on the coastal side of the Sierra Nevada and the Cascade Mountains in Oregon, Washington, northern California, and the coastal regions of Alaska. Type III represents Gulf of Mexico and Atlantic coastal areas where tropical storms bring large 24-h rainfall amounts. Type II is applicable to the remaining United States, Puerto Rico, and the Virgin Islands. Type II and Type III distributions are very similar to each other.

HEC-1 Model Approach

This method uses specified values of precipitation depths for 5-, 15-, and 60-min and 2-, 3-, 6-, 12-, 24-, 96-, 168-, and 240-h durations depending on the duration of the design storm. The model generates a triangular distribution such that the rainfall depth specified for any duration occurs during the central part of the storm. This method has been found to be convenient for cases where site-specific sequences of incremental precipitation cannot be determined.

Soil Losses

Soil losses include the portion of precipitation that is lost due to infiltration, transpiration and interception by vegetation, depression storage, and evaporation. Methods to estimate soil losses that are relatively simple to use and do not require parameters, some of which may be relatively difficult to obtain, are described here.

1. **Constant initial loss (mm) followed by uniform loss rate (mm/h).** This method is suitable for watersheds where calibrated values for the uniform loss and uniform loss rate are available for storm events similar to those for which the surface runoff hydrograph is to be developed. The entire precipitation is lost until the prescribed initial loss is satisfied. Thereafter, precipitation loss occurs at the prescribed uniform rate.

Table 2-14. Time distribution and sequencing of design storms

Duration (h)	Fraction of 24-h rainfall depth			
	Type I	Type IA	Type II	Type III
0.0	0.00000	0.00000	0.00000	0.00000
0.5	0.00871	0.01000	0.00513	0.00500
1.0	0.01745	0.02000	0.01050	0.01000
1.5	0.02621	0.03500	0.01613	0.01500
2.0	0.03500	0.05000	0.02200	0.02000
2.5	0.04416	0.06600	0.02813	0.02519
3.0	0.05405	0.08200	0.03450	0.03075
3.5	0.06466	0.09800	0.04113	0.03669
4.0	0.07600	0.11600	0.04800	0.04300
4.5	0.08784	0.13500	0.05525	0.04969
5.0	0.09995	0.15600	0.06300	0.05675
5.5	0.11234	0.18000	0.07125	0.06419
6.0	0.12500	0.20600	0.08000	0.07200
6.5	0.13915	0.23700	0.08925	0.08063
7.0	0.15600	0.26800	0.09900	0.09050
7.5	0.17460	0.31000	0.10925	0.10163
8.0	0.19400	0.42500	0.12000	0.11400
8.5	0.21900	0.48000	0.13225	0.12844
9.0	0.25400	0.52000	0.14700	0.14575
9.5	0.30300	0.55000	0.16300	0.16594
10.0	0.51500	0.57700	0.18100	0.18900
10.5	0.58300	0.60100	0.20400	0.21650
11.0	0.62300	0.62400	0.23500	0.25000
11.5	0.65550	0.64500	0.28300	0.29800
12.0	0.68400	0.66400	0.66300	0.50000
12.5	0.70925	0.68300	0.73500	0.70200
13.0	0.73200	0.70100	0.77200	0.75000
13.5	0.75225	0.71900	0.79900	0.78350
14.0	0.77000	0.73600	0.82000	0.81100
14.5	0.78625	0.75281	0.83763	0.83406
15.0	0.80200	0.76924	0.85350	0.85425
15.5	0.81725	0.78529	0.86763	0.87156
16.0	0.83200	0.80096	0.88000	0.88600
16.5	0.84625	0.81625	0.89119	0.89838
17.0	0.86000	0.83116	0.90175	0.90950
17.5	0.87325	0.84569	0.91169	0.91938
18.0	0.88600	0.85984	0.92100	0.92800
18.5	0.89825	0.87361	0.92969	0.93581
19.0	0.91000	0.88700	0.93775	0.94325
19.5	0.92125	0.90001	0.94519	0.95031
20.0	0.93200	0.91264	0.95200	0.95700
20.5	0.94225	0.92489	0.95844	0.96336
21.0	0.95200	0.93676	0.96475	0.96944
21.5	0.96125	0.94825	0.97094	0.97523
22.0	0.97000	0.95936	0.97700	0.98075
22.5	0.97825	0.97009	0.98294	0.98598
23.0	0.98600	0.98044	0.98875	0.99094
23.5	0.99325	0.99041	0.99444	0.99561
24.0	1.00000	1.00000	1.0000	1.00000

Source: USDA (1986); Ponce (1989)

2. **Curve Number Method** (USDA 1972, 1985; ASCE 1996). In this method, a curve number (CN) is assigned to each component of the watershed based on soil type, land use, and antecedent moisture condition of the soils. Published soil surveys for most counties in the United States are available in many libraries and local offices of the Natural Resources Conservation Service (NRCS). The soils are classified into four hydrologic soil groups:
   a. Soils having high infiltration rates even when thoroughly wetted and consisting chiefly of deep, well-drained to excessively drained sands or gravels. These soils have a low surface runoff potential and are assigned low curve numbers.
   b. Soils having moderate infiltration rates when thoroughly wetted and consisting chiefly of moderately deep to deep, moderately well- to well-drained soils with moderately fine to moderately coarse textures.
   c. Soils having slow infiltration rates when thoroughly wetted and consisting chiefly of soils with a layer that impedes downward movement of water, or soils with moderately fine to fine texture. These soils have a relatively high runoff potential and are assigned moderately high curve numbers.
   d. Soils having very slow infiltration rates when thoroughly wetted and consisting chiefly of clay soils with a high swelling potential, soils with a permanently high water table, soils with a clay pan or clay layer at or near the surface, and shallow soils over nearly impervious material. These soils have high runoff potential and are assigned high curve numbers.

The soil moisture condition resulting from weather conditions preceding the storm event for which a surface runoff hydrograph is to be developed is referred to as the antecedent moisture condition (AMC). AMCs are divided into three classes:

1. **AMC I:** This represents a condition when soils are dry but not to the wilting point and antecedent precipitation (within 5 days prior to the storm event) is less than about 13 mm for the dormant season and less than about 36 mm for the crop growing season.
2. **AMC II:** This represents average conditions typifying annual floods when antecedent precipitation is 13 to 28 mm for the dormant season and 36 to 53 mm for the growing season.
3. **AMC III:** This represents a condition when heavy rainfall (greater than 28 mm during the dormant season and 53 mm during the growing season) or light rainfall and low temperatures have occurred within 5 days preceding the storm event.

According to the curve number method,

$$Q = (P - I_a)^2 / [P + 0.8 S_a] \qquad (2\text{-}40)$$

where

Q = surface runoff (cm)

P = total rainfall (cm)

S_a = potential maximum retention as defined in Eq. (2-6) (cm)

I_a = initial abstraction (cm) = $0.2 S_a$

If $P \leq I_a$, there is no runoff.

Table 2-15. Typical runoff curve numbers (AMC II and $I_a = 0.2 S_a$)

Land use	Cover Treatment or practice	Hydrologic condition	Hydrologic soil group A	B	C	D
Fallow	Straight row	—	77	86	91	94
Row crops	Straight row	Poor	72	81	88	91
		Good	67	78	85	89
	Contoured	Poor	70	79	84	88
		Good	65	75	82	86
	Contoured and terraced	Poor	66	74	80	82
		Good	62	71	78	81
Small grain	Straight row	Poor	65	76	84	88
		Good	63	75	83	87
	Contoured	Poor	63	74	82	85
		Good	61	73	81	84
	Contoured and terraced	Poor	61	72	79	82
		Good	59	70	78	81
Closed seeded legumes or rotation meadow[a]	Straight row	Poor	66	77	85	89
		Good	58	72	81	85
	Contoured	Poor	64	75	83	85
		Good	55	69	78	83
	Contoured and terraced	Poor	63	73	80	83
		Good	51	67	76	80
Pasture or range	General	Poor	68	79	86	89
		Fair	49	69	79	84
		Good	39	61	74	80
Meadow	General	Good	30	58	71	78
Woods	General	Poor	45	66	77	83
		Fair	36	60	73	79
		Good	25	55	70	77
Farmsteads	General	General	59	74	82	86
Roads	Dirt	General	72	82	87	89
	Hard surface	General	74	84	90	92
Open spaces (lawns, parks, golf courses, cemeteries, etc.)	Grass cover <50%	Poor	68	79	86	89
	Grass cover 50 to 75%	Fair	49	69	79	84
	Grass cover >75%	Good	39	61	74	80
Commercial and business areas (85% impervious)	General	General	89	92	94	95
Industrial districts (72% impervious)	General	General	81	88	91	93
Residential areas	CN reduces with increase in lot size	Lot sizes of 1/8 to 2 acres	46–77	65–85	77–90	82–92
Paved parking lots	General	General	95–98	95–98	95–98	95–98

[a]Close-drilled or broadcast
Source: USDA (1972, 1985).

Curve numbers for AMC II are given in Table 2-15 (USDA 1972, 1985). The corresponding curve numbers for AMC I and AMC III can be estimated using the equations (Hawkins et al. 1985):

$$CN_I \text{ (CNs for AMC I)} = CN_{II}/(2.3 - 0.013\, CN_{II}) \tag{2-41}$$

$$CN_{III} \text{ (CNs for AMC III)} = CN_{II}/(0.43 + 0.0057\, CN_{II}) \tag{2-42}$$

If the watershed includes segments with several types of soils and soil cover complexes, appropriate CNs are assigned to each segment and a weighted CN can be estimated for the entire watershed using the equation

$$CN \text{ (weighted)} = CN(1)A_1 + CN(2)A_2 + \ldots + CN(n)A_n \tag{2-43}$$

where

$CN(1)A_1$, etc. = CN and area, respectively, of different segments

n = total number of segments in the watershed

Snowmelt and Snow Loads

A simple method to estimate the contribution of snowmelt to rainfall excess is to use the degree-day method, as in the equation

$$S_m = C_m(T - T_m) \tag{2-44}$$

where

S_m = snowmelt per day (mm)

T = air temperature (°C) at the midpoint of the zone of snowpack during the time interval for which snowmelt is being calculated

T_m = air temperature at which snow melts (°C)

C_m = melt coefficient per degree-day (mm/°C), usually about 3.20

Often, air temperatures at different elevations within the zone of snowpack are not available. To estimate air temperatures at different elevations from known values at one elevation, an average temperature lapse rate with changes in elevation may be used. A typical value of lapse rate is about 0.60°C per 100-m change in elevation.

If data about rainfall intensity and wind velocity are available, the contribution of snowmelt during rain may be evaluated using Eq. (2-45) or (2-46) (Chow 1964):

- For open (<10% cover) or partly forested (10 to 60% cover) basin areas,

$$S_m = [1.326 + 0.2386kv + 0.0126P]T + 2.29 \tag{2-45}$$

- For heavily forested (>80% cover) areas,

$$S_m = [3.383 + 0.0126P]T + 1.27 \tag{2-46}$$

where

S_m = daily snowmelt (mm)

T_a = mean temperature of saturated air at the 3.0-m level (°C)

v = mean wind velocity (km/h) at the 15.2-m level

P = rate of precipitation in mm/day in open portions of the basin

k = coefficient varying from 0.3 for forested areas to 1.0 for unforested plains

The water resources engineer is often required to compute rain and snow loads on roofs of industrial structures (e.g., nuclear power plant structures) and some residential buildings. If a parapet is provided so that the snow and rainfall are both retained on the roof, then the roof load can be estimated using design depths and unit weights of snow and rainwater. The average initial specific gravity of snow is about 0.10. In some cases, snowfall is followed by small amounts of intermittent rainfall. As rainwater falls on snow, the snow is compacted, and it absorbs water until a threshold is reached when drainage of excess water begins. The average specific gravity of compacted snow is about 0.40. Computational steps to estimate roof loads due to compacted snow on roofs without parapets are listed below:

- Estimate 100-year mean recurrence interval snow load on the ground for the location of interest from available maps (e.g., NBS 1972).
- Obtain the corresponding snow load on the roof by multiplying the snow load on the ground by a basic snow load coefficient of 0.8. This coefficient may be decreased for sloping roofs to account for slide-off of snow and increased to account for accumulation of snow on pitched or curved roofs.
- Obtain the snow depth, h_0 (m), of the computed initial snow load on roof using a specific gravity of 0.10 for snow.
- Obtain initial water equivalent of snow, H_0(m), using the relationship $H_0 = 0.10\, h_0$.
- Obtain the depths of compacted snow, h (m), and its water equivalent, H (m), at the threshold condition using the experimental relationships (USBR 1966):

$$H = 0.4h \quad (2\text{-}47)$$

$$h/h_0 = 1.474 - 0.474 H/H_0 \quad (2\text{-}48)$$

- Estimate roof load of compacted snow for the equivalent water depth, H (m).

Example 2-7: Estimate roof load of compacted snow on a building near Milwaukee, Wisconsin.

Solution:

100-yr mean recurrence interval snow load on the ground = 146.5 kg/m² (NSB 1972)

Corresponding snow load on roof = 0.80 × 146.5 = 117.2 kg/m²

Initial snow depth = h_0 = 117.2/(0.10 × 1000) = 1.172 m

Initial water equivalent, $H_0 = 0.10 h_0 = 0.1172$ m

$h/h_0 = H/(0.4 h_0) = 1.474 - 0.474 H/H_0$

Therefore,

$H = 0.2386$ m

Roof load due to compacted snow = 0.2386 × 1000 = 238.6 kg/m^2

Baseflow

A number of methods have been proposed for the estimation of baseflows (e.g., ASCE 1996). Usually, baseflow is a small component of the flood hydrograph and its estimates are hard to verify. Unless calibrated values are known, a constant baseflow may be found reasonable for most practical purposes. This constant value may be nearly equal to the dry weather flow at the point of computation.

Combining Hydrographs and Routing through Channels

The surface runoff hydrographs for various subwatersheds should be combined at various junction points along the main stream. Some times, hydrographs may have to be lagged before combining and routing. Various methods are available for routing hydrographs through stream channels (e.g., USACE 1991a, 2002; ASCE 1996). Depending on the ease of estimating the required parameters, any one of these methods may be used. Methods to lag a hydrograph by a certain number of time steps, though straightforward, may require input modification for different models. For the HEC-1 model, use of the Muskingum or Straddle-Stagger options may be convenient. In these cases, the channel loss parameters may be set to zero. The routing steps may be set to 1, and the number of ordinates to be averaged may be set to zero (see HEC-1 or HEC-HMS User's Manuals, USACE 1991a, 2002).

Example 2-8: Develop 100-yr flood hydrograph for a 60.44-km^2 watershed, shown in Figure 2-3.

Solution:

1. Divide the watershed into 12 subwatersheds as shown in Figure 2-3. Subwatersheds 1 to 10 have point outlets and unit hydrographs have to be developed for them. Subwatersheds 11 and 12 contribute lateral flows and are analyzed by the kinematic wave method. For this purpose, Subwatershed 11 is further subdivided into 11(a), 11(b), 11(c), and 11(d), and 12 is divided into 12(w) and 12(e).
2. From topographic and soil maps of the watershed, estimate hydraulic and kinematic wave parameters as shown in Tables 2-16(a) and (b).
3. Prepare line diagram of hydrologic network as shown in Figure 2-4.
4. Prepare sequence of rainfall-runoff computations as shown in Table 2-16(c).
5. Determine time distribution of precipitation depths.

6. Use the input developed in the previous steps for the selected model to develop a surface runoff hydrograph for each subwatershed and combine them to develop the surface runoff hydrograph for the entire watershed.
7. Compare estimated peak of the hydrograph for the watershed with estimates using other methods as shown in Table 2-11.

Probable Maximum Flood Hydrograph

The probable maximum flood (PMF) is defined as the hypothetical flood (peak discharge, volume, and hydrograph shape) that is considered to be the most severe reasonably possible, based on comprehensive hydrometeorological application of probable maximum precipitation (PMP) and other hydrological factors favorable for maximum flood runoff, such as sequential storms and snowmelt (USNRC 1977). Reasonably conservative watershed parameters (e.g., time of concentration and soil loss parameters) along with heuristically combined hydrometeorological events are used to develop the PMF hydrograph (Prakash 1978; 1983). These include the magnitude of the principal storm PMP of required duration, time distribution and sequence of incremental precipitation, antecedent moisture condition, and the antecedent storm that is assumed to precede the principal storm.

The antecedent storm may be taken to be 40% of the PMP occurring about 5 days prior to the principal storm, and the antecedent moisture condition may be taken to be AMC-II.

For watersheds within the United States, PMP estimates for various durations can be obtained from relevant publications of the U.S. Department of Commerce (e.g., USDOC

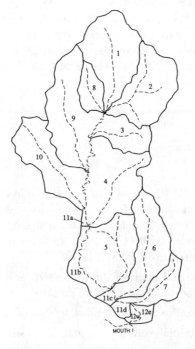

Figure 2-3. Watershed map

Table 2-16(a). Hydraulic parameters of subwatersheds

Subwatershed	Hydraulic parameters					
	D.A. (km^2)	L (km)	H (m)	t_c (h)	t_t (h)	CN
1	7.61	5.49	45.72	1.55	0.93	65
2	6.58	4.43	48.77	1.18	0.71	65
3	3.16	2.72	53.34	0.65	0.39	65
4	8.18	4.73	76.81	1.07	0.64	65
5	7.43	4.12	57.91	1.02	0.61	65
6	6.21	6.24	86.56	1.41	0.85	65
7	2.54	3.35	32.00	1.00	0.60	65
8	1.94	2.80	49.38	0.69	0.42	65
9	7.98	6.55	68.58	1.63	0.98	65
10	6.32	5.49	67.36	1.34	0.80	65

D.A. = drainage area; L = hydraulic length; H = elevation difference between outlet and upstream edge of subwatershed.

1961, 1962, 1965, 1969, 1977, 1978, 1982, 1983, 1988, 1994). In areas where both short-duration local intense thunderstorms and long-duration general storms are common, two estimates of the principal storm PMP must be made. Examples of such areas include those west of the 103d meridian in the United States. Methods to estimate the short-duration (usually 6 h) local storm PMP and long-duration (usually 24 to 96 h) general storm PMP are given in the above-mentioned publications. As an example, the following steps are used to compute the general storm PMP for the Colorado River and Great Basin Drainages using figures and tables in Hydrometeorological Report No. 49 (USDOC 1977):

1. Estimation of 24-h convergence PMP for 26-km^2 area at the location of the drainage area of interest.
2. Reduction of the estimated 24-h convergence PMP for barrier elevation.

Table 2-16(b). Kinematic wave parameters for Subwatersheds 11 and 12

Subwatershed	D.A. (km^2)	Lateral flow parameters				Main channel parameters			
		L	S	n	% area	L	S	W.D.	Z
11(a)	0.21	213.35	0.050	0.30	100	1158.18	0.0026	3.05	2.0
11(b)	1.48	365.74	0.008	0.20	100	4267.00	0.0025	3.05	2.0
11(c)	0.36	609.57	0.005	0.20	100	914.36	0.0023	3.05	2.0
11(d)	0.44	457.18	0.008	0.20	100	1066.75	0.0009	3.05	2.0
12(w)	0.31	304.79	0.015	0.30	32	N.A.	N.A.	N.A.	N.A.
12(e)	0.62	457.18	0.010	0.30	68	1676.32	0.0009	3.66	2.0

L = length (m); S = slope (m/m); W.D. = bottom width of channel approximated by a trapezoid (m); Z = slope of the bank (horizontal:vertical); and N.A. = not applicable.

Table 2-16(c). Sequence of rainfall-runoff computations

1. Develop hydrograph for Subwatershed 1.
2. Develop hydrograph for Subwatershed 2.
3. Develop hydrograph for Subwatershed 8.
4. Combine hydrographs for Subwatersheds 1, 2, and 8.
5. Route through the channel up to the outlet of Subwatershed 3.
6. Develop hydrograph for Subwatershed 3.
7. Combine hydrographs for Subwatersheds 1, 2, 8, and 3.
8. Route through the channel up to the outlet of Subwatershed 9.
9. Develop hydrograph for Subwatershed 9.
10. Combine hydrographs for Subwatersheds 1, 2, 8, 3, and 9.
11. Route through the channel up to the outlet of Subwatershed 10.
12. Develop hydrograph for Subwatershed 10.
13. Combine hydrographs for Subwatersheds 1, 2, 8, 3, 9, and 10.
14. Develop hydrograph for lateral flow from Subwatershed 11a and route combined hydrograph to the outlet of Subwatershed 4 (using kinematic wave approach of the HEC-1 model).
15. Develop hydrograph for Subwatershed 4.
16. Combine hydrographs for Subwatersheds 1, 2, 8, 3, 9, 10, 11a, and 4.
17. Develop hydrograph for lateral flow from Subwatershed 11b and route combined hydrograph to the outlet of Subwatershed 5 (using kinematic wave approach of the HEC-1 model).
18. Develop hydrograph for Subwatershed 5.
19. Combine hydrographs for Subwatersheds 1, 2, 8, 3, 9, 10, 11a, 4, 11b, and 5.
20. Develop hydrograph for lateral flow from Subwatershed 11c and route combined hydrograph to the outlet of Subwatershed 6 (using kinematic wave approach of the HEC-1 model).
21. Develop hydrograph for Subwatershed 6.
22. Combine hydrographs for Subwatersheds 1, 2, 8, 3, 9, 10, 11a, 4, 11b, 5, 11c, and 6.
23. Develop hydrograph for lateral flow from Subwatershed 11d and route combined hydrograph to the outlet of Subwatershed 7 (using kinematic wave approach of the HEC-1 model).
24. Develop hydrograph for Subwatershed 7.
25. Combine hydrographs for Subwatersheds 1, 2, 8, 3, 9, 10, 11a, 4, 11b, 5, 11c, 6, 11d, and 7.
26. Develop hydrograph for lateral flow from Subwatersheds 12w and 12e and route combined hydrograph for all subwatersheds to the outlet of Subwatershed 12 (using kinematic wave approach of the HEC-1 model).

3. Estimation of cumulative convergence PMP for 6-, 12-, 18-, 24-, 48-, and 72-h durations using specified multiplying factors.

4. Estimation of incremental convergence PMP for 6-, 12-, 18-, 24-, 48-, and 72-h durations.

5. Estimation of areally reduced incremental convergence PMP for 6, 12-, 18-, 24-, 48-, and 72-h durations.

HYDROLOGIC ANALYSES

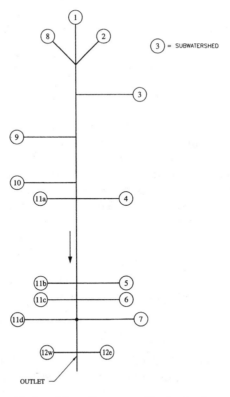

Figure 2-4. Line diagram of hydrologic network

6. Estimation of cumulative drainage area average convergence PMP for 6-, 12-, 18-, 24-, 48-, and 72-h durations.

7. Estimation of drainage area average 24-h orographic PMP index.

8. Estimation of areally and seasonally adjusted 24-h orographic PMP.

9. Estimation of orographic PMP for 6-, 12-, 18-, 24-, 48-, and 72-h durations using specified multiplying factors.

10. Estimation of total PMP for 6-, 12-, 18-, 24-, 48-, and 72-h durations (i.e., sum of convergence and orographic PMPs).

Storm transposition and moisture maximization methods described in the aforementioned publications and information included in World Meteorological Organization publications (e.g., WMO 1986) may be used for watersheds in other parts of the world where regional or site-specific estimates of PMP have not been developed. For situations where mean ($\bar{X}$) and standard deviation (s) of the 24-h annual maximum rainfall depths can be estimated from available site-specific or regional data, preliminary estimate of 24-hr point PMP (P_m) may be made using the relationship (NRC 1985)

$$P_m (24\ h) = \bar{X} + 20\ s \qquad (2\text{-}49)$$

In the absence of reliable information, the empirical relationship enveloping the world's maximum rainfalls of different durations may be used for preliminary estimates of the PMP (USDOC 1961):

$$P_m \text{ (cm)} = 38.9 D^{0.486} \tag{2-50}$$

where D = duration in hours.

The total precipitation depth during the storm is divided into precipitation depths for time increments equal to the unit duration of the unit hydrograph. The incremental values are arranged in descending order of magnitude (i.e., 1, 2, 3, 4, 5, 6, etc.). Several methods to divide the PMP into smaller time increments equal to the unit duration and to sequence them to develop the PMF hydrograph are included in the aforementioned publications and also in the HEC-1 and HEC-HMS models. The following are some commonly used sequences of incremental precipitation:

1. 6, 4, 2, 1, 3, 5
2. 5, 3, 1, 2, 4, 6
3. A hypothetical distribution and sequence included in the HEC-1 model
4. Methods for distribution and sequencing of Standard Project and PMP storms of 24-, 48-, 72-, and 96-h durations included in the HEC-1 model

An optimal sequence, which depends on the shape of the unit hydrograph for the watershed, may be used for cases where extremely conservative PMF peaks are desired (Prakash 1978).

Since the PMF is a conservatively estimated hypothetical event, reasonableness of the estimated PMF peak should be verified using heuristic methods. Where practicable, watershed parameters such as the time of concentration and soil loss parameters may be verified by calibration using observed historic rainfall and flood hydrographs. A few simple empirical methods to estimate peak flows approaching the PMF peak are given here (Crippen 1982).

1. Crippen's Equation

$$Q_m = 577.26 A^{0.8405} \times Z^{-0.751} \tag{2-51a}$$

$$Z = 0.6214 A^{0.5} + 5 \tag{2-51b}$$

2. Creager's Equation

$$Q_m = 130(0.386 A)^B \tag{2-52a}$$

$$B = 0.9358 A^{-0.048} \tag{2-52b}$$

3. Matthai's Equation

$$Q_m = 174.3 A^{0.61} \tag{2-53}$$

where

Q_m = peak flow (m³/s)

A = drainage area (km²)

It should be verified that the estimated PMF peak is close to the range of those computed by the empirical methods. In addition, the estimated PMF peak should be reconciled with any other previously estimated PMF peaks in the region.

Example 2-9: The watershed of a stream is located in a hurricane-prone area. A dam is to be constructed on this stream at a point where the drainage area is 535 km^2. Rainfall analysis of limited data and extrapolations from data for watersheds in similar latitudes resulted in PMP depths shown in Table 2-17(a). Develop a PMF hydrograph for the design of this dam assuming a constant baseflow of 100 m^3/s.

Solution: The watershed is divided into seven subwatersheds as shown in Figure 2-5. The portions of Subwatershed 3 on the left and right sides of the river contribute lateral flow along the stream reach. So, the kinematic wave option of the HEC-1 model is used to develop the hydrograph for this subwatershed. Relevant parameters for each subwatershed are shown in Table 2-17(b). An AMC II curve number of 70 is used for all subwatersheds. The sequence of rainfall runoff computations is shown in Table 2-17(c).

For the sake of simplicity, the hypothetical distribution and sequence of incremental precipitation incorporated in the HEC-1 model is adopted. This requires PMP values for 5- and 15-min durations. In the absence of site-specific information, values for smaller durations may be estimated from the 1-hr value using the ratios shown in Table 2-17(d) (NOAA 1973):

5-min PMP = 0.29 × 109.74 = 31.82 mm

15-min PMP = 0.57 × 109.74 = 62.55 mm

The input developed previously may be used for the selected rainfall-runoff model. In this case, the HEC-1 model is used, which results in a PMF peak of 11,435 m^3/s.

To check the conservatism of the estimated PMF peak, use the empirical equations described previously, as shown below:

1. Crippen's Equation

$$Z = 0.6214 \times 535^{0.5} + 5 = 19.373, \text{ and}$$

$$Q_m = 577.26 \times 535^{0.8405} \times 19.373^{-0.751} = 12,242 \text{ m}^3/\text{s}$$

2. Creager's Equation

$$B = 0.9358 \times 535^{-0.048} = 0.6922, \text{ and}$$

$$Q_m = 130 \, (0.386 \times 535)^{0.6922} = 5,204 \text{ m}^3/\text{s}$$

3. Matthai's Equation

$$Q_m = 174.3 \times 535^{0.61} = 8,046 \text{ m}^3/\text{s}$$

Thus, the estimated value of 11,435 m^3/s is judged to be fairly conservative.

Table 2-17(a). PMP depths

Duration (h)	Rainfall (mm)
1	109.74
2	211.67
3	309.06
6	554.87
12	828.07
24	1092.58
48	1338.40

Risk Analysis and Estimation of Failure Probabilities

ASCE (1988) and Prakash (1992b) provide elaborate risk analysis of dams and other structures. In some situations, the water resources engineer is required to estimate the probability or risk of occurrence of a certain flood during the design life of a structure, which has been designed for a T-yr ($p = 1/T$) flood. For such analyses,

- Probability that at least one (i.e., one or more) failure events will occur in n years:

$$P(\geq 1) = 1 - (1 - p)^n \qquad (2\text{-}54)$$

- Probability that no failure events will occur in n years:

$$P(\text{none}) = (1 - p)^n \qquad (2\text{-}55)$$

Table 2-17(b). Subwatershed parameters

Subwatershed	Area (km²)	CN	Lag time (hrs) and kinematic wave parameters
1	70	70	1.60[a]
2	70	70	2.08[a]
4	24	70	2.14[a]
5	56	70	1.94[a]
6	164	70	7.42[a]
7	106	70	4.58[a]
3 (lateral flow from right side looking downstream)[b]	27	70	L = 4,500 m; S = 0.50 m/m; n = 0.30
3 (lateral flow from left side looking downstream)[b]	18	70	L = 1,500 m; S = 0.40 m/m; n = 0.30
3 (main channel flow)[b]	45	70	L = 8,500 m; S = 0.0133 m/m; n = 0.040; W.D. = 85 m; z = 1.0

[a] Lag time (hrs).
[b] Kinematic wave parameters.
L = hydraulic length; S = land slope; n = roughness coefficient; W.D. = bottom width of channel approximated by a trapezoid; and Z = slope of the bank (horizontal to vertical).

Table 2-17(c). Sequence of rainfall-runoff computations for hurricane-prone watershed

1. Develop hydrograph for Subwatershed 1.
2. Develop hydrograph for Subwatershed 2.
3. Combine hydrographs for Subwatersheds 1 and 2.
4. Lag the combined hydrograph by 2 hours to reflect flow to the end of Subwatershed 3.
5. Develop PMF hydrograph for Subwatershed 3.
6. Combine the hydrograph for Subwatersheds 1 and 2 with that for Subwatershed 3.
7. Lag the combined hydrograph by 1.5 hours to reflect flow to the end of Subwatershed 4.
8. Develop hydrograph for Subwatershed 4.
9. Develop hydrograph for Subwatershed 5.
10. Combine hydrograph for Subwatersheds 1, 2, and 3 with those for Subwatersheds 4 and 5.
11. Lag the combined hydrograph by 3 hours to reflect flow to the end of Subwatershed 6.
12. Develop hydrograph for Subwatershed 6.
13. Develop hydrograph for Subwatershed 7.
14. Combine hydrograph for Subwatersheds 1, 2, 3, 4, and 5 with those from Subwatersheds 6 and 7.

Table 2-17(d). PMP depths for 5- to 30-min durations

Duration (min)	Ratio to 1-hr value
5	0.29
10	0.45
15	0.57
30	0.79

Source: NOAA (1973).

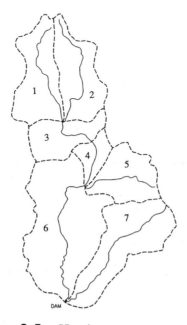

Figure 2-5. Hurricane-prone watershed

Example 2-10: A bridge is designed for the 100-yr flood ($p = 0.01$). What is the probability of (1) no overtopping and (2) one or more overtopping events during the next 20 and 100 yr.

Solution:

- Probability of at least one overtopping event, $P (\geq 1)$, in 20 yr:

    $1 - (1 - 0.01)^{20} = 0.18$ or 18%

- Probability of at least one overtopping event, $P (\geq 1)$, in 100 yr:

    $1 - (1 - 0.01)^{100} = 0.63$ or 63%

- Probability of no overtopping event, P (none), in 20 yr:

    $(1 - 0.01)^{20} = 0.82$ or 82%

- Probability of no overtopping event, P (none), in 100 yr:

    $(1 - 0.01)^{100} = 0.37$ or 37%

Thus, there is an 18% chance that the bridge may be overtopped at least once during the next 20 yr.

In certain cases, it is desirable to estimate the probability of x number of T-yr events occurring in a period of n years. For these computations, Poisson's distribution can be used. Thus, (Haan 1977):

$$p(\lambda,x) = \lambda^x e^{-\lambda}/x! \qquad (2\text{-}56)$$

where

   $x!$ = factorial x

   $\lambda = n/T$

   $p(\lambda,x)$ = probability that x number of T-yr events occur in n years

---

Example 2-11: The risk of damages to a flood control facility is to be assessed for the flood season during which no time may be available for repairs. Estimate the probability that three 10-yr floods would occur in that period of 6 mo when repairs may be difficult.

Solution:

$$n = 0.5 \text{ yr}; T = 10 \text{ yr}; \lambda = 0.5/10 = 0.05; \text{ and } x = 3.$$

Therefore,

$$p(0.05,3) = (0.05)^3 \exp(-0.05)/(3 \times 2 \times 1) = 2 \times 10^{-5}.$$

Benefit-Cost Analysis

Benefit-cost analyses are required to perform comparative evaluation of alternative water resources development projects. An approximate procedure is illustrated here.

Example 2-12: An erosion and flood control project is proposed for a stream passing through an urbanized area. Perform a simplistic benefit-cost analysis and identify the optimum level of flood protection for which the project should be designed. Ignore decision factors other than monetary benefits and costs. Use a discount rate of 6% to convert present-day capital cost to annual values.

Solution: Preliminary estimates were prepared for alternative projects designed to provide different levels of protection using a project life of 50 years. The benefit-cost computations are included in Tables 2-18(a) and (b). The operation and maintenance cost decreases as the cost (size) of the project increases and the indicated costs are adjusted to include price escalation from year to year.

$$\text{Annual cost} = Ci(1 + i)^n / [(1 + i)^n - 1]$$

where

C = present-day capital cost

i = discount rate

The estimated benefit-cost ratios suggest that, based on costs alone, it may be desirable to design the project for 70- to 100-yr return period floods.

Table 2-18(a). Computation of expected damage protection or benefits

Design return period (T, yr)	Probability ($p = 1/T$)	Incremental probability or frequency (Δp)	Damage protection (D, million $)	Incremental expected damage protection ($D\Delta p$, million $)	Total expected damage protection (million $)
≤1	1.00	—	—	—	—
2	0.50	0.50	0.10	0.05	0.05
5	0.20	0.30	0.20	0.06	0.11
10	0.10	0.10	0.30	0.03	0.14
20	0.05	0.05	0.40	0.02	0.16
25	0.04	0.01	0.50	0.005	0.165
30	0.033	0.007	0.60	0.0042	0.1692
40	0.025	0.008	0.80	0.0064	0.1756
50	0.020	0.005	1.00	0.0050	0.1806
70	0.014	0.006	1.50	0.0090	0.1896
100	0.010	0.004	1.7	0.0068	0.1964
150	0.0067	0.0033	2.3	0.0076	0.2040
300	0.0033	0.0034	3.5	0.0119	0.2159
>300	≅0	0.0033	4.0	0.0132	0.2291

Table 2-18(b). Computation of benefit-cost ratio

Design return period (T, yr)	Total expected damage protection (million $)	Present-day capital cost (million $)	Annual cost (million $)	Average annual operation & maintenance cost[a] (million $)	Total annual cost (million $)	Benefit/cost
≤1	—	—	—	—	—	—
2	0.05	1.103	0.07	0.07	0.14	0.36
5	0.11	1.261	0.08	0.06	0.14	0.79
10	0.14	1.576	0.10	0.05	0.15	0.93
20	0.16	1.734	0.11	0.045	0.155	1.03
25	0.165	1.891	0.12	0.04	0.160	1.03
30	0.1692	2.049	0.13	0.034	0.164	1.03
40	0.1756	2.128	0.135	0.030	0.165	1.06
50	0.1806	2.285	0.145	0.025	0.170	1.06
70	0.1896	2.443	0.155	0.020	0.175	1.08
100	0.1964	2.522	0.160	0.017	0.177	1.11
150	0.2040	3.468	0.20	0.01	0.21	0.97
300	0.2159	4.729	0.30	0.01	0.31	0.70
>300	0.2291	7.881	0.50	0.01	0.51	0.45

[a]Includes price escalation from year to year.

Reservoir Operation Studies

Reservoir operation studies are required for sizing reservoirs to meet specified demands for municipal water supply, irrigation, recreation, flood control, hydropower, and instream flows. Several types of hydrologic analyses may be required to perform reservoir operations studies. Commonly required analyses are described here.

Generation of Streamflow Sequences

If the objective of the study is to evaluate the adequacy of available surface water supply on a daily basis, then daily streamflow data have to be collected or synthetically generated. This may be required for run-of-river systems where stream water is used or diverted depending on the quantity available in the river at that point in time, e.g., for diversion dams with little storage capacity. For reservoirs with relatively large conservation storage to absorb diurnal fluctuations in streamflows, monthly streamflow data may be adequate. If reliable data are available, those data must be used. However, in many cases, only limited data are available at the time of reservoir planning. In such cases, streamflow generation models may be used to extend available data and generate daily or monthly streamflow sequences for 50 to 200 years or so.

In some cases, only monthly data are available and sequences of only monthly streamflows can be generated, but daily streamflow data are required for reservoir operation studies. In such cases, daily streamflow data for streams in similar watersheds may be examined. Similarity between watersheds may be judged by size of drainage area, time of concentration, precipitation pattern, climatic region, baseflow patterns (i.e., low streamflow patterns), etc. The percentage of monthly streamflow that occurred during each day of a month in an average year may be estimated for the stream for which daily streamflow data are available. This distribution may be used to disaggregate monthly streamflows into daily values. Note that this distribution is useful only for preliminary planning and must be modified as soon as appropriate data are available.

Deterministic Methods

These involve simulation of continuous hourly, daily, or monthly streamflows using continuous hourly rainfall data for the watershed along with other parameters governing temporal variation of soil moisture conditions, evapotranspiration, percolation, etc. (Prakash and Dearth 1990). Some of the models that can be used for this purpose are the National Weather Service (NWS) River Forecast System (NWS 1998); Hydrologic Simulation Program FORTRAN (HSPF) (USEPA 1991a); Precipitation-Runoff Modeling System (PRMS) (USGS 1983); and Streamflow Synthesis and Reservoir Regulation (SARR) (USACE 1986). The respective user's manuals must be studied for the use of these models. Such analyses may be treated as special studies.

Regression Analysis

Regression analysis is useful for cases where daily or monthly streamflow data with concurrent daily or monthly rainfall, mean air temperature, and other parameters that may affect surface runoff are available for a relatively short period of time (e.g., 5 to 10 years) and data for rainfall, temperature, and other relevant parameters are available for a longer period of time (e.g., 25 to 50 years or more). In these cases, regression equations may be developed using the 5 to 10 year data with streamflow as the dependent variable. The correlation coefficients and standard errors for regressions with untransformed variables and logarithms of variables should be compared and the best-fitting regression equation should be used to generate additional streamflow data, using additional available data for the independent variables.

Example 2-13: For evaluating the feasibility of a water supply project, monthly streamflows have to be estimated for a period of 24 years at an ungauged stream at a point where the drainage area is 38 km². However, only monthly rainfall data are available for the watershed for the above period of 24 years. Monthly streamflow and rainfall data are available for another stream in the vicinity with similar watershed characteristics for a period of 13 years at a point where the drainage area is 40.66 km². The average basin slopes, altitudes, hydrologic soil groups, and mean annual precipitation in the ungauged and gauged watersheds are 0.013 and 0.02 m/m, 68 and 100 m, A to B and A to B, and 500 and 550 mm, respectively.

Solution: The drainage areas for the two streams are different and the watershed characteristics are similar. The following is an approximate approach to generate monthly streamflows:

- Estimate monthly streamflows per square kilometer of drainage area for the gauged stream.

- Develop a linear regression equation between monthly streamflows per square kilometer and monthly rainfall for the gauged stream. For comparison, develop a second linear regression equation between the logarithms of these two variables.
- Use the equation with the higher correlation coefficient to estimate monthly streamflows per unit square kilometer from known monthly rainfall data for the ungauged stream.

Monthly streamflow and rainfall data for the gauged watershed are shown in Table 2-19(a). Liner regression between Q/A and P and log (Q/A) and log (P) resulted in

$$Q/A = -0.00966 + 9.14 \times 10^{-5}, \quad r^2 = 0.71 \tag{i}$$

$$\text{Log } (Q/A) = -6.27588 + 1.732749 \log (P), \quad r^2 = 0.85$$

or

$$Q/A = 0.000000530 \, P^{1.732749} \tag{ii}$$

The second equation is adopted because it has a higher correlation coefficient:

$$r = \sqrt{0.85} = 0.92.$$

The known monthly rainfall values and estimated monthly flows for the ungauged watershed ($A = 38 \text{ km}^2$) using the second equation are shown in Table 2-19(b).

Statistical and Stochastic Methods

These are useful for situations where only daily or monthly streamflow data are available for a limited period of time (e.g., 10 years or more) with little information about con-

Table 2-19(a). Monthly streamflow and rainfall data for gauged watershed

Year	Rainfall P (mm)	Log (P)	Streamflow Q (m³/s)	Q/A (m³/s/km²)	Log (Q/A)
1	45.5	1.6577	0.0453	0.0011	−2.9531
2	351.7	2.5461	0.5379	0.0132	−1.8784
3	253.0	2.4031	0.2831	0.0070	−2.1572
4	323.0	2.5091	0.2690	0.0066	−2.1795
5	393.4	2.5949	0.3398	0.0084	−2.078
6	96.1	1.9829	0.0275	0.0007	−3.1707
7	147.0	2.1671	0.0849	0.0021	−2.6801
8	68.3	1.8346	0.0311	0.0008	−3.1158
9	324.5	2.5112	0.7928	0.0195	−1.7100
10	103.3	2.0139	0.0340	0.0008	−3.0780
11	316.6	2.5005	1.2458	0.0306	−1.5137
12	628.9	2.7986	2.7180	0.0668	−1.1749
13	52.1	1.7166	0.0235	0.00068	−3.2381

A = drainage area in km².

Table 2-19(b). Known monthly rainfall and estimated monthly flows

Year	P (mm)	Q (m³/s)
1	5.33	0.0004
2	112.52	0.0721
3	96.52	0.0553
4	29.21	0.0070
5	193.55	0.1846
6	171.70	0.1500
7	244.09	0.2760
8	156.97	0.1284
9	72.64	0.0338
10	94.23	0.0530
11	33.78	0.0090
12	234.19	0.2569
13	82.80	0.0424
14	152.40	0.1220
15	216.15	0.2236
16	16.76	0.0027
17	27.43	0.0062
18	35.05	0.0096
19	186.94	0.1739
20	28.96	0.0069
21	142.24	0.1083
22	12.45	0.0016
23	166.88	0.1428
24	127.00	0.0890

current precipitation. Sophisticated stochastic modeling may be treated as a special study for which the reader may refer to other publications (e.g., Quimpo 1968; Fiering and Jackson 1971; Yevjevich 1972b, 1982). For generating long sequences of monthly streamflows using monthly streamflow data for a limited period, the HEC-4 model (USACE 1971a) is a convenient tool. This model has the additional capability to fill in flow values for months for which data are missing.

Flow Duration Analysis

Flow duration curve is a plot of streamflows (on y-axis) against percentage of times that flow is equaled or exceeded (on x-axis). It is useful to determine the dependability of daily or monthly streamflows at a given location to meet a specified water demand on a daily or monthly basis. Daily computations may be required for run-of-river diversions to meet hydropower or irrigation requirements. Monthly computations may be required for reservoirs with significant storage so that daily variations in streamflows are not important, so long as the total monthly inflows are maintained. Computational procedure for flow duration curves is illustrated in Example 2-15.

Mass Curve (Rippl) Analysis

Mass curve analysis is useful in determining preliminary size of a reservoir required to satisfy a simplified pattern (i.e., time sequence) of water demand.

Mass curves include two plots on the same graph. The first curve is a plot of cumulative streamflows or inflows (y-axis) against time (x-axis), and the second is a plot of cumulative demand on y-axis against time on x-axis. If the demand curve is a straight line (i.e., demand is constant), then select a point on the inflow curve at the beginning of the longest low-flow period and draw a line tangent to the mass flow curve and parallel to the demand curve starting from this point. The maximum vertical distance between this line and the inflow curve gives the required storage. In certain cases, the demand may be variable with time. For instance, water demand may be low during the initial and final months of the construction of a major project and may be high during the peak construction period. A slightly modified procedure is used for this case. The computational procedure is illustrated in Example 2-14.

Example 2-14: Five years of monthly streamflow data are available at the location of a proposed dam (Table 2-20(a)). The reservoir is proposed to meet a constant demand of 8 m^3/s including evaporation and other losses. Determine the size of conservation storage required to meet the above demand. Assume that the reservoir size is such that diurnal fluctuations in streamflows will not impact the capability of the reservoir to meet the required demand. Also, determine the conservation storage to meet the variable demand shown in column (3) of Table 2-20(b).

Solution: For the case with constant demand, flows, cumulative flows, cumulative demand, flow-demand, and cumulative flow-cumulative demand are shown in columns (2) through (6) of Table 2-20(c). Then,

- Identify the pairs of successive peaks (*P*) and troughs (*T*) in column (6).
- Compute the differences: $P1 - T1$; $P2 - T2$; $P3 - T3$, etc.

The maximum of these differences is the required storage. Note that some peaks and troughs with relatively small differences $(P - T)$ are not included in column (6). The maximum difference is between $P2$ and $T2$ and its value is 39.37 m^3/s-mo. To verify that this storage is adequate, reservoir storage and spills during each month are shown in columns (7) and (8).

The graphical solution for a constant demand of 8 m^3/s is illustrated in Figures 2-6(a) and (b). Figure 2-6(a) includes plots of cumulative flows and demand against time. The peak preceding the longest low-flow period is denoted by *P*. A tangent to the mass curve of flows at *P* drawn parallel to the mass curve of demand is shown in Figure 2-6(b). The largest vertical intercept gives the storage, which can be scaled to be about 39.37 m^3/s-mo.

The flows, demand, flow-demand, cumulative flows, cumulative demand, and cumulative flow-cumulative demand for the case with variable demand are shown in columns (2) through (7) in Table 2-20(b). As in the previous case, successive peaks and troughs are identified, and the differences—$P1 - T1$; $P2 - T2$; $P3 - T3$; and $P4 - T4$—are computed. The maximum difference is between $P2$ and $T2$ and its value is 64.33 m^3/s-mo. This gives the required storage. To verify that this storage is adequate, reservoir storage and spills during each month are shown in columns (8) and (9).

Table 2-20(a). Monthly streamflow data (m³/s)

Year	Jan.	Feb.	March	April	May	June	July	Aug.	Sept.	Oct.	Nov.	Dec.
1	8.77	4.91	5.52	11.13	22.48	11.50	9.51	14.10	20.15	35.92	6.65	4.60
2	24.53	9.47	3.35	4.59	3.85	4.32	9.27	7.04	36.47	6.45	6.15	5.28
3	5.72	5.36	3.61	3.47	5.78	4.28	4.90	1.94	7.39	4.30	15.26	18.97
4	9.55	5.83	4.50	8.22	16.86	12.28	10.75	10.46	9.92	9.01	16.39	16.16
5	6.16	9.46	8.37	7.29	19.20	23.97	26.77	10.38	5.67	4.69	3.91	15.95

Table 2-20(b). Mass curve analysis with variable demand

(1)	(2)	(3)	(4)	(5)	(6)	(7)	(8)	(9)
Month	Flow (m³/s-mo)	Demand (m³/s-mo)	Flow-demand (m³/s-mo)	Cum. flow (m³/s-mo)	Cum. demand (m³/s-mo)	Cum. flow-cum. demand (m³/s-mo)	Storage (m³/s-mo)	Spill (m³/s-mo)
1	0	0	0	0	0	0	64.33	0
2	8.77	4	4.77	8.77	4	4.77	64.33	4.77
3	4.91	5	−0.09	13.68	9	4.68	64.24	0
4	5.52	6	−0.48	19.2	15	4.2	63.76	0
5	11.13	7	4.13	30.33	22	8.33	64.33	3.56
6	22.48	7	15.48	52.81	29	23.81	64.33	15.48
7	11.5	9	2.5	64.31	38	26.31	64.33	2.5
8	9.51	10	−0.49	73.82	48	25.82	63.84	0
9	14.1	8	6.1	87.92	56	31.92	64.33	5.61
10	20.15	6	14.15	108.07	62	46.07	64.33	14.15
11	35.92	8	27.92	143.99	70	73.99	64.33	27.92
12	6.65	6	0.65	150.64	76	74.64	64.33	0.65
13	4.6	3	1.6	155.24	79	76.24	64.33	1.6
14	24.53	4	20.53	179.77	83	96.77	64.33	20.53
15	9.47	5	4.47	189.24	88	101.24 $P1$	64.33	4.47
16	3.35	6	−2.65	192.59	94	98.59	61.68	0
17	4.59	7	−2.41	197.18	101	96.18	59.27	0
18	3.85	7	−3.15	201.03	108	93.03	56.12	0
19	4.32	9	−4.68	205.35	117	88.35	51.44	0
20	9.27	10	−0.73	214.62	127	87.62	50.71	0
21	7.04	8	−0.96	221.66	135	86.66 $T1$	49.75	0
22	36.47	6	30.47	258.13	141	117.13	64.33	15.89
23	6.45	8	−1.55	264.58	149	115.58	62.78	0
24	6.15	6	0.15	270.73	155	115.73	62.93	0
25	5.28	3	2.28	276.01	158	118.01	64.33	0.88
26	5.72	4	1.72	281.73	162	119.73	64.33	1.72
27	5.36	5	0.36	287.09	167	120.09 $P2$	64.33	0.36
28	3.61	6	−2.39	290.7	173	117.7	61.94	0
29	3.47	7	−3.53	294.17	180	114.17	58.41	0

(*continued*)

Table 2-20(b). (*Continued*)

(1)	(2)	(3)	(4)	(5)	(6)	(7)	(8)	(9)
Month	Flow (m³/s-mo)	Demand (m³/s-mo)	Flow-demand (m³/s-mo)	Cum. flow (m³/s-mo)	Cum. demand (m³/s-mo)	Cum. flow-cum. demand (m³/s-mo)	Storage (m³/s-mo)	Spill (m³/s-mo)
30	5.78	14	−8.22	299.95	194	105.95	50.19	0
31	4.28	18	−13.72	304.23	212	92.23	36.47	0
32	4.9	20	−15.1	309.13	232	77.13	21.37	0
33	1.94	15	−13.06	311.07	247	64.07	8.31	0
34	7.39	12	−4.61	318.46	259	59.46	3.7	0
35	4.3	8	−3.7	322.76	267	55.76 *T2*	0	0
36	15.26	6	9.26	338.02	273	65.02	9.26	0
37	18.97	3	15.97	356.99	276	80.99	25.23	0
38	9.55	4	5.55	366.54	280	86.54	30.78	0
39	5.83	5	0.83	372.37	285	87.37	31.61	0
40	4.5	6	−1.5	376.87	291	85.87	30.11	0
41	8.22	7	1.22	385.09	298	87.09	31.33	0
42	16.86	14	2.86	401.95	312	89.95 *P3*	34.19	0
43	12.28	18	−5.72	414.23	330	84.23	28.47	0
44	10.75	20	−9.25	424.98	350	74.98	19.22	0
45	10.46	15	−4.54	435.44	365	70.44	14.68	0
46	9.92	12	−2.08	445.36	377	68.36 *T3*	12.60	0
47	9.01	8	1.01	454.37	385	69.37	13.61	0
48	16.39	6	10.39	470.76	391	79.76	24.0	0
49	16.16	3	13.16	486.92	394	92.92	37.16	0
50	6.16	4	2.16	493.08	398	95.08	39.32	0
51	9.46	5	4.46	502.54	403	99.54	43.78	0
52	8.37	6	2.37	510.91	409	101.91	46.15	0
53	7.29	7	0.29	518.2	416	102.2	46.44	0
54	19.2	14	5.2	537.4	430	107.4	51.64	0
55	23.97	18	5.97	561.37	448	113.37	57.61	0
56	26.77	20	6.77	588.14	468	120.14 *P4*	64.33	0.5
57	10.38	15	−4.62	598.52	483	115.52	59.71	0
58	5.67	12	−6.33	604.19	495	109.19	53.38	0
59	4.69	8	−3.31	608.88	503	105.88	50.07	0
60	3.91	6	−2.09	612.79	509	103.79 *T4*	47.98	0
61	15.95	3	12.95	628.74	512	116.74	60.93	0

P = peak; T = trough.

Estimation of 7-day Average 10-yr Low Flow

Many times, flow diversions from streams are permissible only if flows equal to or more than the 7-day, 10-yr (7Q10) low flows of the stream are left in the stream to meet instream flow requirements. The 7Q10 low flow is also used to evaluate the impacts of wastewater discharges on the water quality of the receiving stream. The 7-day average implies an average of all successive combinations of 7 days of flows in a year, i.e., average of flows from day 1 to day 7, day 2 to day 8, day 3 to day 9, etc. This is also known as the 7-day moving average. The lowest 7-day average flow is computed for each year for which records are available. The mean,

Table 2-20(c). Mass curve analysis with constant demand[a]

(1)	(2)	(3)	(4)	(5)	(6)	(7)	(8)
Month	Flow (m³/s-mo)	Cum. flow (m³/s-mo)	Cum. demand (m³/s-mo)	Flow-demand (m³/s-mo)	Cum. flow-cum. demand (m³/s-mo)	Storage (m³/s-mo)	Spill (m³/s-mo)
1	0	0	0	0	0	39.37	0
2	8.77	8.77	8	0.77	0.77	39.37	0.77
3	4.91	13.68	16	−3.09	−2.32	36.28	0
4	5.52	19.20	24	−2.48	−4.80	33.80	0
5	11.13	30.33	32	3.13	−1.67	36.93	0
6	22.48	52.81	40	14.48	12.81	39.37	12.04
7	11.50	64.31	48	3.50	16.31	39.37	3.50
8	9.51	73.82	56	1.51	17.82	39.37	1.51
9	14.10	87.92	64	6.10	23.92	39.37	6.10
10	20.15	108.07	72	12.15	36.07	39.37	12.15
11	35.92	143.99	80	27.92	63.99	39.37	27.92
12	6.65	150.64	88	−1.35	62.64	38.02	0
13	4.60	155.24	96	−3.40	59.24	34.62	0
14	24.53	179.77	104	16.53	75.77	39.37	11.78
15	9.47	189.24	112	1.47	77.24 *P*1	39.37	1.47
16	3.35	192.59	120	−4.65	72.59	34.72	0
17	4.59	197.18	128	−3.41	69.18	31.31	0
18	3.85	201.03	136	−4.15	65.03	27.16	0
19	4.32	205.35	144	−3.68	61.35 *T*1	23.48	0
20	9.27	214.62	152	1.27	62.62	24.75	0
21	7.04	221.66	160	−0.96	61.66	23.79	0
22	36.47	258.13	168	28.47	90.13 *P*2	39.37	12.89
23	6.45	264.58	176	−1.55	88.58	37.82	0
24	6.15	270.73	184	−1.85	86.73	35.97	0
25	5.28	276.01	192	−2.72	84.01	33.25	0
26	5.72	281.73	200	−2.28	81.73	30.97	0
27	5.36	287.09	208	−2.64	79.09	28.33	0
28	3.61	290.70	216	−4.39	74.70	23.94	0
29	3.47	294.17	224	−4.53	70.17	19.41	0
30	5.78	299.95	232	−2.22	67.95	17.19	0
31	4.28	304.23	240	−3.72	64.23	13.47	0
32	4.90	309.13	248	−3.10	61.13	10.37	0
33	1.94	311.07	256	−6.06	55.07	4.31	0
34	7.39	318.46	264	−0.61	54.46	3.70	0
35	4.30	322.76	272	−3.70	50.76 *T*2	0	0
36	15.26	338.02	280	7.26	58.02	7.26	0
37	18.97	356.99	288	10.97	68.99	18.23	0
38	9.55	366.54	296	1.55	70.54	19.78	0
39	5.83	372.37	304	−2.17	68.37	17.61	0
40	4.50	376.87	312	−3.50	64.87	14.11	0
41	8.22	385.09	320	0.22	65.09	14.33	0
42	16.86	401.95	328	8.86	73.95	23.19	0

(*continued*)

Table 2-20(c). (*Continued*)

Month (1)	Flow (m³/s-mo) (2)	Cum. flow (m³/s-mo) (3)	Cum. demand (m³/s-mo) (4)	Flow- demand (m³/s-mo) (5)	Cum. flow- cum. demand (m³/s-mo) (6)	Storage (m³/s-mo) (7)	Spill (m³/s-mo) (8)
43	12.28	414.23	336	4.28	78.23	27.47	0
44	10.75	424.98	344	2.75	80.98	30.22	0
45	10.46	435.44	352	2.46	83.44	32.68	0
46	9.92	445.36	360	1.92	85.36	34.60	0
47	9.01	454.37	368	1.01	86.37	35.61	0
48	16.39	470.76	376	8.39	94.76	39.37	4.63
49	16.16	486.92	384	8.16	102.92	39.37	8.16
50	6.16	493.08	392	−1.84	101.08	37.53	0
51	9.46	502.54	400	1.46	102.54	38.99	0
52	8.37	510.91	408	0.37	102.91	39.36	0
53	7.29	518.20	416	−0.71	102.20	38.65	0
54	19.20	537.40	424	11.20	113.40	39.37	10.48
55	23.97	561.37	432	15.97	129.37	39.37	15.97
56	26.77	588.14	440	18.77	148.14	39.37	18.77
57	10.38	598.52	448	2.38	150.52 *P3*	39.37	2.38
58	5.67	604.19	456	−2.33	148.19	37.04	0
59	4.69	608.88	464	−3.31	144.88	33.73	0
60	3.91	612.79	472	−4.09	140.79 *T3*	29.64	0
61	15.95	628.74	480	7.95	148.74	37.59	0

^aConstant demand = 8m³/s.
P = peak; T = trough.

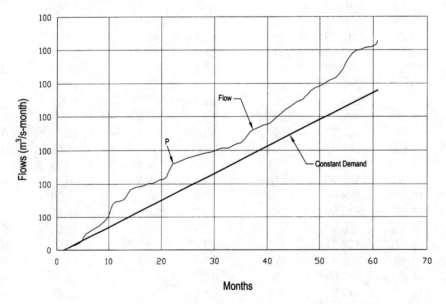

Figure 2-6(a). Mass curves of flow and demand

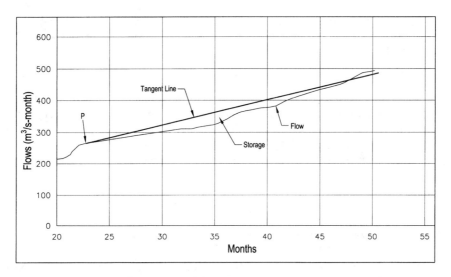

Figure 2-6(b). Mass curve analysis

standard deviation, and skew coefficient of the resulting lowest annual 7-day average low flows are computed, as described in the section of this chapter entitled "Statistical Analysis of Available Data." Using normal, log-normal, or log Pearson Type III distributions, the T-year low flow is computed using the equation,

$$Q_T = \overline{X} - Ks \quad (2\text{-}21)$$

where K is as defined in the section of this chapter entitled "Statistical Analysis of Available Data." Gumbel used the Extreme Value Type III or Bounded Exponential distribution of the smallest values for analyzing annual low streamflows (Yevjevich 1972a, 1997). Analytical estimation of K values for this distribution is relatively complex. For most practical cases, the values may be plotted on plotting papers for Extreme Value Type III distribution and low streamflows for the desired return period obtained by eye judgment.

Example 2-15: Daily streamflow data for a low-flow year for a stream are shown in Table 2-21(a). It is proposed to divert water from this stream to meet a uniform daily demand of 7.0 m³/s with minimum instream flow requirements of 1.0 m³/s. Determine the dependability of this source to meet the water demand of the municipality.

Solution: Develop a flow duration curve for the low-flow period following these computational steps:

- Sort all daily flows in descending order of magnitude.
- Assign a rank number to each daily value (i.e., rank number 1 to the highest flow and so on). If the same flow occurs two or more times, then each value is assigned a different rank in succession.

- Divide the rank of each value by the total number of days for which data are available and multiply the quotient by 100 to obtain the percentage of times that flow is equaled or exceeded.
- Plot flows (on y-axis) against percentage exceedance (on x-axis) as shown in Figure 2-7.

Table 2-21(b) shows 27 selected flows, their ranks, and percentage of times each flow is equaled or exceeded. It may be seen that the dependability of the stream to meet a total daily demand of 8 m³/s is 66.03%. The available streamflow will be short of the demand during 33.97% (100 – 66.03%) of days in a low-flow year.

Table 2-21(a). Daily streamflow data (1 year) (m³/s)

Day	Jan.	Feb.	March	April	May	June	July	Aug.	Sept.	Oct.	Nov.	Dec.
1	14.39	18.75	20.84	11.33	10.27	17.4	5.66	17.15	11.45	7.67	7.5	12.94
2	33.74	6.06	9.79	5.96	6.97	19.17	4.58	2.84	15.08	17.55	5.69	8.49
3	10.8	26	6.79	18.5	13.09	18.31	7.67	2.56	15.22	4.66	5.06	14.19
4	8.93	10.81	5.04	19.01	7.42	11.2	21.19	8.44	21.92	5.35	11.97	14.76
5	8.15	9.23	19.05	10.19	20.36	10.69	7.8	16.19	15.84	2.63	16.45	25.93
6	11.06	5.47	12.61	15.2	26.95	8.14	4.88	4.26	24.07	3	8.77	27.07
7	6.58	8.74	20.02	7.02	21.99	5.14	6.01	4.61	24.91	6.44	9.75	7.77
8	10.98	5.42	17.55	7.78	25.91	4.41	9.66	8.8	16.22	4.03	12.91	6.28
9	9.35	28.34	33.64	3.92	20.06	3.53	8.69	8.4	27.53	7.47	5.58	3.92
10	7.38	25.25	26.26	3.5	16.04	3.93	24.16	11.04	16.81	10.39	7.73	3.42
11	12.78	23.68	16.07	43.92	11.98	2.66	20.44	9.02	10.57	3.32	2.85	31
12	13.26	8.64	13.31	31.42	24.46	2.82	16.05	8.66	11.45	24.25	2.67	11
13	14.9	47.05	7.78	6.93	13.12	4.01	7.72	31.9	7.48	26.6	3.69	8.9
14	7.82	9.94	12.57	9.38	11.38	6	7.14	14.14	8.56	19.56	5.66	9.85
15	5.61	8.65	8.16	9.21	9.35	9.41	11.97	7.55	3.45	17.22	3.26	6.09
16	4.53	7.96	17.51	10.51	7.66	23.06	15.44	5.44	2.76	11.25	26	3.34
17	23.62	6.18	17.09	8.47	6.11	41.72	11.4	4.39	17.9	6.7	21.82	8.16
18	39.37	7.67	41.75	8.06	6.63	10.97	10.81	3.98	38.45	5.17	23.62	3.76
19	66.9	7.34	46.6	12.25	5.95	12.81	19.72	15.26	4.96	7.24	28.77	9.18
20	13.37	14.16	9.64	43	7.81	15.83	9.31	8.35	6.99	4.21	20.93	7.79
21	73.1	8	31.34	17.42	9.84	10.73	9.61	17.3	7.29	11.82	32.67	14.7
22	35.69	8.91	11.18	14.83	14.72	10.26	10.65	33.88	5.12	8.15	26.04	22.19
23	27.85	11.09	58.18	11.23	10.88	10.4	9.47	22.6	9.68	3.69	14.68	9.55
24	9.94	23.71	13.65	12.07	13.14	7.54	3.3	18.98	8.71	7.1	11.06	2.93
25	8.59	17.22	5.77	15.89	15.06	7.12	3.73	19.02	7.02	6.85	9.78	7.79
26	8.8	8.04	4.76	30.15	11.85	6.43	3.35	16.31	3.65	11.51	18.35	8.13
27	4.51	16.37	3.81	12.01	12.56	17.98	2.38	24.04	3.73	11.7	21.21	6.29
28	12.29	10.4	4.57	13.37	10.25	8.37	5.49	13.1	4.12	30.48	20.12	10.96
29	9.17		9.47	10.27	8.72	13.3	7.47	23.67	7.66	12.54	11.57	5.01
30	8.28		4.83	11.97	7.07	7.03	23.13	19.55	23.46	6.05	5.58	5.53
31	13.32		10.12		5.74		19.35	13.36		7.92		8.1

Table 2-21(b). Flow duration table

Q (m³/s)	Rank	% Exceedance
73.10	1	0.27
33.64	15	4.11
26.26	30	8.22
23.67	45	12.33
20.44	60	16.44
18.50	75	20.55
16.81	90	24.66
15.20	105	28.77
13.37	120	32.88
12.57	135	36.99
11.51	150	41.10
10.98	165	45.21
10.27	180	49.59
9.64	195	53.42
8.93	210	57.53
8.49	225	61.64
8.04	240	65.75
8.00	241	66.03
7.67	255	70.14
7.14	270	73.97
6.44	285	78.08
5.69	300	82.19
5.06	315	86.30
4.26	330	90.41
3.65	345	94.52
2.76	360	98.63
2.38	365	100.00

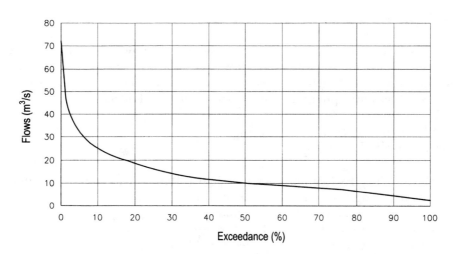

Figure 2-7. Flow duration curve

Example 2-16: Use the daily streamflow data of Example 2-15 to estimate the 7-day average low flow for that year. The estimated 7-day average lowest flows for the previous 15 yr for that stream are given in Table 2-22(a). Estimate the 7-day 10-yr low flow for that stream.

Solution: The steps of computation are as follows:

- Record daily flows given in Table 2-21(a) in a column of a spreadsheet.

- Using statistical functions of a standard software package (e.g., Excel), compute 7-day moving averages and insert in the next column. This gives 359 7-day average values for 365 daily flows. A few selected values are shown in Table 2-22(b).

- Find the minimum of these 359 values, which is found to be 3.786 m^3/s. This is the 7-day average low flow for the year for which data are given in Table 2-21(a).

- Use this value along with the given values for the preceding 15 yr to compute the mean, standard deviation, and skew coefficient for the 16 values. The results are shown in Table 2-22(c).

Using normal distribution as described in the section "Statistical Analysis of Available Data" in this chapter:

$$\text{7-day, 10-yr low flow (7Q10)} = 4.16 - 1.28155 \times 0.95 = 2.94 \text{ m}^3/\text{s}$$

Using log-Pearson Type III distribution:

$$K \text{ (by Eq. (2-22))} = (2/0.341)[\{(1.28155 - (0.341/6)) \times (0.341/6) + 1\}^3 - 1] = 1.312$$

$$\log [7Q10] = 0.609 - (1.312 \times 0.096) = 0.483$$

$$7Q10 = 3.04 \text{ m}^3/\text{s}$$

Thus, a value of 3.0 m^3/s appears reasonable.

Hydrologic Models

With continuing advances in computer technology, there has been a proliferation of hydrologic models that can interact with various graphics and visual packages, e.g., Surfer, CADD, GIS. A number of these models are available from various private vendors. A brief discussion on some commonly used public-domain hydrologic models follows.

Models for Development and Routing of Storm Runoff Hydrographs

- **HEC-1, Flood Hydrograph Package** (USACE 1991a): This model computes surface runoff hydrographs for given rainfall and snowmelt events with known hydraulic characteristics of watersheds, combines hydrographs of subwatersheds at prescribed locations, and routes them through reservoirs and channels in a given network. It can also perform simplified dam-break analysis and economic analysis for flood damages.

Table 2-22(a). 7-day average annual low flows

Year	Flow (m³/s)
1	4.111
2	3.212
3	5.313
4	6.005
5	2.891
6	2.976
7	3.505
8	3.987
9	3.831
10	5.892
11	4.905
12	4.303
13	4.550
14	3.525
15	3.761
16	a

[a] Value to be determined by moving average analysis of given data.

Table 2-22(b). Computation of 7-day average low flow

Daily flow (m³/s)	7-day moving average (m³/s)
14.39	
33.74	
10.80	
8.93	
8.15	
11.06	
6.58	13.379
10.98	12.891
9.35	9.407
—	—
—	—
—	—
5.14	12.864
4.41	11.009
3.53	8.774
3.93	6.720
2.66	5.500
2.82	4.376
4.01	3.786
6.00	3.909
9.41	4.623
—	—
—	—
—	—

Table 2-22(c). Statistical parameters of 7-day average annual low flows

Parameter	Untransformed 7-day average low flows (m³/s)	Logarithms of 7-day average low flows
Number of values	16.00	16.000
Mean	4.16	0.609
Standard deviation	0.95	0.096
Skew coefficient	0.72	0.341

- **HEC-HMS, Hydrologic Modeling System** (USACE 2002): This model is a successor to the HEC-1 model. It is an interactive model and has the capabilities that are available in the HEC-1 model. In addition, it includes:
  - A distributed runoff model for use with distributed precipitation data, such as data available from weather radar.
  - A continuous soil-moisture accounting model used to simulate the long-term response of a watershed to wetting and drying.
  - An automatic calibration package that can estimate certain model parameters and initial conditions, given observations of hydrologic conditions in the watershed.
  - Link to a database management system that permits data storage, retrieval, and connectivity with other analysis tools available from the USACE Hydrologic Engineering Center and other sources.
- **TR-20, Program for Project Formulation Hydrology** (USDA 1983a): This model performs generally the same computations as the HEC-1 model. It has the capability to develop runoff hydrographs, route them through channels and reservoirs, and combine or separate them at confluences. It is designed to perform multiple analyses in a single run so that various alternatives can be evaluated in one run.
- **TR-55, Urban Hydrology for Small Watersheds** (USDA 1986): This model includes simplified procedures to compute storm runoff volume, peak rate of discharge, hydrographs, and storage volumes required for flood-retarding reservoirs in small watersheds. Available support functions include computation of CN, time of concentration, and travel time through a subwatershed.
- **SWMM, Storm Water Management Model** (USEPA 1989a): This model analyzes both water quantity and quality associated with a single storm event or continuous long-term climatic patterns in urban or rural watersheds. It can simulate flows through storm sewers and natural channels, perform storage and subsurface flow routing, and simulate movement of runoff and pollutants from ground surface to the receiving water body through pipe and channel networks and storage treatment units.

Although choice among the aforementioned models should be made based on specific study requirements, it may be convenient to use the HEC-1 or HEC-HMS models in most cases.

Continuous Flow Simulation Models

- **NWSRFS, National Weather Service River Forecast System Model** (NWS 1998): This model includes a snow accumulation and ablation model, soil-moisture accounting model, rainfall-runoff model, and a reservoir routing model to produce a continuous flow hydrograph using continuous (hourly and daily) climatic data.

- **PRMS, Precipitation-Runoff Modeling System** (USGS 1983): PRMS simulates mean daily flows and can generate shorter time hydrographs using continuous climatic data. The model includes soil moisture accounting, channel routing, reservoir routing, sediment transport computations, and groundwater flow computations.

- **HSPF, Hydrologic Simulation Program-FORTRAN** (USEPA 1991a). This model performs basin-scale analysis for one-dimensional channels, including soil-moisture accounting, overland flow, routing through channels, sediment transport, and movement of several water quality constituents using continuous climatic data.

- **SSARR, Streamflow Synthesis and Reservoir Regulation Model** (USACE 1986): SSARR performs year-round surface runoff simulations for flood forecasting and reservoir operations using continuous climatic data. It includes models for watershed hydrology and river and reservoir routing, snow accumulation and melting, infiltration, and interception.

These models are fairly complex and require large volumes of data. Studies involving these models should be treated as special studies.

Other Useful Models

- **HEC-4, Monthly Streamflow Simulation** (USACE 1971a): This model develops sequences of monthly streamflows for desired number of years (e.g., 50 to 100 years or more) using available monthly streamflow data for a smaller number of years (e.g., 10 to 15 years or more). It reconstitutes any missing streamflows within the available data, estimates the statistical characteristics of available or reconstituted data, and generates sequences of streamflows having the same statistical characteristics. This model is fairly simple to use and provides useful data for reservoir operations.

- **HEC-3, Reservoir System Analysis for Conservation and HEC-5- Simulation of Flood Control and Conservation Systems** (USACE 1981, 1991b): These models perform storage routing computations for a system of reservoirs for hydropower generation, water supply, navigation, recreation, flow augmentation, and other seasonal or monthly uses. For HEC-3, the input includes monthly streamflows obtained from models such as HEC-4 or data collected from other sources. HEC-5 can use any time interval from one minute to one month, and it also allows multiple time intervals within a single simulation. Economic computations can be made for hydropower benefits and flood damage evaluation. Reservoir operations can be performed to minimize downstream flooding, evacuate flood control storage as quickly as possible, provide for low-flow requirements, and meet water supply and hydropower requirements. These models are useful for reservoir planning.

- **HEC-FFA, HEC Flood Flow Frequency Analysis and FREQ** (USACE 1995): HEC-FFA performs frequency computations of given annual maximum floods using Water Resources Council Guidelines for Determining Flood Flow Frequency (USWRC 1981). FREQ is a graphics-based LP Type III flood frequency estimation model. It allows the user to calculate unbiased frequency factors and confidence limits for estimates. It also allows alteration of skew coefficients.

CHAPTER 3

HYDRAULIC ANALYSIS

Classification of Flows

The subject of hydraulics includes the study and analysis of the movement of water through natural and man-made watercourses including open channels and pipes. The flow of water is classified in several categories:

1. **Free Surface and Pressure Flows.** Free surface flow occurs when the top of the water surface is at atmospheric pressure. Pressure flow occurs when the water surface is confined by a solid boundary (e.g., flow in closed conduits and through certain culverts and bridges under flooding conditions).
2. **Uniform and Nonuniform Flows.** Uniform flow occurs when the depth of flow is the same at every cross section of the channel and the water surface is parallel to the channel bed. Nonuniform or varied flow occurs when flow enters or leaves the channel along its length and the depth of flow varies from cross section to cross section.
3. **Steady and Unsteady Flows.** The flow is steady when the depth of flow does not change with time and unsteady when it does change from time to time. Steady flow may be uniform, gradually varied, or rapidly varied (e.g., occurrence of hydraulic jump). Unsteady uniform flow is practically impossible; however, unsteady flow may be gradually varied (e.g., in a flood wave) or rapidly varied (e.g., in a bore).
4. **Laminar and Turbulent Flows.** Laminar flow is dominated by viscosity and turbulent flow by inertial or gravitational forces. Whether the flow is laminar or turbulent is indicated by the Reynolds number, R_e, defined as

$$R_e = VR/\nu \tag{3-1}$$

where

V = velocity of flow (m/s)

R = hydraulic radius (m) = A/P

A = cross-sectional area of flow (m²)

P = wetted perimeter (m)

ν = kinematic viscosity of water (m²/s)

If $R_e < 500$, the flow is laminar; for $500 < R_e < 12{,}500$, flow is transitional from laminar to turbulent flow; and for $R_e \gg 12{,}500$, flow is turbulent. The upper limit may vary significantly depending on specific flow conditions.

5. **Subcritical, Critical, and Supercritical Flows.** When gravity forces dominate, flow is characterized by the Froude number, F, defined as:

$$F = V/\sqrt{(gD)} \qquad (3\text{-}2)$$

where

g = gravitational acceleration = 9.81 m²/s

D = hydraulic depth (m) = A/T

T = top width of flow (m)

Flow is subcritical if $F < 1$; critical if $F = 1$; and supercritical if $F > 1$.

Steady Uniform Flow in Open Channels

Uniform flow in open channels or conduits is governed by the following:

1. Manning's Equation

$$V = (1/n)\, R^{2/3} \sqrt{S} \qquad (3\text{-}3)$$

2. Chezy's Equation

$$V = C\sqrt{(R\,S)} \qquad (3\text{-}4)$$

3. Darcy-Weisbach Equation (generally used for pipe flows)

$$h = f\,l\,V^2/(2\,g\,d) \qquad (3\text{-}5)$$

4. Hazen-Williams Equation (mostly used for pipe flows)

$$V = 0.84917\, C_H R^{0.63} S^{0.54} \qquad (3\text{-}6)$$

where

S = energy slope (m/m)

l = length of pipe (m)

h = head loss in the pipe (m)

d = pipe diameter (m) = $4R$

n = Manning's roughness coefficient

C = Chezy's coefficient

f = Darcy-Weisbach friction coefficient

C_H = Hazen-Williams coefficient

The coefficients in Eqs. (3-3) to (3-6) are interrelated:

$$C = R^{1/6}/n \quad \text{or} \quad n = R^{1/6}/C = R^{1/6}\sqrt{f}/\sqrt{(8\,g)} \qquad (3\text{-}7)$$

$$f = 8\,g\,n^2/R^{1/3} \quad \text{or} \quad n = R^{1/6}\sqrt{f}/\sqrt{(8\,g)} \qquad (3\text{-}8)$$

$$C_H = 1.17762\,R^{0.037}/(n\,S^{0.04}) \quad \text{or} \quad n = 1.17762\,R^{0.037}/(C_H\,S^{0.04}) \qquad (3\text{-}9)$$

If grain-size distribution of channel bed material or armor material is known, then Manning's roughness coefficient may be estimated using Strickler's Equation (Simons and Senturk 1976, 1992):

$$n = d_{90}^{1/6}/26 \qquad (3\text{-}10)$$

where d_{90} = particle size (m) than which 90% of the material is finer by weight. Values estimated by Eq. (3-10) may be too low for many field situations. Typical values for field conditions are shown in Table 3-1 (Chow 1959).

Design of Nonerodible Channels

Nonerodible channels include concrete-lined, rock-cut, and armored channels. Usually, such channels are required where ground slope along the channel alignment is steep, available right-of-way is limited, anticipated bed and bank erosion for an unarmored channel is unacceptable, or the channel has to pass through exposed rock. The procedure for the design of such channels is iterative, and the computational steps are listed below:

1. Determine design discharge, Q (m³/s) (e.g., peak flood flow for flood control or diversion channel, or full supply discharge for an irrigation or power canal).

2. Plot ground profile along the channel alignment, mark required invert elevations at the upstream and downstream ends, and determine feasible bed slope for the channel to minimize excavation. Channel bed may not be placed on fill.

3. Estimate maximum permissible velocity, V (m/s), for the type of lining or armor material. Concrete-lined or rock-cut channels may withstand about 5 to 15 m/s depending on the abrasive resistance of concrete or rock; those lined with grass may withstand 1.0 to 2.5 m/s depending on the type of soils and grass. Preliminary estimates of maximum permissible velocity for armor material may be estimated using procedures described in the section of this chapter entitled "Design of Erodible Channels."

4. Estimate Manning's roughness coefficient, n (Table 3-1), and hydraulic radius using:

$$V = (1/n)\,R^{2/3}\sqrt{S} \qquad (3\text{-}3)$$

Table 3-1. Typical values of Manning's roughness coefficient

Type of channel	Manning's roughness coefficient, n
Clean, straight and uniform, excavated in earth	0.016–0.020
Straight and uniform with short grass and some weeds, excavated in earth	0.022–0.033
Winding and sluggish earthen channel with no vegetation	0.023–0.030
Winding and sluggish earthen channel with grass and some weeds	0.025–0.033
Smooth and uniform rock-cut channel	0.025–0.040
Jagged and irregular rock-cut channel	0.035–0.050
Unmaintained channel with dense weeds as high as flow depth	0.050–0.120
Unmaintained channel with dense brush and high stages	0.080–0.140
Natural channel with clean straight bank, full stage, no rifts or deep pools	0.025–0.033
Natural channel with straight bank, full stage, no rifts or deep pools, some weeds and stones	0.030–0.040
Natural winding, clean channel with some pools and shoals	0.033–0.045
Natural winding, clean channel with some pools and shoals, lower stages, more ineffective slopes and sections	0.040–0.055
Natural winding, clean channel with some pools and shoals, some weeds and stones	0.035–0.050
Natural winding, clean channel with some pools and shoals, lower stages, more ineffective slopes and stony sections	0.045–0.060
Sluggish river reaches, weedy or with very deep pools	0.050–0.080
Very weedy stream reaches	0.075–0.150
Floodplains, pasture, no brush	0.025–0.050
Floodplains, cultivated areas	0.025–0.050
Floodplains with brush	0.035–0.160
Floodplains with trees	0.050–0.160
Concrete-lined channel	0.011–0.016
Gunited channel	0.016–0.025
Channel with vegetal lining	0.030–0.50

Source: Chow (1959).

5. Estimate cross-sectional area of flow:

$$A = Q/V \qquad (3\text{-}11)$$

6. Estimate suitable side slope for the channel banks. Typical values are shown in Table 3-2. For earthen channels, side slopes should be tested by slope stability analysis.

7. Assume water depth, D (m), and estimate bottom width, B (m), for the desired trapezoidal channel:

$$A = BD + zD^2 \qquad (3\text{-}12)$$

Table 3-2. Suitable side slopes for nonerodible channels

Type of channel	Side slope (z Horizontal:1 Vertical)
Concrete lined	Vertical walls (0:1) to 2:1 or flatter
Rock-cut	0:1 to practical limit of rock cut
Armored with stone	1.5:1 or flatter
Shotcrete or gunited	Existing rock face to 1:1 or flatter
Soil cement	2:1

Source: Chow (1959).

8. Estimate:

$$P = [B + D\,2\,\sqrt{(z^2 + 1)}] \tag{3-13}$$

9. Estimate:

$$R = A/P \tag{3-14}$$

If this value of R is not nearly equal to the value estimated in step (4), assume a new value of D and repeat steps (7) to (9) until the two estimates of R are nearly equal.

The design steps for other channel shapes are similar except that different expressions will be required for A and P.

Example 3-1: Design a concrete-lined section for a 2-km stream reach with $S = 0.0025$ m/m, upstream of an existing drop structure. The existing channel is approximately 55 m wide. The channel downstream of the drop structure is lined and has sufficient capacity such that the proposed lining will not impact flooding and erosion conditions in the downstream reach. Use $Q = 500$ m^3/s. Permissible velocity for concrete lining is 5.0 m/s and $n = 0.015$.

Solution: From Eq. (3-3), $5.0 = (1/0.015) R^{2/3} \sqrt{(0.0025)}$. So, $R = 1.837$ m, $A = 500/5.0 = 100$ m^2. Since concrete lining will be installed on earth slopes, a suitable side slope of $2H:1V$ is adopted. Computations with trial values of D are shown in Table 3-3.

To minimize potential for excessive flood velocities in localized areas, the channel may be designed to have a bottom width of about 46 m, side slopes of $2H:1V$, and a total depth of 2.65 m, which will provide a freeboard of about 0.65 m above the anticipated flood level.

Velocity in flow sections of open channels varies in both vertical and horizontal directions. Different equations are used to estimate velocities at specified depths for smooth and rough channels. For turbulent flows, a channel may be treated as smooth if $kU_*/\nu < 5$; rough if $kU_*/\nu > 70$; and transitional if $5 < kU_*/\nu < 70$, where U_* = friction or shear velocity (m/s) = $\sqrt{(gRS)}$ and k = equivalent roughness height (m). For riprap-lined channels, $k = d_{65}$. For movable bed channels, k will normally be much larger than d_{65} because of boundary irregularities and bed forms and may vary from 0.03 to 0.91 m. If Chezy's C or Manning's n are known, then k for rough channels may be estimated by

$$C = R^{1/6}/n = 32.6 \log (12.2\, R/k) \tag{3-15}$$

Table 3-3. Trial computations for concrete-lined channel

Trial value of D (m)	B (m) = $(A - zD^2)/D$ (Eq. (3-12))	P (m) (Eq. (3-13))	R (m) = A/P
1.75	53.64	61.47	1.63
2.50	35.0	46.18	2.16
2.0	46.0	54.94	1.82
2.027	45.28	54.345	1.84
2.023	45.3855	54.4327	1.837

For turbulent flow in smooth channels,

$$U/U_* = 5.5 + 5.75 \log (U_* y/\nu) = 5.75 \log (9 U_* y/\nu) \qquad (3\text{-}16)$$

where U = velocity at depth, y, above channel bottom (m/s).

For turbulent flow in rough channels,

$$U/U_* = 8.5 + 5.75 \log (y/k) = 5.75 \log (30 y/k) \qquad (3\text{-}17)$$

In transition between smooth and rough channels, the constant 8.5 in Eq. (3-17) varies with kU_*/ν (Simons and Senturk 1976, 1992). For laminar flow in a rectangular open channel, velocity distribution in vertical direction is given by (Chow 1959):

$$U = (\gamma S/\mu)(yD - y^2/2) \qquad (3\text{-}18)$$

where

γ = unit weight of water (kg/m^3)

D = depth of flow (m)

μ = dynamic viscosity of water (kg-s/m^2)

Laminar flow may occur if velocity and depth of flow are relatively small.

Depth-averaged velocities at different locations along a channel cross section may be assumed to be proportional to $y^{2/3}$, in which y is depth (m) of water at that location. Distribution of depth-averaged velocity along channel cross sections should preferably be obtained from HEC-RAS (USACE 1998), HEC-2 (USACE 1991c), or other similar models.

Design of Erodible Channels

Maximum Permissible Velocity

The bed and banks of erodible channels experience erosion and deposition under varying flow conditions. The stability of such channels is governed by the character of the bed and bank materials and hydraulics of flow. A relatively simple and practical method for the de-

Table 3-4. Typical values of Manning's roughness coefficients and maximum permissible velocities

		Maximum permissible velocity (m/s)		
Material	n	Clear water	Water transporting colloidal silt	Water transporting noncolloidal silts, sands, gravels, or rock fragments
Fine sand (colloidal)	0.02	0.46	0.76	0.46
Sandy loam (noncolloidal)	0.02	0.53	0.76	0.61
Silt loam (noncolloidal)	0.02	0.61	0.91	0.61
Alluvial silt (noncolloidal)	0.02	0.61	1.07	0.61
Ordinary firm loam	0.02	0.76	1.07	0.69
Volcanic ash	0.02	0.76	1.07	0.61
Fine gravel	0.02	0.76	1.52	1.14
Stiff clay (very colloidal)	0.025	1.14	1.52	0.91
Graded loam to cobbles (noncolloidal)	0.03	1.14	1.52	1.52
Alluvial silt (colloidal)	0.025	1.14	1.52	0.91
Graded silt to cobbles (colloidal)	0.030	1.22	1.68	1.52
Coarse gravel (noncolloidal)	0.025	1.22	1.83	1.98
Cobbles and shingles	0.035	1.52	1.68	1.98
Shales and hard pans	0.025	1.83	1.83	1.52

Source: SCS (1954); Simons and Senturk (1976, 1992); Chow (1959).

sign of erodible channels is based on maximum permissible velocities for the bed and bank materials. Typical values of Manning's roughness coefficient and maximum permissible velocities for different types of soils are given in Table 3-4 (SCS 1954; Simons and Senturk 1976, 1992; Chow 1959).

Suitable side slopes for erodible channels are indicated in Table 3-5 (Chow 1959).

Table 3-5. Suitable side slopes for erodible channels

Material	Side slope ($H:V$)
Stiff clay	0.5:1 to 1:1
Firm compacted clay or soils having clay, silt, and sand mixtures	1.5:1
Silt, loam, and sandy soils	2:1
Sandy loam, porous clay, and fine sands	3:1

Source: Chow (1959).

Nonsilting, Nonscouring Velocity

To avoid silt deposition in channels (e.g., irrigation canals), design of erodible channels may sometimes require that channel velocities be above a certain minimum permissible (nonsilting) velocity. The following are commonly used equations to estimate nonsilting, nonscouring velocities:

1. Kennedy's nonsilting, nonscouring velocity:

$$V_0 \text{ (m/s)} = CD^{0.64} \quad (3\text{-}19)$$

where $C = 0.37$ for extremely fine soils; 0.55 for fine silty, sandy soils; 0.60 for coarse, light, sandy soils; 0.66 for sandy, loamy soils; and 0.71 for coarse silts or hard soil debris (Chow 1959; Singh 1967).

2. Lacey's regime velocity:

$$V_0 \text{ (m/s)} = (Qf^2/140)^{1/6} \quad (3\text{-}20)$$

where

f = Lacey's silt factor = $1.76\sqrt{d}$

d = mean particle size (mm) (Davis and Sorensen 1970; Zipparro and Hansen 1993)

Example 3-2: Design an earthen (unlined) channel for the stream reach of Example 3-1. The channel bed contains graded silt and loam to cobbles. Use side slopes of $2H{:}1V$.

Solution: This channel may not be well maintained in the future and some weeds may grow with time. So, use $n = 0.035$. The existing channel bed slope is steep (0.0025 m/m). To reduce channel bed slope, provide a 4.1-m drop structure at the upstream end or one 2-m drop structure at the upstream end and another 2.1-m drop structure 1.0 km downstream from the upstream end. This is determined after several trials with different heights of drop structures such that the resulting bed slope provides no more than the maximum permissible velocity with a bed width not much larger than the existing bed width of the channel.

Average bed slope with drop structures = $(2{,}000 \times 0.0025 - 4.1)/2{,}000 = 0.00045$ m/m.

During flood events, the channel will transport noncolloidal silts, sands, gravels, and rock fragments. Thus,

Permissible velocity = $V = 1.52$ m/s (Table 3-4) and $A = 500/1.52 = 328.95$ m².

$1.52 = (1/0.035)R^{2/3}\sqrt{(0.00045)}$. So, $R = 3.97$ m $= (BD + 2D^2)/[B + \{2\sqrt{(5)}D\}]$. By trial and error, $D = 4.605$ m and $B = 62.223$ m. To minimize the potential for excessive flood velocities in localized areas, the channel may be designed to have a bottom width of about 63 m with a total depth of 5.25 m, which will provide a freeboard of about 0.65 m above the anticipated flood level. With drop structures, the channel excavation will be more, the channel bed will be lower, and flood levels may be somewhat lower.

HYDRAULIC ANALYSIS

Analysis of sediment stability (i.e., potential for erosion and deposition) is required to design stable channels, evaluate erosion protection requirements, and assess the potential for exposure of contaminated sediments already deposited on channel beds. Commonly used equations to evaluate sediment stability include the following:

1. **Shields Shear Stress Equation** (Simons and Senturk 1976, 1992). This method compares the Shields shear stress for initiation of motion with the shear stress at the channel bed due to flow in the channel.

$$\text{Shear resistance of sediments} = a \, (\gamma_s - \gamma) \, d \tag{3-21}$$

$$\text{Shear stress at canal bed due to flow} = \gamma \, R \, S \tag{3-22}$$

where

γ_s = unit weight of sediments (kg/m^3)

γ = unit weight of water (kg/m^3)

d = particle size (m)

R = hydraulic mean radius (m)

S = energy gradient (m/m)

a = a coefficient that is a function of U_*d/ν

U_* = shear velocity (m/s) = $\sqrt{(g \, R \, S)}$

ν = kinematic viscosity of water (m^2/s)

Approximate values of a for given values of U_*d/ν are given in Table 3-6. Values of a for $U_*d/\nu < 0.2$ have not been estimated. The trend of available experimental data suggests that the coefficient "a" increases with decreasing U_*d/ν. For most open-channel flows, a value of 0.047 is suggested.

2. **Meyer-Peter-Muller Equation** (USBR 1984). The modified Meyer-Peter-Muller bed load transport equation for beginning of transport of individual particles is

$$d = D \, S / [57.9\{n/(d_{90}^{1/6})\}^{3/2}] \tag{3-23}$$

where

d = nontransportable particle size (m)

D = mean water depth (m)

n = Manning's roughness coefficient

d_{90} = particle size in millimeters than which 90% of the bed material is finer by weight

3. **Einstein-Strickler-Manning Equations** (Simons and Senturk 1976, 1992; USACE 1991d; Singh 1967). This method is based on Manning's equation to estimate channel roughness, Strickler's equation to estimate grain roughness, and Einstein's equation

Table 3-6. Approximate values of coefficient a in Shields equation

$U_* d/v$	a
0.25	0.45
0.3	0.39
0.4	0.29
0.5	0.22
0.6	0.19
1.0	0.12
2.0	0.06
3.0	0.047
4.0	0.040
6.0	0.035
8.0	0.034
10	0.033
20	0.034
30	0.035
40	0.037
60	0.040
100	0.045
200	0.052
500	0.06
800	0.06
1,000	0.06

Source: Adapted from Shields diagram in Simons and Senturk (1976).

for initiation of motion. According to this equation, particle size for initiation of motion is given by

$$d = 18.18 \, R'S \tag{3-24}$$

where R' = hydraulic radius associated with grain roughness. Approximate values of R' may be estimated by (Singh 1967):

$$R'/R = (n'/n)^{3/2} \tag{3-25}$$

where n' = grain roughness coefficient given by Strickler's equation (Simons and Senturk 1976, 1992):

$$n' = (d_{50})^{1/6}/25.67 \text{ (if } d_{50} \text{ is in meters)} \tag{3-26}$$

or

$$n' = (d_{50})^{1/6}/81.2 \text{ (if } d_{50} \text{ is in millimeters)}$$

where d_{50} = particle size than which 50% of channel bed material is finer by weight.

4. **Camp's Equation** (ASCE 1976). This method is based on the Shields equation for initiation of motion, tractive stress at the canal bed, and Manning's equation to estimate energy slope. According to this equation, average channel velocity for initiation of motion, v_0, is given by

$$v_0 = (1/n) \ R^{1/6} \ [\sqrt{\{a \ (s-1) \ d\}}] \tag{3-27a}$$

$$v_0 = \sqrt{[(8 \ g/f) \ a \ (s-1) \ d]} \tag{3-27b}$$

where

s = specific gravity of sediment particles

f = Darcy-Weisbach friction factor

d = particle size (m)

For the coefficient a, Camp suggests a value of 0.04 for initiation of motion and 0.8 for significant scour irrespective of the energy slope. For flat slopes where $U_*d/\nu < 0.3$, larger values of the coefficient a may be used for initiation of motion (see Table 3-6).

Equations (3-21) to (3-27) are based on limited laboratory or field data for noncohesive sediments and practical experience. Particle sizes smaller than about 0.009 mm (e.g., clays) may exhibit cohesive resistance, which is not accounted for in these equations. Estimates of erodible particle sizes obtained from these equations for given hydraulic conditions may be significantly different. It is advisable to estimate erosion potential using more than one equation and select reasonable sizes of erodible particles by judgment.

Example 3-3: Estimate the potential for sediment erosion from the bed of a controlled channel under normal flow conditions when $R = 3.4$ m, $S = 0.00001$, and Manning's roughness coefficient, n, = 0.025. The d_{50} and d_{90} of channel bed material are 0.3 and 0.8 mm, respectively. Use $\nu = 1.13 \times 10^{-6}$ m^2/s, $\gamma = 1{,}000$ kg/m^3, and $\gamma_s = 2{,}650$ kg/m^3. The channel is wide, so assume $R \approx D$ (mean water depth).

Solution: Using Manning's equation, average channel velocity = $(1/0.025) \times (3.4)^{2/3} \times \sqrt{0.00001} = 0.286$ m/s.

Use Shields, Meyer-Peter-Muller, Einstein-Strickler-Manning, and Camp's equations to estimate particle sizes that can be eroded under given hydraulic conditions.

1. Shields Equation

    $U_* = \sqrt{(g \ RS)} = \sqrt{(9.81 \times 3.4 \times 0.00001)} = 0.018$ m/s.

    As a first trial, for $d = 0.3$ mm, $U_*d/\nu = 0.018 \times 0.3 \times 10^6/(1{,}000 \times 1.13) = 4.78$.

    Use $a = 0.04$ (Table 3-6).

    τ (canal bed) = $\gamma \ RS = 1{,}000 \times 3.4 \times 0.00001 = 0.034$ kg/m^2.

    τ (resistance) = $0.04 \ (\gamma_s - \gamma) \ d = 0.04 \times 1650 \times d$.

Equating the tractive and resistive shear stresses, $d = 0.00051$ m $= 0.51$ mm. With $d = 0.51$ mm, $U_*d/\nu = 8.12$ and the adopted value of a (i.e., 0.04) is conservative.

2. Meyer-Peter-Muller Equation

   $d = 3.4 \times 0.00001/[57.9 \{0.025/(0.8)^{1/6}\}^{3/2}] = 0.00014$ m $= 0.14$ mm.

3. Einstein-Strickler-Manning Equation

   $n' = (0.3)^{1/6}/81.2 = 0.010; R' = 3.4 \times (0.010/0.025)^{1.5} = 0.87$ m.

   $d = 18.18 \times 0.87 \times 0.00001 = 0.00016$ m $= 0.16$ mm.

4. Camp's Equation

   $s = (2{,}650)/1{,}000 = 2.65$.

   $0.286 = (1/0.025) \times (3.4)^{1/6} \times \sqrt{(0.04 \times 1.65\, d)}$

   So, $d = 0.00051$ m $= 0.51$ mm.

Two of the estimates of erodible particle size are less than and the other two are greater than the d_{50} of channel bed material. This suggests that there is potential for sediment movement. Note that the estimates are for initiation of motion and not for significant scour.

Design of Vegetated Channels

From environmental, aesthetic, and economic considerations, it is often desirable to evaluate the feasibility of vegetative lining for erosion control on channel banks. A commonly used method for the design of vegetated channels is based on the use of permissible velocities, *n-VR* chart or table, and retardance classes for different types of vegetal linings defined by the U.S. Soil Conservation Service (SCS 1954; Chow 1959; Barfield et al. 1981). *VR* represents the product of velocity and hydraulic radius, and *n* is Manning's roughness coefficient. Grasses have been divided into five retardance classes (i.e., A, B, C, D, and E) based on type, height, and condition of grasses. Guidelines for determining retardance class of a vegetal lining are included in Tables 3-7 and 3-8 (SCS 1954).

Permissible (nonscouring) velocities for different types of grass covers are shown in Table 3-9 (SCS 1954). Velocities exceeding 1.52 m/s should be used only where good covers and proper maintenance of vegetal lining can be ensured.

The values of Manning's *n* and the product *VR* for different retardance classes are shown in Tables 3-10(a) and (b) (SCS 1954).

After a retardance class has been assigned to the desired type of vegetal lining, the steps of design computations are as follows:

1. Select permissible velocity, *V*, for the type of vegetal lining from Table 3-9.
2. Assume a trial value of Manning's *n* and read the corresponding *VR* from Table 3-10(a) or (b).
3. Compute the value of $R = VR/V$.
4. Use Manning's formula to estimate $VR = (1/n)\, R^{5/3} \sqrt{S}$.
5. If this value of *VR* is not nearly equal to the value obtained in step (2), then try another value of *n* until the two estimates closely match.

Table 3-7. Retardance classes for different vegetal covers

Retardance class	Vegetal cover	Condition
A	Weeping lovegrass	Excellent stand, tall, (average 76 cm)
	Yellow bluestem *Ischaemum*	Excellent stand, tall, (average 91.5 cm)
B	Kudzu	Very dense growth, uncut
	Bermuda grass	Good stand, tall, (average 30.5 cm)
	Native grass mixture (little bluestem, blue grama, and other long and short midwest grasses)	Good stand, unmowed
	Weeping lovegrass	Good stand, tall, (average 61 cm)
	Lespedeza sericea	Good stand, not woody, tall (average 48 cm)
	Alfalfa	Good stand, uncut (average 28 cm)
	Weeping lovegrass	Good stand, mowed (average 33 cm)
	Kudzu	Dense growth, uncut
	Blue grama	Good stand, uncut (average 33 cm)
C	Crabgrass	Fair stand, uncut (25.4 to 122 cm)
	Bermuda grass	Good stand, mowed (average 15 cm)
	Common *lespedeza*	Good stand, uncut (average 28 cm)
	Grass-legume mixture-summer (orchard grass, redtop, Italian ryegrass, and common *lespedeza*)	Good stand, uncut (15 to 20 cm)
	Centipede grass	Very dense cover (average 15 cm)
	Kentucky bluegrass	Good stand, headed (15 to 30.5 cm)
D	Bermuda grass	Good stand, cut to 6 cm height
	Common *lespedeza*	Excellent stand, uncut (average 11.5 cm)
	Buffalo grass	Good stand, uncut (7.5 to 15 cm)
	Grass legume mixture-fall, spring (Orchard grass, redtop, talian ryegrass, and common *lespedeza*)	Good stand, uncut (10 to 13 cm)
	Lespedeza sericea	After cutting to 5 cm height. Very good stand before cutting.
E	Bermuda grass	Good stand, cut to 3.8 cm height
	Bermuda grass	Burned stubble

Source: SCS (1954).

6. Estimate the area of flow using $A = Q/V$.
7. Determine acceptable side slopes, z, for the trapezoidal channel section, and assume a reasonable trial bed width, B.
8. Estimate channel depth from $A = BD + zD^2$.

Table 3-8. Guide for selection of retardance class

Stand	Average length of grass (cm)	Degree (class) of retardance
Good	>76	A
	28–61	B
	15–25	C
	5–15	D
	<5	E
Fair	>76	B
	28–61	C
	15–25	D
	5–15	D
	<5	E

Source: SCS (1954).

Table 3-9. Permissible velocities for channels lined with vegetation

		Permissible velocity (m/s)	
Grass cover	Slope range (percent)	Erosion-resistant soils	Easily eroded soils
Bermuda grass	0–5	2.44	1.83
	5–10	2.13	1.52
	>10	1.83	1.22
Buffalo grass	0–5	2.13	1.52
Kentucky bluegrass	5–10	1.83	1.22
Smooth brome	>10	1.52	0.91
Blue grama			
Grass mixture	0–5	1.52	1.22
	5–10	1.22	0.91
	>10	NR	NR
Lespedeza sericea	0–5	1.07	0.76
Weeping lovegrass	>5	NR	NR
Yellow bluestem			
Kudzu			
Alfalfa			
Crabgrass			
Common *lespedeza*[a]	0–5	1.07	0.76
Sudangrass[a]	>5	NR	NR

[a]Annuals used on mild slopes or as temporary protection until permanent covers are established.
NR = Not recommended.
Source: SCS (1954).

Table 3-10(a). VR and n values for different retardance classes

Product of velocity and hydraulic radius, VR (m²/s)	Manning's n values for retardance class				
	A	B	C	D	E
0.093	0.29	0.15	0.085	0.06	0.036
0.186	0.175	0.10	0.058	0.046	0.030
0.279	0.13	0.08	0.048	0.040	0.029
0.372	0.11	0.07	0.042	0.038	0.027
0.464	0.092	0.063	0.039	0.037	0.026
0.557	0.085	0.059	0.038	0.035	0.025
0.743	0.075	0.052	0.035	0.034	0.025
0.929	0.07	0.048	0.034	0.032	0.025
1.858	0.06	0.04	0.032	0.030	0.024

Source: SCS (1954).

Table 3-10(b). n and VR values for different retardance classes

Manning's n	Product of velocity and hydraulic radius, VR (m²/s), for retardance class				
	A	B	C	D	E
0.024	—	—	—	—	1.858
0.030	—	—	—	1.858	0.195
0.040	—	1.858	0.446	0.297	0.068
0.050	—	0.836	0.248	0.145	0.036
0.060	1.858	0.520	0.172	0.092	0.021
0.070	0.929	0.362	0.128	0.064	0.014
0.080	0.641	0.279	0.100	0.046	0.010
0.090	0.502	0.223	0.084	0.037	—
0.10	0.409	0.186	0.072	0.031	—
0.15	0.232	0.104	0.040	0.015	—
0.20	0.156	0.064	0.027	—	—
0.30	0.086	0.033	—	—	—
0.375	0.037	—	—	—	—

— indicates that the product VR is not viable for that value of n and retardance class.
Source: SCS (1954).

Example 3-4: Estimate the dimensions of a channel to be excavated in easily erodible soils and lined with Bermuda grass that will be maintained at an average height of about 6 cm. The channel is to carry a discharge of 3 m³/s with a bed slope of 0.02 and side slopes of 2.5H:1V.

Solution: From Tables 3-7 and 3-8, the retardance class is D, and from Table 3-9, permissible maximum nonscouring velocity for $S = 2.0\%$ is 1.83 m/s.

Assume a trial $n = 0.033$. From Table 3-10(a) or (b), $VR = 0.836$ m²/s. So $R = 0.836/1.83 = 0.4568$ m.

Using Manning's equation,

$$VR = (1/n)\, R^{5/3}\, \sqrt{S} = (1/0.033)\, (0.4568)^{5/3}\, \sqrt{0.02} = 1.16.$$

Similarly, other trial values of Manning's n give the values shown below:

(1)	(2)	(3)	(4)
n	VR (from Table 3-10)	R	$VR = (1/n)\, R^{5/3}\, \sqrt{S}$
0.034	0.743	0.406	0.926
0.037	0.464	0.254	0.388
0.035	0.557	0.304	0.556

For $n = 0.035$, values in columns (2) and (4) are close. So select $R = 0.304$ m.

$$\text{Cross-sectional area} = A = 3/1.83 = 1.64 \text{ m}^2 = BD + zD^2 \qquad \text{(i)}$$

$$\text{Wetted perimeter} = P = A/R = 1.64/0.304 = 5.386 = B + [2\sqrt{(z^2 + 1)}]\,D = B + [2\sqrt{(2.5^2 + 1)}]\,D = B + 5.385\,D \qquad \text{(ii)}$$

Here, D = water depth and B = bed width of the channel. By trial and error, estimate the value of D, which gives the required area $A = 1.64$ m².

D (trial value)	B [from Eq. (ii)]	A [from Eq. (i)]
0.39	3.286	1.662
0.38	3.34	1.630
0.383	3.323	1.64

To achieve a channel velocity less than the permissible nonscouring value of 1.83 m/s, and to allow a freeboard of 0.32 m, use $B = 3.4$ m and $D = 0.7$ m.

Flow Through Bends

Due to centrifugal forces, water surface elevation on the outer (convex) side of a channel bend tends to be higher and that on the inner side lower than that at the channel center. This results in shifting of maximum velocity from the center line of the channel. Erodible channels tend to scour toward the outer side of bends. The rise in water surface elevation (superelevation) may be estimated by the following equation (USACE 1994):

$$\Delta y = (C V^2 B)/gr \qquad (3\text{-}28)$$

where

Δy = superelevation = difference in water surface elevation (m) between theoretical water surface at the center line and at the outer side of the bend

V = mean channel velocity (m/s)

B = channel width at elevation of center-line water surface (m)

r = radius of bend (m) at channel center line

C = 0.5 for subcritical and 1.0 for critical and supercritical flows

To minimize secondary flows and distortion of velocity distribution at bends, r should be greater than 3 times the channel width.

In some cases, flood diversion channels in erodible soils may be purposely designed with a specified curvature away from an existing or proposed structure to minimize the potential of channel migration toward the structure.

It has been found that a sand bed channel tends to develop a meandering pattern when

$$S Q^{1/4} \leq 0.0007 \qquad (3\text{-}29)$$

The same channel may tend to develop a braided pattern when

$$S Q^{1/4} \geq 0.0041 \qquad (3\text{-}30)$$

and is in transitional stage between the values given by Eqs. (3-29) and (3-30) (Simons and Senturk 1976, 1992).

Freeboard

Freeboard is the distance between the design water surface elevation and the top of channel bank or levee. Suggested values of freeboards are given below (USACE 1994):

- Rectangular channels: 0.61 m
- Trapezoidal sections of concrete- or riprap-lined channels: 0.76 m
- Earth levees: 0.91 m

Water Surface Profiles

Steady flow analysis to determine water surface profiles for channels whose cross sections and discharges change from location to location are performed using computer models, such as HEC-2 (USACE 1991c), HEC-RAS (USACE 1998), and WSP-2 (USDA 1976). These analyses are required to estimate flow velocities and water depths at different locations along and across the channel, height and alignment of levees, waterways of bridges and bridge backwater, and delineation of floodplains, etc. Floodplain analysis involves water surface profile computations for floods of different return periods (e.g., 10-, 25-, 50-, 100-, and 500-yr floods) with and without encroachments. Hydraulic analysis with encroachments provides information required to limit developments within the floodplain and involves identification of imaginary encroachment boundaries on either side of the stream bank, such that the resulting rise in water surface elevation is within a prescribed maximum (e.g., 0.3 m) or the available unobstructed width of flow is not less than a prescribed minimum. The flow section within such imaginary encroachments is called "floodway." Options to perform backwater analysis with different types of encroachments are included in the aforementioned models. In addition, these models include analysis of flow through bridges, junctions of tributaries, and bifurcation or split flows. The basic input data include channel cross sections along the study reach; discharges at different cross sections; and loss coefficients for friction, contraction, expansion, and bridge piers and abutments. Although there are no limits on the number of cross sections required for a particular study, it is desirable to have a minimum of 6 to 10 cross sections for the study reach. Additional cross sections must be included in the vicinity of structures (e.g., bridges and culverts, weirs and drop structures, channel junctions, and groins). The cross sections should be oriented nearly perpendicular to the anticipated flow lines. To meet this condition, sometimes the cross sections may have to be curved or crooked.

Some environmental mitigation projects for major streams (e.g., Missouri and Mississippi rivers) require habitat restoration on the floodplains by diverting a small portion of the streamflow onto the floodplain through ungated chutes or diversion channels. These chutes are intended to create wetlands, riffles and pools, and meandering watercourses through the floodplains, generally along the alignment of historic chutes that may have been blocked after channelization. The diverted flow returns to the main river at the chute outlet. Split flow analysis for such projects involves estimation of flows that would be diverted by a chute with specified dimensions (i.e., width and invert elevations of the ungated inlet and outlet structures, and cross sections and bed slope of the chute) under different streamflow conditions. This analysis can be accomplished by trial and error using hydraulic models, such as HEC-RAS and HEC-2 (USACE 1998, 1991c). The computational steps are as follows (Chow 1959):

1. Select a discharge, Q, passing through the main channel upstream of the inlet.

2. Assume that the flow, Q, is divided into Q_m (flow through the main channel) and Q_c (flow through the chute) at the inlet, such that $Q = Q_m + Q_c$ (see Figure 3-1). Thus, Q_m is the main channel flow through the reach from the location of the chute inlet to that of the chute outlet. At the location of the outlet, the flow, Q_c, diverted through the chute returns to the main channel; thereafter, flow through the main channel becomes Q.

3. Compute water surface profile for flow, Q_m, through the main channel from chute outlet to chute inlet and for flow, Q, starting from a specified downstream station to the chute outlet. Find the resulting energy grade elevation at the inlet and the water surface elevation at the outlet of the chute.

4. Compute water surface profile for the flow, Q_c, through the chute, starting with the computed water surface elevation at the outlet, and find the energy grade elevation at the inlet.

HYDRAULIC ANALYSIS

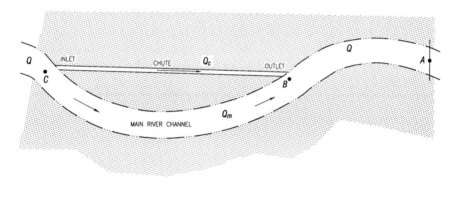

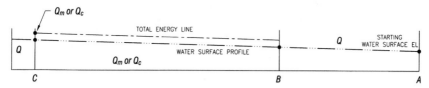

Figure 3-1. Split flow schematic

5. If the two energy grade elevations at the inlet do not match, then repeat steps (2) through (4) with modified Q_m until the two energy grade elevations are reasonably close.

In planning stages, a large number of trial computations are required to determine acceptable chute dimensions and diverted flows under different flow conditions in the river. However, usually flow diversion through the chute, Q_c, is a small fraction of the total river flow, Q (less than 10%), so that the energy grade and water surface elevations through the main channel with a flow, Q_m, are not significantly different from those with the total flow, Q. For such cases, simplified computational steps are as follows:

1. Compute water surface elevations at the outlet and energy grade elevations at the inlet for the main channel for different river flow conditions, Q.

2. For each river flow condition, assume several values for diverted flow and perform water surface profile computations for the chute for those flows, Q_c, starting with computed water surface elevation at the outlet (step (1)).

3. Select diverted flow, Q_c, which results in energy grade elevation at the inlet, which is nearly equal to that computed in step (1).

Example 3-5: A trapezoidal chute—with B = 30.5 m, side slopes 3H:1V, bed slope = 0.00021, and invert elevation = 245.63 m—is proposed to divert flows from a major stream with an average annual flow of 1,060 m³/s. The energy grade elevation in the river at chute inlet for average annual flow conditions is 246.32 m. Estimate flow diverted by the chute under average annual flow conditions. Use $n = 0.035$.

Solution: Since the chute has a constant cross section and bed slope, use Manning's formula to estimate energy elevation in the chute at inlet.

By trial and error with several trial values of Q_c, $y = 0.685$ m; $A = 22.30$ m²; $P = 34.83$ m; $R = 0.6402$ m; $V = 0.3075$ m/s; and $Q_c = 6.857$ m³/s for chute flow. Energy grade elevation at chute inlet = $245.63 + 0.685 + (0.3075^2)/2\,g = 246.32$ m, which is equal to the given energy grade elevation at the inlet for average annual river flow. So, flow diverted by the chute = 6.857 m³/s.

Critical Flow

Critical flow is defined by minimum specific energy. Specific energy, E (m), at a flow section is given by

$$E = y + V^2/2g \qquad (3\text{-}31)$$

where

  y = flow depth (m)
  V = velocity (m/s)

For a rectangular open channel,

$$E = y + q^2/(2g\,y^2) \qquad (3\text{-}32)$$

where q = discharge per unit width of channel (m²/s). To obtain minimum specific energy, set $dE/dy = 0$. Thus,

$$y_c = (q^2/g)^{1/3} \qquad (3\text{-}33)$$

$$E\,(\text{minimum}) = 3/2\,(y_c) \qquad (3\text{-}34)$$

$$V_c = \sqrt{(g\,y_c)} \qquad (3\text{-}35)$$

where

  y_c = critical depth (m) and
  V_c = velocity for critical flow (m/s)

At critical flow, Froude number = $F = V/\sqrt{(g\,y)} = 1$. When $F > 1$, flow is supercritical and subcritical when $F < 1$.

For a prismatic channel with given discharge, Q (m³/s),

$$E = y + Q^2/(2g\,A^2) \qquad (3\text{-}36)$$

To obtain minimum specific energy, set $dE/dy = 0$. Thus,

$$Q^2/g = A^3/T \qquad (3\text{-}37)$$

or

$$V^2/2\,g = D/2 \qquad (3\text{-}38)$$

where

$T = dA/dy$ = top width (m)

D = hydraulic depth = A/T

Eq. (3-37) can be used to derive equations for critical discharge for various channel shapes. For example, for a rectangular channel,

$$Q_c = B\sqrt{(g y^3)} \qquad (3\text{-}39)$$

For a triangular channel,

$$Q_c = z y^{2.5} \sqrt{(g/2)} \qquad (3\text{-}40)$$

For a trapezoidal channel,

$$Q_c = (B y + z y^2)^{1.5} \sqrt{\{g/(B + 2z y)\}} \qquad (3\text{-}41)$$

For a parabolic channel,

$$Q_c = T\sqrt{\{(8/27)\, g y^3\}} \qquad (3\text{-}42)$$

where

z = side slope

B = bottom width

A critical flow section may constitute a control section for both subcritical and supercritical flows. A control section is one where a definite relationship can be established between stage and discharge. At this section, flow is controlled in such a way that flow conditions on one side of it (i.e., either upstream or downstream) do not affect those on the other side. In the case of subcritical flow, flow control is at the downstream end; in the case of supercritical flow, flow control is at the upstream end.

Hydraulics of Weirs and Spillways
Broad-Crested and Sharp-Crested Weirs

If the crest of a weir is sufficiently wide to prevent the nappe from springing free at the upstream corner, the weir is classified as broad-crested (Rouse 1950). The weir behaves as broad-crested if the head above the crest, H, is less than 1.5 B, where B = width of the crest along the direction of flow. If $H > 1.5\, B$, the nappe of the free overfall becomes detached, and the weir behaves similar to a sharp-crested weir.

If the upstream edge of a broad-crested weir is well rounded to minimize contraction, and if friction along its width is negligible or the crest is sloped to overcome loss of head due to friction, then flow occurs at critical depth, y_c. In a free overfall, y_c occurs at 3 y_c to 4 y_c upstream of the brink (downstream corner) of the crest, and $y_c = 1.4\, y_b$, where y_b = water depth at brink. At the critical section,

$$y_c = (q^2/g)^{1/3} \qquad (3\text{-}33)$$

$$H = y_c + V_c^2/2g = y_c + (q/y_c)^2/2g = y_c + y_c/2 = 3/2\, y_c = 2.098\, y_b \qquad (3\text{-}43)$$

$$q = [\sqrt{(8\, g/27)}]\, H^{3/2} = 5.17\, y_b^{3/2} \qquad (3\text{-}44)$$

Thus, for a broad-crested weir, the theoretical maximum discharge is given by

$$Q = 1.70\, L\, H^{3/2} = 5.17\, L\, y_b^{3/2} \qquad (3\text{-}45)$$

The head, H, must be measured about $2.5H$ upstream from the upstream edge of the broad-crested weir; for most practical cases, the weir coefficient, 1.70, should be reduced to account for friction losses on the crest. This coefficient has been found to vary with the head, H, and width of the broad-crested weir (Brater et al. 1996). Flow over a broad-crested weir is not appreciably affected by submergence if $y_T < (2/3\, H + P)$, where y_T = tailwater depth and P = height of weir above channel bottom.

Flow over a sharp-crested weir is given by

$$Q = 2/3\, C_d\, L\, [\sqrt{(2\, g)}]\, H^{3/2} = C\, L\, H^{3/2} \qquad (3\text{-}46)$$

The weir coefficient, C, varies from about 1.76 to 2.54 with the ratio H/P and end contractions due to sharp edges at the ends of the weir. Flow over a sharp-crested weir is not appreciably affected by submergence if $y_T < (0.25\, H + P)$. For larger submergence,

$$Q_1/Q = [1 - (H_1/H)^{3/2}]^{0.385} \qquad (3\text{-}47)$$

where

Q_1 = discharge with submergence

H_1 = tailwater depth above weir crest

Ogee Crest

If the upper and lower nappes of flow over a sharp-crested weir are aerated (i.e., both nappes are at atmospheric pressure), then the discharge coefficient is given by (Chow 1959)

$$C = 1.805 + 0.221 H/P \qquad (3\text{-}48)$$

If the lower nappe is not aerated, it results in undesirable hydraulic performance and negative pressures on the downstream face. To minimize negative pressures, ogee profiles have been developed to conform to the shape of the lower nappe as far as practicable:

$$X^n = K H_d^{n-1}\, Y \qquad (3\text{-}49)$$

where

X and Y are coordinates of the profile with origin at the highest point of the crest

H_d = design head including velocity head due to approach flow

K and n are given in Table 3-11

Table 3-11. Coefficients and exponents of ogee profile equation

Slope of upstream face of spillway	K^a	n^a	K^b	n^b
Vertical	2.000	1.872	2.141	1.837
1H:3V	2.000	1.851	2.141	1.817
2H:3V	1.901	1.802	2.058	1.772
1H:1V	1.852	1.780	2.058	1.761

Source: USBR (1987).
[a]Neglecting velocity of approach head.
[b]Velocity of approach head = 0.20 H_d.

Based on model tests, geometric shapes have been developed to provide a smooth upstream edge to the ogee crest in the form of two arcs with different radii. A typical ogee crest shape is illustrated in Figure 3-2 (USBR 1987).

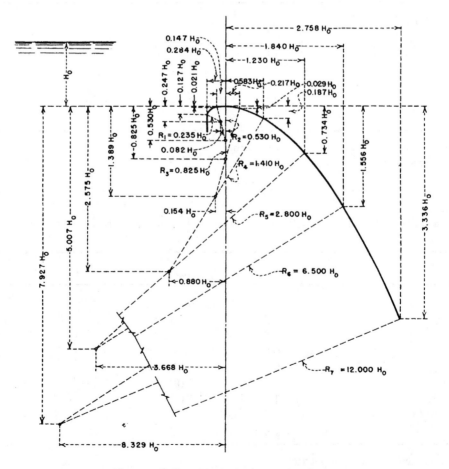

Figure 3-2. Typical ogee crest shape
Source: USBR (1987).

The discharge for an ogee crest is given by

$$Q = CLH_o^{3/2} \qquad (3\text{-}50)$$

where

$C = 2.181$ to 2.225 if $H_o = H_d$

$P/H_d > 2.50$

Experimental curves are available to estimate C for other values of H_o and P/H_d, H_o being other than the design head (USBR 1987). Selected values of C for different values of P/H_d are included in Table 3-12(a).

Approximate values of discharge coefficients for other than the design head of the ogee crest are included in Table 3-12(b) (USBR 1987).

Table 3-12(a). Discharge coefficients for vertical-faced ogee crest with varying values of P

P/H_d	C
0	1.7
0.5	2.098
1.0	2.142
1.5	2.159
2.0	2.17
2.5	2.176
3.0	2.181

H_d = design head including velocity of approach.
Source: USBR (1987).

Table 3-12(b). Discharge coefficients for vertical-faced ogee crest for other than design head

H_o/H_d	C_a/C
0.05	0.8
0.2	0.85
0.4	0.9
0.6	0.94
0.8	0.97
1.0	1.0
1.2	1.025
1.4	1.05
1.6	1.07

C_a = discharge coefficient for head H_o.
Source: USBR (1987).

Flow along Steep Slopes

A commonly encountered case of supercritical flow involves flow along downstream faces of spillways or sloping aprons downstream of drop structures. Estimates of the velocity and depth of flow along and at the toe of these steep slopes is required to design energy dissipation measures for such structures. This can be done using the energy equation

$$H + z = y + q^2/(2\,g\,y^2) + q^2\,n^2\,L/y^{3.33} \tag{3-51}$$

where

H = head above the crest including velocity of approach (m)

z = height of crest above toe (m)

L = length of slope (m)

q = unit discharge (m²/s) = $C\,H^{1.5}$

C = discharge coefficient of crest

y = depth of flow at toe (m)

Eq. (3-51) is solved by trial and error. Knowing y, velocity at the toe = $V = q/y$ can be estimated.

For steep slopes varying from $0.6H:1V$ to $0.8H:1V$, velocity at the toe also may be estimated from experimental curves available in the literature (e.g., Chow 1959; Peterka 1978). Selected values are shown in Table 3-13.

Hydraulic Jump

A hydraulic jump occurs at the transition from supercritical to subcritical flow. Commonly encountered cases include supercritical flow down the steep slope of a spillway or drop

Table 3-13. Toe velocities for spillways with downstream face slopes of $0.6H:1V$ to $0.8H:1V$

Z (m)	Toe velocity (m/s)					
	H = 0.76 m	H = 2.29 m	H = 3.05 m	H = 6.1 m	H = 9.1 m	H = 12.2 m
12	12.2	13.4	13.7	13.7	13.7	13.7
24	15.2	18.6	18.9	19.8	19.8	19.8
37	15.8	21.3	22.9	24.4	24.4	24.4
49	15.8	23.2	24.7	27.4	28.0	28.0
61	15.8	24.1	26.2	29.9	31.7	32.0
73	15.8	24.4	27.4	32.0	33.8	34.4
85	15.8	24.7	28.0	33.5	36.3	37.2
98	15.8	24.7	28.3	35.4	38.1	39.6
110	15.8	24.7	28.3	36.6	39.6	41.5
122	15.8	24.7	28.3	37.2	41.5	43.6

Source: Peterka (1978).

structure or supercritical flow through a gate opening entering a mild slope channel on the downstream side. For hydraulic jumps in horizontal rectangular channels (see Figure 3-3),

$$y_1 y_2 (y_1 + y_2) = 2 q^2/g \qquad (3\text{-}52a)$$

$$y_2/y_1 = 1/2 \, [\sqrt{(1 + 8 F_1^2)} - 1] \qquad (3\text{-}52b)$$

$$H_L = (y_2 - y_1)^3/(4 y_1 y_2) \qquad (3\text{-}53)$$

$$E_1 = y_1 + V_1^2/2g = y_1 + q^2/(2 g y_1^2) \qquad (3\text{-}54a)$$

$$E_2 = y_2 + V_2^2/2g = y_2 + q^2/(2 g y_2^2) \qquad (3\text{-}54b)$$

where

y_1 = pre-jump supercritical water depth

y_2 = post-jump subcritical water depth

V_1, V_2 = pre-jump and post-jump velocities, respectively

F_1 = pre-jump Froude number = $V_1/\sqrt{(g y_1)}$

H_L = energy loss through jump

q = discharge per unit width of channel

E_1, E_2 = pre-jump and post jump specific energy, respectively

Post-jump depth, y_2, may be estimated from published experimental plots between y_2/y_1 and F_1 for different apron slopes (e.g., Chow 1959). In many practical situations, the unit discharge, q, and surface water elevations and velocities upstream and downstream of the drop are known from water surface profile computations. A trial and error procedure may be adopted to determine hydraulic jump parameters for such cases. The parameters are non-

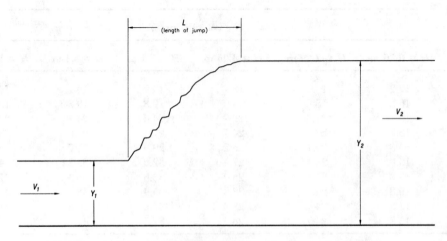

Figure 3-3. Hydraulic jump in a horizontal channel

dimensionalized by setting $x = y_1/y_c$; $y = y_2/y_c$; $l = H_L/y_c$; $m = E_1/y_c$; and $n = E_2/y_c$, where y_c = critical depth = $(q^2/g)^{1/3}$. Note that x varies from 0 to 1. With any selected value of x, the other non-dimensional parameters — m, y, n, and z — can be computed recognizing that

$$m = x + 1/(2\ x^2) \tag{3-55a}$$

$$x\ y\ (x + y) = 2 \tag{3-55b}$$

$$n = y + 1/(2\ y^2) \tag{3-55c}$$

$$l = (y - x)^3/(4\ x\ y) \tag{3-55d}$$

Selected values of y, x, m, n, and l are included in Table 3-14. One set of plots can be prepared between y_1 and E_1 for various values of q. This set can be expanded to include plots between y_2 and E_2. This becomes the specific energy plot. A typical specific energy curve is illustrated in Figure 3-4. A second set of plots can be prepared between H_L and E_2 for various values of q. These curves, called Blench curves, are useful for location of stilling basin floors in practical applications (Singh 1967). A typical Blench curve is illustrated in Figure 3-5.

Location of hydraulic jump is required for the design of aprons and stilling basins below drop structures, spillways, and sluice gates. Typical configuration of a drop with sloping apron is illustrated in Figure 3-6. If available tailwater depth is higher than the post-jump depth, the hydraulic jump will form on the sloping apron, constituting the downstream face of a drop structure or spillway. If the tailwater depth is lower than the post-jump depth, then the hydraulic jump will form on the horizontal floor connecting the drop to the downstream channel or downstream of the sluice gate. The location of hydraulic jump on horizontal aprons with low friction, where flow is subcritical, may be unstable and may move downstream with relatively small changes in discharge or tailwater depth. On the other hand, hydraulic jump on a sloping apron may be confined to a well-defined zone for the full range of flow conditions. Therefore, from a practical standpoint, it is desirable that the hydraulic jump forms above or at the toe of the sloping apron for most flow conditions (see Figure 3-6).

The length of jump, L, on a horizontal floor may be estimated from values given in Table 3-15 (USBR 1987).

Computational steps for the location of hydraulic jump on a horizontal apron or floor with mild slope (Figure 3-7(a)) are as follows (Chow 1959):

1. Compute water depth, y, on the sloping apron or immediately downstream of the sluice gate and tailwater depth, y_2', in the downstream channel.

2. Compute critical depth, y_c.

3. Compute mild slope water surface profile starting with y_c at the downstream end up to y at the upstream end.

4. For each depth, y_1, on the computed water surface profile, compute F_1 and the corresponding post-jump depth, y_2, and plot these post-jump depths (Figure 3-7(a)). The jump will form on the mild slope if $y_2' < y_2$.

5. Locate the point of intersection of the post-jump profile and tailwater depth line.

Table 3-14. Selected values of x, y, m, n, and l

$y = y_2/y_c$	$x = y_1/y_c$	$m = E_1/y_c$	$n = E_2/y_c$	$l = H_L/y_c$
1.0	1.0	1.5	1.5	0
1.1	0.906256	1.515047	1.513223	0.001824
1.2	0.82361	1.56071	1.547222	0.013488
1.3	0.750343	1.638419	1.595858	0.042561
1.4	0.685125	1.750323	1.655102	0.095221
1.5	0.626893	1.899175	1.722222	0.176953
1.6	0.574773	2.088256	1.795313	0.292944
1.6664	0.543177	2.237854	1.846458	0.391396
1.7	0.528031	2.321321	1.87301	0.448311
1.8	0.486042	2.602566	1.954321	0.648245
1.81	0.482082	2.633514	1.96262	0.670894
1.82	0.478165	2.664995	1.970948	0.694047
1.83	0.474288	2.697013	1.979303	0.71771
1.84	0.470452	2.729573	1.987684	0.741889
1.85	0.466656	2.76268	1.996092	0.766588
1.86	0.462899	2.79634	2.004525	0.791814
1.87	0.459182	2.830556	2.012984	0.817573
1.88	0.455503	2.865335	2.021467	0.843868
1.89	0.451863	2.900681	2.029974	0.870708
1.9	0.44826	2.9366	2.038504	0.898096
2.0	0.414214	3.328427	2.125	1.203427
2.1	0.383486	3.783431	2.213379	1.570052
2.2	0.35571	4.307359	2.303306	2.004054
2.3	0.330562	4.906319	2.394518	2.511801
2.4	0.307758	5.586763	2.486806	3.099957
2.5	0.287043	6.355485	2.58	3.775485
2.5078	0.285508	6.419362	2.587303	3.832059
2.6	0.268193	7.219615	2.673964	4.54565
2.7	0.251012	8.18661	2.768587	5.418022
2.8	0.235324	9.264254	2.863776	6.400479
2.9	0.220974	10.46065	2.959453	7.501202
3.0	0.207825	11.78423	3.055556	8.728679
3.1	0.195755	13.24374	3.152029	10.09171
3.2	0.184657	14.84821	3.248828	11.59938
3.3	0.174434	16.60703	3.345914	13.26111
3.4	0.165003	18.52986	3.443253	15.08661
3.5	0.156287	20.6267	3.540816	17.08588
3.6	0.148219	22.90783	3.63858	19.26925
3.7	0.140739	25.38386	3.736523	21.64733
3.8	0.133793	28.06568	3.834626	24.23106
3.9	0.127335	30.96452	3.932873	27.03165
4.0	0.12132	34.09188	4.03125	30.06063

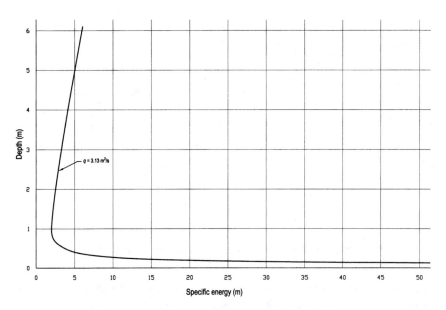

Figure 3-4. Typical specific energy curve

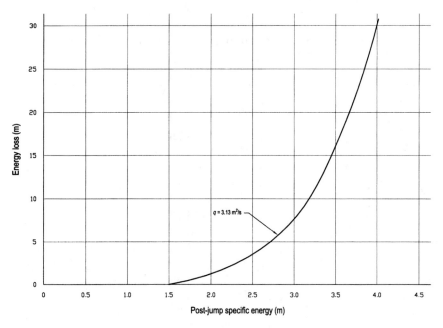

Figure 3-5. Typical Blench curve

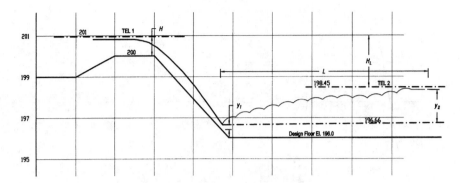

Figure 3-6. Drop with sloping apron

6. Estimate trial length of jump, L, corresponding to the post-jump depth at the aforementioned point of intersection.
7. Mark a horizontal intercept, L, between the tailwater depth and post-jump depth profiles. This is the location of the jump (A). If the post-jump depth at this location is significantly different from that used for estimation of the trial length of jump, then recompute trial jump length and mark that intercept to refine the jump location.

Table 3-15. Length of jump on horizontal floor

F	L_j/y_2
1.5	3.6
2	4.4
3	5.25
4	5.8
5	6.0
6	6.15
7	6.15
8	6.15
9	6.15
10	6.15
11	6.10
12	6.05
13	6.02
14	6.0
15	5.92
16	5.85
17	5.8
18	5.7
19	5.6
20	5.45

Source: USBR (1987).

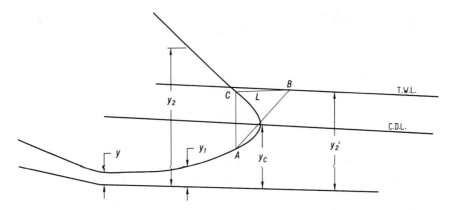

Figure 3-7(a). Hydraulic jump on mild slope

The hydraulic jump will form on the steep apron if $y_2' > y_2$. Computational steps for this case (Figure 3-7(b)) are as follows (Chow 1959):

1. Compute water depth, y, on the sloping apron or immediately downstream of the sluice gate and tailwater depth, y_2', in the downstream channel.

2. Compute critical depth, y_c, and plot the critical depth line (CDL) above the steep apron.

3. Compute steep slope water surface profile starting with y_c at the upstream end up to y_2' at the downstream end.

4. For each depth y_1 at the water surface on the steep apron, compute the post-jump (sequent) depth y_2 and plot the sequent depth curve.

5. Locate the point of intersection of the post-jump profile and tailwater depth line.

6. Estimate trial length of jump, L, corresponding to the post-jump depth at the above-mentioned point of intersection.

7. Mark a horizontal intercept, L, between the tailwater depth and post-jump depth profile. This is the location of the jump (A). If the post-jump depth at this location

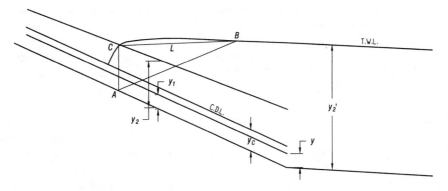

Figure 3-7(b). Hydraulic jump on steep slope

is significantly different from that used for estimation of the trial length of jump, then recompute trial jump length and mark that intercept to refine the jump location.

Where water surface profile computations for a known q are available both upstream and downstream of the drop structure, and hydraulic jump is desired to be located on a horizontal floor at or upstream of the toe of the sloping apron, simplified computational steps for practical cases are as follows:

- Estimate total energy upstream of the drop, TEL1 = upstream water surface elevation + $V_a^2/2g$, where V_a = velocity of approaching flow. Neglect friction losses on the sloping apron.

- Estimate total energy downstream of the drop, TEL2 = tailwater elevation + $V^2/2g$, where V = velocity in downstream channel.

- Estimate head loss = H_L = TEL1 − TEL2.

- With known q and H_L, estimate E_2 using Blench curves or computational steps used to develop those curves (described previously in this section).

- With known q and E_2, estimate y_2 from specific energy plots or computational steps used to develop those curves (described previously in this section).

- Draw a line at depth E_2 below TEL2, and locate the point where this line intersects the sloping apron. If the horizontal apron is located at or below this elevation, the hydraulic jump is likely to form at or above the toe of the sloping apron (Figure 3-6).

- If the length of the sloping apron is large, then y_1 and V_1 at the estimated location of the horizontal apron may be estimated using the procedures given in the section entitled "Flow along Steep Slopes." Then, TEL1 is computed using these values and the remaining computational steps are repeated to obtain revised location of the horizontal apron.

Design elevation of the horizontal floor or stilling basin downstream of a drop, spillway, or sluice gate requires testing for various discharges (e.g., 10, 25, 50, 75, and 100% of the design discharge). The horizontal apron or apron toe should be located below the location of hydraulic jump for the full range of anticipated discharges.

Example 3-6: Two design conditions for a stilling basin include q = 1.7 m²/s and 35.1 m²/s, respectively. Water surface profile computations for the lower flow indicated TEL1 = 201 m, tailwater elevation = 198.33 m, and TEL2 = 198.45 m. For the higher flow, TEL1 = 208.18 m, tailwater elevation = 206.05 m, and TEL2 = 206.22 m. Determine the location and approximate length of the horizontal floor.

Solution: Using Blench curves, E_2 can be estimated from known values of q and H_L = TEL1 − TEL2. Then, y_2 corresponding to the estimated E_2 can be read for given q from the specific energy plots. The elevation of the horizontal floor should be at or below the elevation TEL2 − E_2.

An alternative trial and error method of computations is illustrated in Table 3-16. A trial value of x is assumed and the corresponding value of y is computed from Eq. (3-55b). This enables computation of y_1, y_2, E_1, and E_2. The trials are continued until $E_1 - E_2$ is close to the known value of H_L.

Computations for the two given discharge conditions suggest that a 40-m-long horizontal apron should be provided at elevation 196.0 m, approximately (see Figure 3-6).

Unsteady Flow

One of the most common unsteady flow problems encountered in practice is the analysis of dam-break flood waves. For detailed analysis of dam-break flood waves, the NWS Dam-Break model (Fread 1988) or its equivalents should be used. Several simple, approximate equations have been developed to estimate the maximum flow, travel time, and flood depth resulting from dam-break. For some practical cases, these simpler methods also may be useful.

For a full depth, partial width breach,

$$Q_{max} = (8/27)\, bY_0\, (B/b)^{1/4}\, \sqrt{(g\, Y_0)} \qquad (3\text{-}56)$$

For a partial depth, full width breach,

$$Q_{max} = (8/27)\, BY\, (Y_0/Y)^{1/3}\, \sqrt{(g\, Y)} \qquad (3\text{-}57)$$

For a partial depth, partial width breach,

$$Q_{max} = (8/27)\, bY\, (B/b)^{1/4}\, (Y_0/Y)^{1/3}\, \sqrt{(g\, Y)} \qquad (3\text{-}58)$$

where

Q_{max} = peak flow (m³/s)

B = width of channel (m)

b = width of breach (m)

Table 3-16. Hydraulic jump calculations for location of horizontal floor below sloping aprons

Parameter	Lower flow conditions	Higher flow conditions
q (m²/sec)	1.7	35.1
H_L (m)	201.0 − 198.45 = 2.55	208.18 − 206.22 = 1.96
y_c (m)	0.6654	5.00
$l = H_L/y_c$	3.8323	0.392
x (for known l from Table 3-14)	0.2855	0.5432
y (Eq. (3-55b))	2.5078	1.6664
y_1 (m)	0.2855 × 0.6654 = 0.19	2.72
y_2 (m)	1.67	8.33
E_1 (m)	4.27	11.19
E_2 (m)	1.72	9.23
$F_1 = q/\sqrt{(g\, y_1^3)}$	6.55	2.498
Elevation of horizontal floor (m)	198.33 − 1.67 = 196.66	206.05 − 8.33 = 197.72
Length of horizontal floor (m)	6.2 × 1.67 = 10.4	4.8 × 8.33 = 40

Y = depth of water above bottom of breach (m)

Y_0 = depth of water behind dam (m) (USACE 1977)

For instantaneous and complete failure of dam and flow through a rectangular channel, water depth, Y (m), at time, t (s), and distance, x (m), from the dam (neglecting friction) is given by the following equation (Chow 1959; Rouse 1950):

$$x = 2\,t\sqrt{(g\,Y_0)} - 3\,t\sqrt{(g\,Y)} \qquad (3\text{-}59)$$

Empirical equations to estimate dam-break peak flows based on historical dam failure data are (USBR 1987)

$$Q_{max} = 19.13\,(Y_0)^{1.85} \qquad (3\text{-}60\text{a})$$

The Soil Conservation Service equation for dam-break peak flow is (USDA 1981)

$$Q_{max} = 16.58\,(Y_0)^{1.85} \qquad (3\text{-}60\text{b})$$

In practice, preliminary estimates should be made using several different methods, and a reasonably conservative value should be selected by judgment. Another convenient method to estimate dam-break flood flows, travel times, and water depths in a prismatic river channel for instantaneous and complete dam failure is to use Sakka's charts (USACE 1974).

If it is possible to estimate cross-sectional data for the downstream valley, storage-elevation data for the reservoir, length and top elevation of the dam, time of breach development, reservoir elevation at which failure starts, and bottom width, bottom elevation, and side slopes of potential breach, then the HEC-1 model (USACE 1991a) may be used to compute peak flows and flood levels at different times and locations in the downstream channel. Comparative evaluations of HEC-1 and NWS Dam-Break models have indicated that the two produce comparable results for peak flows and maximum stages in the downstream valley (Tschantz and Mojib 1981). It is advisable to use the NWS dam-break model (Fread 1988) for a more sophisticated analysis.

Overland Flow along Slopes

Storm-related overland flow along slopes is usually unsteady, spatially varied, and difficult to analyze. Empirical methods to estimate flow rates along slopes of solid waste piles and tailings impoundments include Izzard's method for laminar flow and Horton's method for turbulent flow (Chow 1959). Overland flows along steep slopes subjected to significant erosion may be assumed to be turbulent. For such cases, Horton's equation for uniform rate of rainfall excess is

$$q_L = 0.000278\,i\,\tanh^2\,[6.06\,t\{i/(c\,L)\}^{0.5}\,S^{0.25}] \qquad (3\text{-}61)$$

where

q_L = flow at the lower end of an elementary strip in l/s per square meter of area contributing surface runoff

i = rate of rainfall excess (mm/h)

t = time from beginning of rainfall excess (h)

Table 3-17. Typical values of roughness factor, c

Type of surface	c
Bare, compacted soil	0.10
Poor grass cover or moderately rough bare surface	0.20
Average grass cover	0.40
Dense grass cover	0.80

Source: Chow (1959).

S = surface slope along direction of flow (m/m)

L = length of elementary strip (m)

c = a roughness factor; values are shown in Table 3-17

Assuming area contributing surface runoff is an elementary strip of length, L, and unit width, overland flow per unit width of the slope is given by

$$q = q_L L \tag{3-62}$$

If the flow is found to be laminar (i.e., $iL < 3{,}871$), Izzard's equation should be used (Chow 1959). Horton's equation gives overland flow hydrograph at the end of the elementary strip due to rainfall excess continuing at a uniform rate for an indefinite period of time. Overland flow hydrograph for a finite duration of rainfall, t_0, can be estimated using the principle of superposition. For conservative estimates of overland flow along steep slopes, rainfall excess may be assumed to be nearly equal to rainfall.

Usually, overland flow at the toe of the slope is collected by a lateral ditch (parallel to the embankment length) and conveyed to nearby streams. The flow near the upstream end of the ditch is that at the bottom of the first elementary strip. Flow in the ditch increases with distance from its upstream end as more and more elementary strips discharge in it (i.e., flow in the ditch is spatially varied). Momentum of the incoming flow from successive elementary strips impedes the flow in the ditch. A relatively simple method to design the ditch is to provide a steep (supercritical) bed slope of about 0.05 and estimate ditch dimensions assuming critical flow at the downstream end (Figure 3-8).

Example 3-7: Develop an overland flow hydrograph at the end of a 61-m-long strip of unit width along a slope with $S = 0.014$ and $c = 0.198$, assuming a uniform rainfall intensity, $i = 243.17$ mm/h, during severe storms of 6- and 12-min durations. The length of the embankment is 1,000 m. Design a lateral ditch along the toe of the embankment slope to carry overland flow to a nearby stream. To be conservative and for the sake of simplicity, assume that the timings of flows from the elementary strips reaching the ditch are such that the peaks can be combined.

Solution: In this case, $i/(cL) = 20.137$; using Eq. (3-61), $q_L = 0.0676 \tanh^2[9.354\,t]$; and $q = 61\,q_L = 4.1236 \tanh^2[9.354\,t]$. Computations with linear superposition are illustrated in Table 3-18. Column (3) represents a hydrograph for continuous rainfall. For a 6-min duration rainstorm, hydrograph ordinates of column (3) are lagged by 0.10 h (column 4), and the difference in column (5) represents the corresponding hydrograph. For a 12-min duration rainstorm, hydrograph ordinates are lagged by 0.20 h, and the difference in column (7) represents the corresponding hydrograph.

The peak of the hydrograph for a 6-min rainstorm is q_{max} = 0.317 l/s, and that for a 12-min rainstorm is 0.629 l/s.

Q = total discharge at the downstream end of the ditch during a 12-min duration storm = $0.629 \times 1,000/1,000 = 0.629$ m³/s. Provide a trapezoidal ditch with $B = 0.3$ m. One side slope of this ditch is 0.014 or 71.428H:1V, and the other is 2H:1V.

$$A = 0.3\,y + (2 + 71.428)y^2/2$$

$$T = 0.3 + (2 + 71.428)y$$

Use critical flow equation, $Q = 0.629 = \sqrt{(gA^3/T)} = 3.132\,(0.3\,y + 36.714\,y^2)^{1.5}/(0.3 + 73.428\,y)^{0.5}$. By trial and error, $y = 0.14$ m. Using a freeboard of 0.2 m, provide a 0.34-m-deep ditch. The bed width may be uniformly reduced in the upstream direction to have a minimum of 0.15 m at the upstream end.

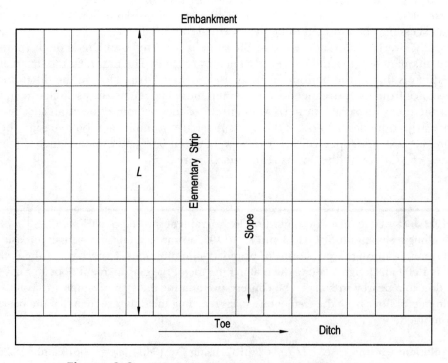

Figure 3-8. Overland flow along embankment slope

Table 3-18. Overland flow computations using Horton's method

			6-min duration rainfall		12-min duration rainfall	
t (h)	tanh (9.354 t)	q (l/s)	q (lagged)	q (diff)	q (lagged)	q (diff)
(1)	(2)	(3)	(4)	(5) = (3) − (4)	(6)	(7) = (3) − (6)
0.1	0.0997	0.041	0	0.041	0	0.041
0.2	0.197	0.161	0.041	0.12	0	0.161
0.3	0.291	0.35	0.161	0.189	0.041	0.309
0.4	0.38	0.595	0.35	0.245	0.161	0.435
0.5	0.462	0.881	0.595	0.285	0.35	0.531
0.6	0.537	1.189	0.881	0.309	0.595	0.594
0.7	0.604	1.506	1.189	0.317	0.881	0.626
0.8	0.664	1.818	1.506	0.312	1.189	0.629
0.9	0.716	2.116	1.818	0.297	1.506	0.61
1	0.762	2.39	2.116	0.276	1.818	0.574
1.1	0.800	2.642	2.392	0.251	2.116	0.527
1.2	0.834	2.866	2.642	0.223	2.39	0.474
1.3	0.862	3.062	2.866	0.196	2.642	0.42
1.4	0.885	3.232	3.062	0.170	2.866	0.366
1.5	0.905	3.378	3.232	0.146	3.062	0.316

Soil Erosion on Slopes

Overland flow along steep slopes results in sheet, rill, and gully erosion. Sheet and rill erosion is described in the next section of this chapter. Gully erosion on steep slopes is described in this section. Empirical equations have been developed to estimate the maximum depth and top width of a gully. Computational steps for a conservative estimate are as follows (Johnson 1999):

1. Estimate drainage area, A (m²), tributary to the gully:

$$A = 0.276 \, [L \cos \theta]^{1.636} \qquad (3\text{-}63)$$

where

L = length of slope (m)

θ = angle of slope with horizontal = arc tan (S), S being the slope (m/m)

2. Estimate average annual precipitation depth, P (m), and total average precipitation over the period of analysis (i.e., t yr). Mean annual precipitation at nearby precipitation gauges may be obtained from published records (e.g., Gale Research Company 1985).

3. Estimate average runoff to rainfall ratio, C, for the site. Average annual runoff in the site vicinity also may be obtained from published records (e.g., Gerbert et al. 1989 and Water Resources Data for different states annually published by the U.S. Geological Survey).

4. Estimate total volume of runoff, V (m³), expected at the toe of the slope during the period of analysis:

$$V = CPtA \tag{3-64}$$

5. Estimate maximum depth, D_{max} (m), of gully incision:

$$D_{max} = GLS \tag{3-65}$$

If clay content of slope material < 15%,

$$G = 1/[2.25 + \{0.789\ V/(H^3)\}^{-0.55}] \tag{3-66}$$

If 15% < clay content of slope material < 50%,

$$G = 1/[2.80 + \{0.197\ V/(H^3)\}^{-0.70}] \tag{3-67}$$

If clay content of slope material > 50%,

$$G = 1/[3.55 + \{0.76\ V/(H^3)\}^{-0.85}] \tag{3-68}$$

where H (m) = $L \sin \theta$ = vertical height of slope.

6. Estimate top width, W (m), of gully at the point of deepest incision:

$$W = (D_{max}/0.61)^{1.149} \tag{3-69}$$

7. Estimate distance, D_L (m), of D_{max} along the slope measured from the crest:

$$D_L/D_{max} = 0.713\ [V S/(L^3)]^{-0.415} \tag{3-70}$$

Because of the empirical nature of these equations, the estimates must be used only for qualitative analysis.

Example 3-8: Estimate potential for gully erosion on the slope of a solid waste disposal pile during periods of 10 and 200 yr after remediation. $\theta = 14°$; $L = 62$ m; clay content of slope material < 15%; $P = 0.5$ m/yr; and $C = 0.20$.

Solution: $S = \tan 14° = 0.25$ and $H = 15$ m. Using Eqs. (3-63) and (3-64), $A = 224.8$ m² and V (10 yr) = $0.20 \times 0.50 \times 10 \times 224.8 = 224.8$ m³ and V (200 yr) = 4,496 m³. Using Eq. (3-66), G (10 yr) = 0.137 and G (200 yr) = 0.31.

Using Eq. (3-65), D_{max} (10 yr) = $0.137 \times 62 \times 0.25 = 2.12$ m.

D_{max} (200 yr) = $0.31 \times 62 \times 0.25 = 4.81$ m.

Using Eq. (3-69), W (10 yr) = 4.18 m and W (200 yr) = 10.73 m.

Using Eq. (3-70), D_L (10 yr) = 48.4 m and D_L (200 yr) = 31.67 m.

The estimated values of D_{max} are such that there is potential for exposure of the waste material.

An empirical equation to estimate the rate of gully head advancement is (Thompson 1964)

$$R = 0.0065\ A^{0.49}\ S^{0.14}\ P^{0.74}\ E \tag{3-71}$$

where

> R = rate of gully head advancement in meters per unit time, the time period is that represented by the rainfall variable, P
>
> P = summation (mm) of those rainfall depths during the period of analysis that equaled or exceeded 12.7 mm in 24 h
>
> A = drainage area above gully head (ha)
>
> S = slope of approach channel above gully head (%)
>
> E = soil factor defined as clay content (i.e., particle sizes of 0.005 mm or smaller) in percentage by weight in the soil profile through which gully head is likely to advance

A simpler empirical equation developed by the Soil Conservation Service is the following: (USDA 1966)

$$R = 0.363\ A^{0.46}\ P^{0.20} \tag{3-72}$$

The rates given by these equations may be significantly different. Judgment and field verification, where possible, must be used to estimate the potential for gully head advancement and design structural or nonstructural measures for its control.

Sediment Yield Analysis

Average annual sediment yield of a watershed can be approximated using the Universal Soil Loss Equation (USLE):

$$T = 2.243\ R\ K\ LS\ CP \tag{3-73}$$

where

> T = soil loss (t/ha)
>
> R = rainfall factor
>
> K = soil erodibility factor
>
> LS = length slope factor
>
> C = soil cover factor
>
> P = conservation practice factor

These parameters may be estimated from published tables and charts (e.g., Barfield et al. 1981; USEPA 1977).

Empirical expressions also have been developed to estimate some of these parameters. To estimate average annual soil loss,

1. For regions where Type I storm (see "Design Storm Duration and Depth" in Chapter 2) is applicable,

$$R = 0.0134\ P^{2.2}_{2,6} \tag{3-74}$$

2. For regions where Type II storm is applicable,

$$R = 0.0219\ P^{2.2}_{2,6} \tag{3-75}$$

where $P_{2,6}$ = 2-yr, 6-h storm precipitation (mm).

To estimate soil loss for a single storm event,

1. For regions where Type I storm (see "Design Storm Duration and Depth" in Chapter 2) is applicable,

$$R = 0.0122\ P^{2.2}/D^{0.6065} \tag{3-76}$$

2. For regions where Type II storm is applicable,

$$R = 0.0156\ P^{2.2}/D^{0.4672} \tag{3-77}$$

where

P = total storm precipitation (mm)

D = storm duration (h)

The values of K for different types of soils in different regions may be obtained from local and state offices of the Natural Resource Conservation Service. Typical values are shown in Table 3-19 (SCS 1978; Barfield et al. 1981).

$$LS = (\lambda/22.13)^m\ [(430\ x^2 + 30\ x + 0.43)/6.613] \tag{3-78}$$

where

$x = \sin \theta$

θ = angle of ground slope

λ = slope length (m)

m = 0.2 for $x \leq 0.01$,

 = 0.3 for $0.01 < x \leq 0.03$,

 = 0.4 for $0.03 < x \leq 0.05$,

 = 0.5 for $0.05 < x \leq 0.12$, and

 = 0.6 for $x > 0.12$ (USEPA 1980).

Usually, C varies from 0.003 for 95 to 100% ground cover with brush to 1.0 for no ground cover, and P = 1.0 where no soil conservation practices are adopted or for land uses other than cropping. For contour cropping, P may vary from 0.50 for land slopes of 2 to 7% to 1.0 for land slopes of 25 to 30% (USEPA 1977). An updated computerized version of the soil loss equation is designated as the revised universal soil loss equation (RUSLE). The basic equation is the same as the USLE. RUSLE is a lumped process-type model based on the analysis of a large mass of experimental data to define the various factors of RUSLE. It includes detailed methods to calculate factors such as the cover management factor, C, for situations

Table 3-19. Typical values of soil erodibility factor, K

Soil type	K
Silt loam, silty clay loam, very fine sandy loam	0.37
Clay, clay loam, loam, silty clay	0.32
Fine sandy loam, loamy very fine sand, sandy loam	0.24
Loamy fine sand, loamy sand	0.17
Sand	0.15

Source: SCS (1978).

where experimental data are inadequate to define these parameters (Renard et al. 1991; Foster et al. 1996).

The Modified Universal Soil Loss Equation (MUSLE) may be used to estimate soil loss during a storm event (Williams 1975):

$$Y = 11.79 \, (Q \, q)^{0.56} \, K \, LS \, CP \tag{3-79}$$

where

Y = soil loss (t)

Q = runoff volume during storm (m^3)

q = peak flow (m^3/s)

All the sediment estimated by USLE or MUSLE may not reach the stream draining surface runoff from the watershed because of deposition along the overland flow path from the location of sediment erosion to the stream. To estimate the net amount of sediment reaching the stream, the above-mentioned estimates should be multiplied by the sediment delivery ratio, SDR (USEPA 1988a):

$$\text{SDR} = 0.77 \, L_d^{-0.22} \tag{3-80}$$

where L_d = overland flow distance (m). An alternative equation for SDR is (USACE 1989)

$$\text{SDR} = 0.30 \, A^{-0.2} \tag{3-81}$$

where A = drainage area (km^2)

Because of the subjective nature of the parameters of the USLE, MUSLE, and SDR, the reasonableness of the estimates should be verified by several alternative methods and, if possible, by field measurements. One alternative method is to use the Dendy and Boulton equations (USACE 1989):

$$T = 101.25 \, R^{0.46} \, [1.54 - 0.26 \log A] \text{ if } R \leq 50.8 \text{ mm} \tag{3-82}$$

$$T = 685.83 \, [\exp(0.002165 \, R) \, \{1.54 - 0.26 \log A\}] \text{ if } R > 50.8 \text{ mm} \tag{3-83}$$

where

T = sediment yield (t/km^2/yr)

R = mean annual surface runoff (mm)

The values of R may be obtained from Gerbert et al. (1989) or flow records of nearby streams published in USGS Water Resources Data for each year.

Computer models, e.g., SEDIMOT-2 (Wilson et al. 1984) have been developed to compute storm runoff hydrographs and sediment yield from the watershed using USLE or MUSLE and sediment delivery ratio, and to route the runoff and sediment hydrographs through a sedimentation basin. These models estimate deposition in the sedimentation basin, based on size fractions and fall velocities of sediments eroded from the watershed and detention time in the basin, and compute sediment load in the outflow.

An empirical equation to estimate the quantity of sediment reaching a reservoir is (USBR 1987)

$$Q_s = 1098 \, A^{-0.24} \tag{3-84}$$

where Q_s = sediment yield (m³/km²/yr). To use this equation to verify the estimates based on USLE or Dendy-Boulton equations, Q_s should be increased by about 5 to 25% to account for bed-load transport (USBR 1987; Simons and Senturk 1976) and by the inverse of the SDR to estimate soil loss from the watershed.

Another approximate method to verify the reasonableness of these estimates is to use available streamflow and sediment load data for other locations on the stream or another stream with similar flow and sediment transport characteristics. The observed daily discharge (m³/s) and suspended sediment load (t/day) are plotted on a log-log paper. A straight line is fitted through the plotted points. This straight line or its equation may be used to estimate the suspended load from known daily discharges at the location of interest. To use this method to verify the reasonableness of the previous estimates, the estimated suspended load should be adjusted for bed load (i.e., multiplied by 1.05 to 1.25) and divided by the SDR. If site-specific data are not available, preliminary estimates should be made using several approximate methods and reasonable values selected by judgment.

Example 3-9: The measured daily suspended sediment load and corresponding streamflows for a creek are shown in Table 3-20. The bed load is estimated to be 5% of the suspended load, and the SDR is estimated to be 0.18. Estimate sediment yield from the watershed during the snowmelt period when the average streamflow is 0.25 m³/s.

Solution: Regression of the logarithms of Q_s (t/day) and Q (l/s) as dependent and independent variables, respectively, gives

$$\log Q_s = -4.182 + 1.783 \log (Q)$$

or

$$Q_s = 0.0000658 \, Q^{1.783}, \text{ with } r^2 = 0.72$$

For $Q = 250$ l/s, $Q_s = 1.24$ t/day. Adjusting for bed load (i.e., multiplying by 1.05) and for SDR (i.e., dividing by 0.18) gives approximate average sediment yield of the watershed as 7.23 t/day. This must be verified by estimates obtained by other methods.

An approximate equation to estimate total sediment transport capacity of a stream is (Johnson 1999; Simons et al. 1981)

$$q_s = [(3.281)^{b+c-2}] \, a \, D^b \, V^c \tag{3-85}$$

Hydraulic Analysis

where

q_s = sediment transport rate in m²/s

D = average depth of flow (m)

V = average velocity (m/s)

a, b, and c are coefficients shown in Table 3-21

Eq. (3-85) may be applicable to Froude number = 1 to 4; V = 1.98 to 7.92 m/s; bed slope = 0.005 to 0.040; and 0.062 mm $\leq d_{50} \leq$ 15 mm.

Table 3-20. Suspended sediment load and streamflow data

Q_s (t/d)	Q (l/S)	Q_s (t/d)	Q (l/S)	Q_s (t/d)	Q (l/S)	Q_s (t/d)	Q (l/S)	Q_s (t/d)	Q (l/S)
0.15	36.81	0.11	73.61	0.13	67.95	95.26	1101.40	0.02	70.78
0.22	45.30	0.34	70.78	0.02	70.78	2.54	543.60	0.04	70.78
0.07	42.47	1.81	368.07	0.05	65.12	1.54	206.68	0.02	70.78
0.39	45.30	0.41	65.12	0.11	124.58	0.91	110.42	0.16	50.96
0.44	65.12	0.25	56.63	0.34	266.14	0.24	87.77	0.01	53.79
0.53	87.77	0.20	65.12	7.35	540.77	0.17	82.11	0.01	53.79
0.08	93.43	0.06	67.95	12.70	637.04	0.03	79.28	0.06	73.61
1.18	235.00	0.14	113.25	5.17	597.40	0.08	84.94	0.20	178.37
0.31	198.19	0.15	130.24	8.44	654.02	1.09	263.31	0.41	70.78
0.68	127.41	3.99	569.09	7.62	679.50	0.41	235.00	0.11	31.14
0.17	42.47	0.58	396.38	19.05	1044.70	0.20	121.74	0.18	42.47
0.12	39.64	0.41	198.19	15.42	959.80	0.05	70.78	0.22	113.25
0.06	42.47	0.29	133.07	2.90	685.17	0.07	59.46	0.32	76.44
0.28	45.30	0.07	65.12	0.81	266.14	0.91	50.96	1.72	362.40
0.33	90.60	0.15	73.61	0.05	62.29	0.02	70.78	0.37	268.97
0.16	84.94	0.06	82.11	0.20	79.28	0.06	56.63	0.23	82.11
0.08	135.90	0.08	104.76	0.04	70.78	0.40	42.47	0.08	56.63
0.33	342.58	0.22	251.98	0.02	70.78	0.22	59.46	0.27	56.63
1.45	274.63	15.42	317.10	0.05	59.46	0.08	67.95	0.13	50.96
0.31	93.43	0.31	73.61	0.44	65.12	0.44	118.91	0.01	33.98
0.12	65.12	0.18	84.94	0.03	104.76	0.40	59.46	0.03	53.79
0.20	53.79	0.05	59.46	0.54	203.85	0.05	28.31	0.09	53.79
0.32	48.13	0.21	67.95	1.54	314.27	0.16	36.81	0.18	31.14
0.03	53.79	0.05	62.29	1.18	311.44	0.05	39.64	0.15	53.79
0.11	50.96	0.14	59.46	1.72	311.44	0.06	42.47	0.46	56.63
0.49	67.95	0.06	67.95	9.98	869.20	0.02	53.79	0.11	104.76
4.35	515.29	0.05	62.29	31.75	922.99	0.02	33.98	0.25	209.51
1.91	232.16	3.72	537.94	94.35	1223.10	0.05	48.13	3.27	467.16
0.28	48.13	2.00	549.27	16.33	1330.70	1.54	319.93	0.46	184.03
0.18	36.81	3.63	311.44	152.41	1647.80	4.81	566.25	0.27	121.74
0.44	50.96	0.15	99.09	1959.60	1843.20	0.51	209.51	0.09	65.12
0.30	45.30	0.06	56.63	470.84	1616.70	0.15	99.09	0.12	67.95

Table 3-21. Coefficients in sediment transport capacity equation

d_{50} (mm)	a	b	c
\multicolumn{4}{c}{$G = $ gradation coefficient $= 0.5\,[(d_{84}/d_{50}) + (d_{50}/d_{16})]$}			

d_{50} (mm)	a	b	c
	$G = 1.0$		
0.1	3.30×10^{-5}	0.715	3.30
0.25	1.42×10^{-5}	0.495	3.61
0.5	7.60×10^{-6}	0.280	3.82
1.0	5.62×10^{-6}	0.060	3.93
2.0	5.64×10^{-6}	−0.140	3.95
3.0	6.32×10^{-6}	−0.240	3.92
4.0	7.10×10^{-6}	−0.300	3.89
5.0	7.78×10^{-6}	−0.340	3.87
	$G = 20$		
0.25	1.59×10^{-5}	0.510	3.55
0.5	9.80×10^{-6}	0.330	3.73
1.0	6.94×10^{-6}	0.120	3.86
2.0	6.32×10^{-6}	−0.090	3.91
3.0	6.62×10^{-6}	−0.196	3.91
4.0	6.94×10^{-6}	−0.270	3.90
	$G = 3.0$		
0.5	1.21×10^{-5}	0.360	3.66
1.0	9.14×10^{-6}	0.180	3.76
2.0	7.44×10^{-6}	−0.020	3.86
	$G = 4.0$		
1.0	1.05×10^{-5}	0.210	3.71

Source: Johnson (1999).

Wind Erosion

Estimates of soil loss due to wind erosion are required to evaluate the potential for exposure of tailings or solid wastes protected by soil covers. Soil loss due to wind erosion is expressed in a functional form (USDA 1982, 1983b):

$$E = f(I, K, C, L, V) \qquad (3\text{-}86)$$

where

E = average annual soil loss tons/acre/yr (t/ha/yr))
f = a function of indicated variables

I = soil erodibility to wind stress, usually expressed by dividing soils into wind erodibility groups (WEGs) based on soil types (e.g., sand, loam, silt, clay, clods, and stones)

K = surface roughness factor, usually expressed by an index of 1.0 for smooth surface and 0.5 for ridged surface

C = climatic factor (expressed as a percentage) developed for each month for different locations based on average wind velocity and precipitation–evaporation index

L = unsheltered distance, usually obtained from available charts for different widths of soil surfaces, distances along prevailing wind direction, and angles of deviation of prevailing wind

V = vegetative cover, usually expressed in terms of weight of flat small grain residue per unit area

Vegetative cover, V, is estimated from charts that express a given amount of flat grain stubble of a given crop (e.g., sorghum, cotton, or corn) in a field into an equivalent amount of flat small grain residue (R) from the previous crop. The charts and tables developed for different states should be used for specific studies. Technical notes, including these charts and tables for different states, are available from local offices of the NRCS (USDA). To facilitate wind erosion predictions, the wind erosion equation also has been computerized (e.g., USNRC 1982).

Example 3-10: Estimate annual soil loss rate due to wind erosion for a site in Wyoming where WEG (reflecting variable I of Eq. (3-86)) = 2; pile surface is unridged (reflecting variable K); C = 30% (for critical month at the location of the site in Wyoming); V = 500 lb/acre (560.4 kg/ha); and L = 2,800 ft (853.4 m). Usually, the critical month would be that in which the most severe wind erosion is expected.

Solution: Relevant values of E for unridged surface in WEG2 with C = 30% for Wyoming are included in Table 3-22. Using Table 3-22 for Wyoming (USDA 1982), E = 26 tons/acre (58.3 t/ha) per year.

Sediment Transport Analysis

Sediment transport analysis includes estimates of aggradation, degradation, and deposition in rivers and reservoirs and can be performed using computer models such as the HEC-6 model (USACE 1991d). Input for this model includes the following:

1. Hydrographs of flow in the main channel and tributaries

2. Channel cross sections with roughness coefficients and discharges at each, similar to the HEC-2 and HEC-RAS models (USACE 1991c, 1998)

3. Grain size distribution of channel bed material at each cross section

4. Suspended loads of inflowing water in the main channel and tributaries and their grain size distributions

In most field situations, detailed site-specific information is not available and input data have to be assimilated based on limited data and judgment. In some cases, preliminary esti-

Table 3-22. Soil loss in t/acre (t/ha) on unridged surface, WEG = 2, C = 30%, in Wyoming

Unsheltered distance across field along prevailing wind direction (ft (m))	Flat small grain residue [lbs/acre (kg/ha)]				
	1,000 (1120.8)	750 (840.6)	500 (560.4)	250 (280.2)	0
100	3.3	6.8	11	14.5	19
(30.5)	(7.4)	(15.2)	(24.7)	(32.5)	(42.6)
200	5.5	10	15.7	20	25.5
(61)	(12.3)	(22.4)	(35.2)	(44.8)	(57.2)
500	8.5	15	21.9	27	33.5
(152.4)	(19.1)	(33.6)	(49.1)	(60.5)	(75.1)
1,000	10	17	24.5	30	37.8
(304.8)	(22.4)	(38.1)	(54.9)	(67.2)	(84.7)
2,000	10.6	17.7	25.8	32	39.9
(609.6)	(23.8)	(39.7)	(57.8)	(71.7)	(89.4)
3,000	10.9	18	26	32.7	40
(914.4)	(24.4)	(40.3)	(58.3)	(73.3)	(89.7)
4,000	11	18	26	33	40
(1,219)	(24.7)	(40.3)	(58.3)	(74.0)	(89.7)
5,000	11	18	26	33	40
(1,524)	(24.7)	(40.3)	(58.3)	(74.0)	(89.7)

Source: Adapted from charts given in USDA (1982).

mates have to be made with limited data if contaminated sediments eroded from a site would reach a specific downstream location or settle mostly within the stream reach upstream of that location. Computational steps for such preliminary analyses include the following:

1. Estimate expected maximum, average, and minimum flow velocities and water depths in the study reach.

2. Estimate average d_{15}, d_{50}, and d_{95} of the sediment likely to be eroded from the site.

3. Estimate fall velocities for known particle sizes from charts (e.g., USBR 1971) or using Rubey's equation (Low 1989):

$$w = [\{(2/3)\ g\ (G_s - 1)\ d^3 + 36\ \nu^2\}^{0.5} - 6\ \nu]/d \qquad (3\text{-}87)$$

where

w = fall velocity (m/s) of particles of size, d (m)

ν = kinematic viscosity (m²/s)

G_s = specific gravity of particles

4. Estimate minimum, average, and maximum travel distances, L (m), of particle of size, d (m), using the relationship $L \cong D\ V/w$, where D = average water depth (m) in the study reach and V = flow velocity (m/s).

Example 3-11: The maximum velocity and depth of flow in a stream flowing by a contaminated site are estimated to be 1.52 m/s and 4.6 m, respectively. The average d_{15} of the contaminated sediments is about 0.05 mm. A reservoir is located about 6 km from the site. Estimate if any significant portion of the contaminated sediments is likely to reach the reservoir. Use $G_s = 2.65$ and $\nu = 0.000001131$ m²/s.

Solution:

$w = [\sqrt{\{(2/3) \times 9.81 \times 1.65 \times (0.00005)^3 + 36 \times (0.000001131)^2\}}$

$- 6 \times (0.000001131)]/0.00005 = 0.00197$ m/s.

$L \text{ (maximum)} = 4.6 \times 1.52/(0.00197) = 3{,}550$ m.

This suggests that potential is remote for any significant portion of contaminated sediments reaching the reservoir.

To control the amount of sediment transported by a stream from reaching downstream facilities, sedimentation basins are provided. A major portion of the annual sediment load of steams is transported during storm events. Sedimentation basins for flood peak attenuation and sediment entrapment during storm events can be designed using computer models such as SEDIMOT-2 (Wilson et al. 1984).

A preliminary estimate of the trap efficiency of a sedimentation basin for sediments of mean size, d (m), may be obtained using the empirical equation (USBR 1971; Barfield et al. 1981; Vetter 1940)

$$W/W_0 = \exp(-wL/q) = \exp(-wA/Q) \qquad (3\text{-}88)$$

where

W_0 = weight of sediment entering basin (kg)

W = weight of sediment leaving basin (kg)

L = length of basin (m)

w = fall velocity (m/s) of particles of size d (m)

q = discharge per meter width of basin (m²/s)

A = surface area of basin (m²)

Q = basin inflow or outflow (m³/s)

Sedimentation or settling basins also are required to trap sediments transported by the inflowing water on a continuous basis (e.g., settling basins are required to trap sediments from water used for groundwater recharge systems or water recycling systems of industrial facilities). Such basins may be sized using the following equation (USBR 1971; Johnson 1999):

$$P = [(1 - \exp(-X)] \qquad (3\text{-}89a)$$

$$X = (1.055\, L\, w)/(V \times D) \qquad (3\text{-}89b)$$

where

P = fraction of sediments of size, d (m), deposited over the length of the basin

V = average flow velocity in the inflowing channel (m/s)

w = fall velocity (m/s) of particles of size, d (m)

D = water depth in the basin (m)

Example 3-12: Estimate the size of a settling basin for a channel carrying 6.5 m³/s with average flow velocity of 0.15 m/s and average annual sediment load of 20,000 m³/yr, with particle size distribution given in Table 3-23. Use $G_s = 2.65$ and $\nu = 0.00000094$ m²/s.

Solution: For first trial, assume a basin with $L = 300$ m and $D = 3$ m; then, $X = 1.055$ $(300\ w)/(0.45) = 703\ w$. The computations for trap efficiency of the settling basin are shown in Table 3-24.

The trap efficiency of the basin is $13{,}181/20{,}000 = 0.66$.

If the basin is to accommodate only one year of sediment load with a sediment depth of 0.3 m, then its average width is $13{,}181/(300 \times 0.3) = 146$ m. If this size of basin is not acceptable, computations should be repeated for other values of L.

Open-Channel Dispersion

Near-Field Dispersion

Dispersion and mixing zone analysis is required to evaluate impacts of pollutant discharges in streams through point and nonpoint sources. For single-port discharges, such analyses can be conducted using models such as CORMIX (USEPA 1996a). Near-field dispersion includes dilution resulting from discharges through jets and diffusers, which is affected by the momentum and buoyancy of the jet discharge and geometry of the outfall. Computational steps for a preliminary analysis for simple horizontal round jets, similar to a pipe outlet, are (Fischer et al. 1979) listed below:

- Estimate the momentum flux of jet (m⁴/s²):

$$M = Q\,V_j \tag{3-90}$$

Table 3-23. Particle size distribution of annual sediment load

% finer	Particle size (mm)
100	0.5
98	0.25
83	0.125
46	0.062
33	0.031
22	0.016
13	0.008
7	0.004

HYDRAULIC ANALYSIS

Table 3-24. Computations of trap efficiency

(1) D(mm)	(2) Fraction in total sediment load	(3) w (m/s)	(4) X = 703 w	(5) P = 1 − e^{−X}	(6) Average P	(7) Inflowing sediment load (m³/yr) (col. (2) × 20,000)	(8) Deposition (m³/yr) (col. (6) × col. (7))
0.5		0.0619	43.52	1.0			
	0.02				1.0	400	400
0.25		0.0305	21.44	1.0			
	0.15				1.0	3,000	3,000
0.125		0.0116	8.15	0.9997			
	0.37				0.96	7,400	7,104
0.062		0.0036	2.53	0.92			
	0.13				0.69	2,600	1,794
0.031		0.00088	0.619	0.46			
	0.11				0.306	2,200	673
0.016		0.000235	0.165	0.152			
	0.09				0.096	1,800	173
0.008		0.000058	0.0408	0.04			
	0.06				0.025	1,200	30
0.004		0.0000146	0.0103	0.0102			
	0.07				0.005	1,400	7
Total	1.0					20,000	13,181

- Estimate the characteristic length:

$$l_c = \text{characteristic length (m)} = Q/M^{1/2} \quad (3\text{-}91)$$

- Estimate the maximum jet velocity at distance, x, from jet orifice:

$$w_m = 7\, M\, l_c/(Q\, x) \quad (3\text{-}92)$$

- Estimate:

$$C_m/C_0 = 5.6\, l_c/x \quad (3\text{-}93)$$

- Estimate mean dilution (s_{mean}) at distance, x, from jet orifice:

$$s_{\text{mean}} = (\text{volume flux at distance } x)/(\text{volume flux at jet orifice}) = \mu/Q = 0.25\, x/l_c \quad (3\text{-}94)$$

where

Q = discharge through jet orifice (m³/s)
V_j = jet velocity (m/s)
μ = volume flux at distance x (m³/s)

C_m = maximum concentration at distance x

C_0 = initial concentration

For a single round buoyant (i.e., density of wastewater is less than the density of the receiving water) vertical port, it is found that flow becomes similar to a plume within a short distance from the jet orifice. Computational steps for a preliminary analysis are (Fischer et al. 1979) as follows:

- Estimate the momentum flux of jet (m⁴/s²):

$$M = Q V_j \qquad (3\text{-}90)$$

- Estimate the characteristic length:

$$l_c = \text{characteristic length (m)} = Q/M^{1/2} \qquad (3\text{-}91)$$

- Estimate the characteristic length for the buoyant jet:

$$l_m \text{ (m)} = M3/4/\sqrt{B} \qquad (3\text{-}95)$$

- Estimate the buoyancy flux:

$$B = [(\rho_a - \rho)/\rho] \, g \, Q \qquad (3\text{-}96)$$

where

ρ_a = density of receiving water

ρ = density of effluent

For $y \gg l_m$, where y = water depth above the port, flow is similar to a plume, and for $y \ll l_m$, flow is similar to a jet.

- Estimate the Richardson number, R_0, for a round jet:

$$R_0 = l_c/l_m = Q\sqrt{B}/M^{5/4} \qquad (3\text{-}97)$$

- Estimate dimensionless depth:

$$\zeta = 0.25 \, (y/l_c) \, (R_0/0.557) \qquad (3\text{-}98)$$

- If $y > l_m$, estimate volume flux in a plume at depth y above port:

$$\bar{\mu} = \zeta^{5/3} \qquad (3\text{-}99)$$

- If $y < l_m$, estimate volume flux in a jet at depth y above port:

$$\bar{\mu} = \zeta \qquad (3\text{-}100)$$

- Estimate mean dilution at depth y:

$$\mu/Q = \bar{\mu} \ (0.557/R_0) \quad (3\text{-}101)$$

Multiport diffusers consist of several closely spaced ports. Wastewater discharges exiting through these ports tend to merge to form a vertical plume within a short distance from the locations of the ports. Steps to perform a preliminary analysis for dilution in the plume are (Fischer et al. 1979) listed below:

- Estimate discharge, q, per unit length of diffuser; initial pollutant concentration, C_0, above that of receiving water; densities of discharge, ρ_a, and receiving water, ρ (kg/m³); $\Delta\rho = \rho_a - \rho$; and $g' = g\Delta\rho/\rho$
- Estimate:

$$(C - C_a)/(C_0 - C_a) = 2.63 \ q^{2/3}/(g'^{1/3} \ y) \quad (3\text{-}102)$$

where

C = diluted concentration at height y above diffuser ports

C_a = background concentration in ambient water

For cases where analysis of multiport diffuser is not practicable, the plane vertical source for far-field dispersion analysis may be assumed to be equal to the length of the diffuser, L, and height equal to 30% of the depth of the diffuser (Chin 1985).

Far-Field Dispersion

Far-field dispersion includes dilution due to flow turbulence in the channel away from the outfall where jet momentum is not significant. Some models for simulation of water quality of rivers and reservoirs include QUAL2E (USEPA 1987a), WQRRS (USACE 1978), and WASP4 (USEPA 1988b). For preliminary analysis of dispersion of contaminants released into a stream, simpler and approximate analytical or quasi-analytical equations may be useful. Generally, regulatory agencies specify an acute or chronic regulatory mixing zone (ARMZ or CRMZ) of surface water bodies within which normal water quality standards are not applied. It is that region of the surface water body downstream of the point of wastewater discharge where physical mixing occurs in all directions until the constituents in the discharge achieve uniform concentrations in the receiving water. Complete mixing is assumed to occur at a stream cross section where concentrations at all points within the cross section are within 5% of the mean value for that cross section. Most water quality standards specify the use of 7-day, 10-yr average low flow (7Q10) to evaluate water-quality–related impacts on surface water. The size of the mixing zone should be kept to a minimum and should allow safe passage, protection, and propagation of aquatic organisms. In addition, contamination in the mixing zone should not be acutely toxic. For a point source, mixing lengths, L_{mix}, may be estimated by (Fischer et al. 1979)

$$L_{mix} \text{ (complete mixing)} = 0.1 \ u \ W^2/D_y \quad (3\text{-}103)$$

if the source is located at center of stream cross section, or by

$$L_{mix} \text{ (complete mixing)} = 0.4 \ u \ W^2/D_y \quad (3\text{-}104)$$

if the source is located on the bank of the stream

where

u = flow velocity

W = width of stream

D_y = coefficient of lateral dispersion

Three analytical models to simulate far-field dispersion in streams are described here. The first of these models uses (Prakash 1977, 1999)

$$C(x, y, z)/C_0 = [Q_0/\{4(z_2 - z_1)\sqrt{(\pi x u D_y)}\}]$$

$$\Sigma \Sigma [\{\text{erf}(E1) - \text{erf}(E2)\}\{\exp(E3)\}] \qquad (3\text{-}105a)$$

where

$$E1 = [z - mD - (-1)^m z_1]/[\sqrt{(4 D_z x/u)}] \qquad (3\text{-}105b)$$

$$E2 = [z - mD - (-1)^m z_2]/[\sqrt{(4 D_z x/u)}] \qquad (3\text{-}105c)$$

$$E3 = -[y - nW - (-1)^n y_0]^2/[(4 D_y x/u)] \qquad (3\text{-}105d)$$

where

erf (x) = error function of x

Σ (first) is from $m = -\infty$ to ∞

Σ (second) is from $n = -\infty$ to ∞

$C(x, y, z)$ = concentration at the point (x, y, z)

C_0 = source concentration

Q_0 = rate of discharge of contaminated water

D_y, D_z = transverse and vertical dispersion coefficients, respectively

D = depth of flow

x, y, z = distances in longitudinal, transverse, and vertical directions, respectively, with origin at the center of the channel

z_1, z_2 = vertical coordinates of the vertical line source

y_0 = y-coordinate of the point where the vertical line source is located

The channel downstream of the source is represented by an equivalent average rectangular cross section. The bed, banks, and water surface of the channel are assumed to be no-flux boundaries. The effects of these boundaries are accounted for by the method of images. For steady-state transport, the contribution of longitudinal dispersion is neglected. The flow in the stream is assumed to be uniform and steady, and the contribution of secondary currents is assumed to be accounted for by the transverse dispersion coefficient.

Eq. (3-105a) is applicable to a vertical line source located at $x = 0$, where x is taken to be positive in the downstream direction. If the source is a vertical plane with finite length,

L, parallel to the direction of flow (i.e., x-direction), then it is assumed to consist of a number of vertical sources of infinitesimal length placed adjacent to each other. The contribution of these vertical sources is obtained by linear superposition:

$$C'(x, y, z, L) = \Sigma\, C(x + i\Delta x, y, z) \tag{3-106}$$

where

$C(x, y, z, L)$ = concentration at a point (x, y, z) due to a plane vertical source of finite length, L

$C'(x + i\Delta x, y, z)$ = concentration at $(x + i\Delta x, y, z)$ due to a vertical line source located at $x = 0$, given by Eq. (3-105a) with Q_0 replaced by Q_0/N

Δx = infinitesimal length of a vertical source = L/N

N = number of vertical sources of infinitesimal length in which the source length, L, is divided

For continuous point source discharge, the convective-dispersion equation in steady, uniform flow (advection) in the longitudinal direction (i.e., along the direction of river flow) becomes

$$C(x, y, z)/(C_0\, Q_0) = [1/\{4\pi x \sqrt{(D_y D_z)}\}]$$
$$[\exp(-\lambda x/u)] \Sigma\Sigma \exp[-\{y - nB - (-1)^n y_0\}^2/(4 D_y x/u)\}] \times$$
$$\exp[-\{z - mD - (-1)^m z_0\}^2/(4 D_z x/u)] \tag{3-107}$$

where

C_0 = concentration in the discharge at outfall

Q_0 = rate of outfall discharge

$C(x, y, z)$ = concentration at the point (x, y, z)

D_y = transverse dispersion coefficient

u = velocity

D_z = vertical dispersion coefficient

λ = decay coefficient (t^{-1})

y_0 = y-coordinate of the point source or outfall

z_0 = z-coordinate of the point source or outfall

x = distance downstream from the outfall, which is located at $x = 0$

Σ (first) is from $n = -\infty$ to ∞

Σ (second) is from $m = -\infty$ to ∞

The origin of coordinates is located at the center of the river at mid-depth at $x = 0$. For mass conservation in some cases, 100 or more terms of Eq. (3-107) may have to be used in the computations. Therefore, it is advisable to perform the computations using a Fortran program or spreadsheet. Discharges from several outfalls located at various points on the riverbank may be simulated using the principle of superposition; i.e.,

$$C(x, y, z, T) = \Sigma\, C_i Q_i\, [C\{x + i\Delta x_i, y, z\}] \quad (3\text{-}108)$$

where

$C(x, y, z, T)$ = concentration at point (x, y, z) due to discharges at T outfalls

subscript $i = 1$, refers to the most downstream

$i = 2$, refers to the next upstream outfall, and so on

Δx_i = distance of outfall number i from the most downstream outfall

$C\{x + i\Delta x_i, y, z\}$ = value (right-hand side) computed from Eq. (3-107) using $[x + i\Delta x_i]$ for x

Σ is from $i = 1$ to total number of outfalls, T

This model may be used to estimate concentrations at various points along the width and depth of the river at the downstream edge of the mixing zone. This model assumes the following:

1. There is steady, uniform flow in the river reach. The steady-state assumption implies continuous discharge from the outfalls. Uniform flow implies a constant river cross section in the entire river reach. A representative average rectangular cross section is estimated for the river reach within the mixing zone. Irregularities in the river cross section are likely to enhance dispersion and dilution. Therefore, the assumption of a uniform rectangular cross section may be conservative.

2. River flow is one-dimensional (i.e., along the river course) with no secondary flows in other directions. This assumption is also conservative because secondary flows tend to enhance mixing or increase transverse dispersion.

The second analytical model to simulate far-field dispersion in streams (USEPA 1987b) uses

$$C(x, y)/C_0 = [1/\sqrt{(4\pi D_y t)}\,[\int \exp(E4)\, dy' + \int \exp(E4)\, dy' + \int \exp(E4)\, dy' + \int \exp(E4)\, dy' + \ldots] \quad (3\text{-}109)$$

where

the limits of the first, second, third, and fourth integrals are from a to b, c to d, e to f, and g to h, respectively

$E4 = -(y - y')^2/(4 D_y t)$

$a = -B/2 - y_s$

$b = B/2 - y_s$

$c = -B/2 + y_s$

$d = B/2 + y_s$

$e = -B/2 - 2W - y_s$

$f = B/2 - 2W - y_s$

$g = -B/2 + 2W + y_s$

$h = B/2 + 2W + y_s$

$t = x/u$

B = length of horizontal line source located across the stream

y_s = distance from center line of source to the stream bank, which is assumed to be located at $y = 0$

This model simulates only longitudinal advection and transverse dispersion and internally computes the transverse dispersion coefficient using the relationship $D_y = 0.6\, u_* D$, where u_* = shear velocity = $\sqrt{(g D S)}$. It assumes a horizontal line source with a finite width across the stream and uses the method of images to account for the effects of the channel boundaries. Each of the integrals in Eq. (3-109) represents the contribution of an image of the source and the image pattern repeats up to infinity.

The third model (USNRC 1976) simulates a point source and uses

$$C(x, q)/C_0 = (Q_0/Q)\,[1 + 2\,\Sigma\,\{\exp(E5) \cos(n\,\pi\,q_s/Q)\cos(n\,\pi\,q/Q)\}] \quad (3\text{-}110)$$

where

Σ is from 1 to ∞

$E5 = -(n^2\,\pi^2\,D^2\,D_y\,u\,x)/Q^2$

Q = total stream flow

q = stream flow from the bank (assumed to be at $y = 0$) to a distance y where $C(x, q)$ is computed, such that as $y \rightarrow W$, $q \rightarrow Q$

y_s = distance of point source from the bank

q_s = stream flow between $y = 0$ and $y = y_s$

This model also simulates only longitudinal advection and transverse dispersion.

Example 3-13: Contaminated water seeps through a 3.35-m-deep seepage zone along the bank of a perennial stream. Estimated rates of seepage through different lengths of the seepage zone along the stream are given in Table 3-25. Average width and depth of the receiving stream are 244 m and 3.35 m, respectively, and average velocity corresponding to the 7Q10 flow is 0.62 m/s. Assume $D_y = 0.16$ m²/s and uniform distribution of concentration along depth. Estimate the length of complete mixing.

Solution: For uniform distribution of concentrations along depth, assume a relatively high value of D_z (e.g., 0.16 m²/s). The source may be simulated as a vertical plane extending from $x = 0$ to $x = -2{,}936$ m (x being positive in downstream direction); $z_1 = -1.675$ m; $z_2 = 1.675$ m; and $y_0 = -122$ m (origin being at the center of channel cross section). For mass conservation, a large number of images (i.e., 1,000) is used in a Fortran program. Dividing the seepage zone into 10 smaller segments and using Eqs. (3-105a) and (3-106), estimated concentrations at different locations are shown in Table 3-26.

At $x = 320{,}000$ m, concentrations throughout the width of the channel are found to be 0.012 mg/l. Thus, the theoretical length required for complete mixing may be about 320,000 m. Using estimated concentrations at additional points along the channel width, mixing zone with its tail defined by a specified concentration (e.g., 0.012 mg/l) can be sketched. A commonly used criterion is to define the limit of complete mixing or mixing zone at the cross section where concentrations at all points are within 5% of the mean value for that cross section (Fischer et al. 1979). The cross section where maximum deviation from the mean concentration is within about 5% is about 25,000 m downstream from the source. Theoretically, this large length may be required because D_y is relatively small.

In certain cases, constituent concentrations in surface waters may be affected by processes such as biodegradation and volatilization. The contribution of these processes may be estimated by introducing a multiplying factor in Eq. (3-105a), such that

$$C(x, y, z, \lambda) = C(x, y, z) \exp(-\lambda\, x/u) \qquad (3\text{-}111)$$

where

$C(x, y, z)$ is given by Eq. (3-105a)

$C(x, y, z, \lambda)$ = concentration at point (x, y, z), including the contribution of biodegradation and volatilization

$\lambda = \lambda_1 + \lambda_2$

λ_1 = biodegradation rate constant (t^{-1})

λ_2 = volatilization rate constant (t^{-1})

Biodegradation rate constants in an aerobic environment for different constituents may be obtained from the literature (e.g., USEPA 1985). Suggested first-order biodegradation rate constants for selected constituents for screening level analyses are given in Table 3-27 (USEPA 1985).

Table 3-25. Rates of seepage of contaminated water

Length of seepage zone (m)	Rate of seepage (l/s/ m)	Concentration (mg/l)
0–564	0.0155	250
564–1,203	0.008	250
1,203–1,961	0.0011	250
1,961–2,701	0.0065	250
2,701–2,936	0.0053	250

Table 3-26. Concentration distribution in stream

Distances from downstream edge of seepage zone, x (m)	Concentration (mg/l)		
	Distances along stream width from bank where seepage occurs		
	0 (m)	70 (m)	120 (m)
30	0.19	0.002	0.00004
300	0.11	0.003	0.00007
3,000	0.05	0.014	0.0016
15,000	0.025	0.018	0.010
320,000	0.012	0.012	0.012

The volatilization rate constant is a function of Henry's law constant and is inversely proportional to the depth of water in which the constituent is dissolved. Henry's law constant (K_H) for selected chemicals is given in Table 3-28. For others, K_H may be estimated by

$$K_H \cong [(\text{vapor pressure in mm of Hg}) \times (\text{molecular weight in gm/mole})] / [(\text{water solubility in mg/l}) \times 760] \quad (3\text{-}112)$$

Table 3-27. First-order biodegradation rate constants and Henry's law constants for selected chemicals

Constituent	Suggested biodegradation rate constant, λ_1, for screening level analyses (day^{-1})[a]	Henry's law constant (atm-m^3/mole)[b]
Heptachlor	0	8.19×10^{-4}
Carbon tetrachloride	>0.5 (use 0.5)	2×10^{-2}
1,1-dichloroethylene	>0.5 (use 0.5)	3.4×10^{-2}
1,2-dichloroethylene-*cis*	0.05 to 0.5 (use 0.05)	7.58×10^{-3}
1,2-dichloroethylene-*trans*	0.05 to 0.5 (use 0.05)	6.56×10^{-3}
Trichloroethylene	0.05 to 0.5 (use 0.05)	9.10×10^{-3}
Tetrachloroethylene	0.05 to 0.5 (use 0.05)	2.59×10^{-2}
Benzene	>0.5 (use 0.5)	5.59×10^{-3}
Chlorobenzene	>0.5 (use 0.5)	3.72×10^{-3}
1,2-dichlorobenzene	0.05 to 0.5 (use 0.05)	1.93×10^{-3}
1,3-dichlorobenzene	0.05 to 0.5 (use 0.05)	3.59×10^{-3}
1,4-dichlorobenzene	0.05 to 0.5 (use 0.05)	2.89×10^{-3}
Ethylbenzene	>0.5 (use 0.5)	6.43×10^{-3}
Toluene	>0.5 (use 0.5)	6.37×10^{-3}

[a]Source: USEPA (1985).
[b]Source: Maidment (1993).

Table 3-28. Typical values of volatilization rate constants

Henry's law constant, K_H (atm-m³/mole)	Volatilization rate constant, λ_2, for mixed water depth of 1 m (day^{-1})
1	4.8
10^{-1}	4.8
10^{-2}	4.7
10^{-3}	4.2
10^{-4}	1.8

Source: USEPA (1985).

Typical values of volatilization rate constant for a water depth of 1 m are given in Table 3-28 (USEPA 1985). Values of λ_2 for water depths other than 1 m may be estimated by dividing the values in Table 3-28 by the water depth in meters.

Example 3-14: Estimate the concentration in Example 3-13 at a distance of 15,000 m from the location of seepage and 70 m from the bank along the river width. The constituent is trichloroethylene (TCE), and contributions of biodegradation and volatilization are to be included.

Solution: Using a Fortran program or spreadsheet to perform the computations of Eq. (105a) at $x = 15,000$ m and at a distance of 70 m from river bank, $C(x, y, z) = 0.018$ mg/l with $u = 0.62$ m/s and water depth = 3.35 m. Also, the constituent is uniformly distributed over depth. From Table 3-25, for TCE, $\lambda_1 = 0.05$ day^{-1}. From Table 3-28, for TCE with $K_H = 9.10 \times 10^{-3}$, $\lambda_2 \cong 4.6$ day^{-1} for water depth of 1 m and $4.6/3.35 = 1.373$ day^{-1} for a water depth of 3.35 m. So, $\lambda = 0.05 + 1.373 = 1.423$ day^{-1}.

Also, $x/u = 15,000/(0.62 \times 86,400) = 0.28$ day. Therefore, $C(x, y, z, \lambda) = 0.018 \exp(-1.423 \times 0.28) = 0.012$ mg/l.

Estimation of Dispersion Coefficients

Some equations to estimate longitudinal dispersion coefficient, D_x, are

$$D_x/(U_* D) = 5.915 \, (W/D)^{0.62} \, (U/U_*)^{1.428} \quad \text{(Seo and Cheong 1995)} \quad (3\text{-}113)$$

$$D_x = 5.93 \, U_* D \quad \text{(USNRC 1976)} \quad (3\text{-}114)$$

where

D = average water depth

U_* = shear velocity

W = average channel width

Experimental data for some streams are included in Table 3-29 (Fischer et al. 1979; Graf 1995).

Table 3-29. Longitudinal dispersion coefficients for selected streams

Stream	Depth (m)	Width (m)	Mean velocity (m/s)	Longitudinal dispersion coefficient, D_x (m²/s)
Sacramento River, California	4.0		0.53	15.0
South Platte River, Colorado	0.46		0.66	16.2
Missouri River	2.7	200	1.55	1,500
Clinch River, Tennessee	0.85	47	0.32	14.0
Clinch River, Virginia	0.58	36	0.21	8.1
Powell River, Tennessee	0.85	34	0.15	9.5
Copper Creek, Virginia	0.40	19	0.16	9.9
Colorado River, Lees Ferry to Nautiloid Canyon	8.2	71.6	0.91	109–164

Source: Fischer et al. (1979); Graf (1995).

Some equations to estimate transverse dispersion coefficient, D_y, are

$$D_y = 0.23 \; U_* \, D \; \text{(USNRC 1976)} \tag{3-115}$$

$$D_y = 0.15 \; U_* \, D \; \text{(Fischer et al. 1979)} \tag{3-116}$$

$$D_y = 0.60 \; U_* \, D \; \text{with variation of 50\% (USEPA 1991c)} \tag{3-117}$$

$$\log (D_y / U_* \, D) = -2.698 + 1.498 \log (W/D) \; \text{(Bansal 1971)} \tag{3-118}$$

Experimental data for some streams are included in Table 3-30 (Fischer et al. 1979). Some equations to estimate vertical dispersion coefficient, D_z, are

$$D_z = 0.067 \; U_* \, D \; \text{(Fischer et al. 1979)} \tag{3-119}$$

$$\log (D_z / \nu) = -8.08 + 1.89 \log (U_* \, D / \nu) \; \text{(Bansal 1971)} \tag{3-120}$$

Table 3-30. Transverse dispersion coefficients for selected streams

Stream	Depth (m)	Width (m)	Mean velocity (m/s)	Transverse dispersion coefficient, D_y (m²/s)
Potomac River, Maryland	0.73–1.74	350	0.29–0.58	0.013–0.058
Missouri River, downstream of Cooper Nuclear Station, Nebraska	4	210–270	5.4	1.1
Missouri River, near Blair, Nebraska	2.7	200	1.75	0.12

Source: Fischer et al. (1979).

A relatively high D_z (nearly equal to D_y) may be used to simulate conditions where the contaminant is well mixed over the depth of the stream and concentration variations in the vertical direction are not significant.

Dispersion coefficients should be estimated using several methods and reasonable values selected by judgment. Where practical, values should be verified by field measurements.

Dissolved Oxygen Concentrations in Streams

Wastewater discharges in streams result in increases in their biochemical oxygen demand (BOD), which consumes dissolved oxygen (DO). DO concentrations in river waters below 4 to 5 mg/l are detrimental to aquatic biota. The hydraulics and processes of BOD exertion and DO consumption in rivers are dependent on the following:

1. Temporal (time-dependent) and spatial variations in streamflows, water depths, and velocities
2. Spatial variations in river geometry (i.e., shapes of stream cross sections and existence of river bends)
3. Temporal variations in climatic factors (e.g., wind velocity, temperature, atmospheric pressure, rain, snow, and sunlight)
4. Temporal and spatial variations in ambient DO in the river
5. River conditions (e.g., areal extent of ice cover, water temperatures below the ice cover and in ice-free areas)
6. Geometry of the outfalls (e.g., point, line, or nonpoint configuration) and velocity of discharge from the outfalls (i.e., high or low velocities that may affect initial mixing)
7. Extent of photosynthesis, respiration, and sediment oxygen demand in the riverine environment

Some commonly used models to simulate these processes include QUAL2E (USEPA 1987a), WASP4 (USEPA 1988b), and WQRRS (USACE 1978). For preliminary estimates of DO concentrations in streams resulting from BOD discharges through point sources (e.g., outfalls), a relatively simple approach based on the Streeter-Phelps formulation may be used. For steady, uniform flow conditions, the Streeter-Phelps equation predicts DO deficit at locations downstream from a point where initial DO deficit and BOD concentrations are known. DO deficit is defined as

$$D = C_s - DO_0 \qquad (3\text{-}121)$$

where

D = DO deficit (mg/l)

C_s = DO saturation concentration (mg/l) at the water temperature of the river and at the altitude of the river at the location of interest

DO_0 = ambient DO concentration in river water (mg/l)

The Streeter-Phelps equation for D at a point downstream from a given location is (USEPA 1985, 1987a; USACE 1978)

$$D = [K_L L_0/(K_a - K_L)] [\exp(-K_L x/u) - \exp(-K_a x/u)] + D_0 \exp(-K_a x/u) \qquad (3\text{-}122)$$

where

K_a = re-aeration coefficient (day^{-1})

K_L = BOD decay coefficient (day^{-1})

D = DO deficit at x (mg/l)

x = distance downstream (m)

u = flow velocity (m/day)

D_0 = initial DO deficit at $x = 0$ (mg/l)

L_0 = initial BOD (mg/l) at $x = 0$

The Streeter-Phelps model assumes that

1. The contributions of photosynthesis, respiration, and sediment oxygen demand are negligible. Usually, these contributions are small if there is no algal growth in the river reach of interest.

2. In the river reach of interest, streamlines are parallel to the direction of flow, and there is no DO or BOD interchange between two parallel streamlines. This is the basic assumption of the Streeter-Phelps formulation. Once steady, uniform flow conditions are assumed, this may be the resultant situation. This may be conservative because it implies no dilution of BOD or DO enhancement due to transverse mixing (by dispersion) in the entire stream reach.

Using Eq. (3-122), the DO deficit or DO concentrations at various locations in the downstream river reach may be estimated. A plot of DO concentrations (y-axis) against river distances is called the oxygen sag curve. The critical (lowest) value of DO concentration or maximum value of DO deficit is given by

$$D_{max} = D_0 \exp(-K_a t_c) + \{K_L L_0/(K_a - K_L)\}\{\exp(-K_L t_c) - \exp(-K_a t_c)\}$$
$$= [K_L L_0/K_a] [\exp(-K_L t_c)] \qquad (3\text{-}123a)$$

where t_c = critical time when D_{max} occurs and is given by

$$t_c = [1/(K_a - K_L)] \ln [(K_a/K_L) - D_0 K_a (K_a - K_L)/(L_0 K_L^2)] \qquad (3\text{-}123b)$$

The distance where D_{max} occurs is given by

$$X = u \, t_c \qquad (3\text{-}124)$$

DO analysis requires that the initial BOD concentration at $x = 0$ be known. A widely used measure of BOD is the 5-day BOD (or BOD_5), which is defined as the BOD that has been exerted (oxidized) after $t = 5$ days of exertion/oxidation, i.e.,

$$BOD_5 = BOD_0 [1 - \exp(-K_L t)] \quad (3\text{-}125)$$

where

BOD_0 = initial BOD (mg/l) at $t = 0$

t = time in days taken to be 5 to estimate BOD_5

Typical values of K_L at 20°C vary from about 0.2 to 0.3 day^{-1} with a commonly used value of 0.23 day^{-1}. Its variation with respect to temperature may be estimated by

$$K_L \text{ (at } T°C) = K_L \text{ (at } 20°C) \times (1.135)^{T-20} \quad (3\text{-}126)$$

The re-aeration coefficient may be estimated by O'Connor's formulation (USEPA 1985), i.e.,

$$K_a \text{ (at } 20°C) = 3.932 \, u^{0.5}/Y^{1.5} \quad (3\text{-}127)$$

where

u = velocity (m/s)

Y = water depth (m)

Values at other temperatures may be estimated by

$$K_a \text{ (at } T°C) = K_a \text{ (at } 20°C) \times (1.024)^{T-20} \quad (3\text{-}128)$$

The DO saturation concentration may be taken from tabulated values (e.g., USEPA 1985) or approximated by

$$C_s \text{ (at } T°C \text{ and at 0-m altitude)} = 14.65 - 0.41022 \, T + 0.00791 \, T^2 - 0.00007774 \, T^3 \quad (3\text{-}129)$$

Values at altitudes, E, other than 0 may be estimated by

$$C_s \text{ (at altitude } E \text{ m)} = C_s \text{ (at altitude 0 m)} \times [1 - 0.00011656 \, E] \quad (3\text{-}130)$$

The values estimated by Eqs. (3-127) and (3-128) are for 0% ice cover. A multiplying factor of 0.05 may be used to estimate values for 100% ice cover (USEPA 1987a). Values for other percentages of ice cover may be approximated by linear interpolation. Typical values of C_s at atmospheric pressure and $T = 0°$ to 35°C are given in Table 3-31 (USEPA 1985).

The BOD discharged from an outfall is diluted by advection and dispersion in the river water. Usually, water quality criteria may not be met within a relatively small mixing zone in the immediate vicinity of the outfall. According to USEPA (1991c), there may be up to two types of mixing zones applicable to aquatic life criteria: chronic and acute. In the zone

Table 3-31. Solubility of oxygen in water at atmospheric pressure

T (°C)	C_s (mg/l)
0	14.6
5	12.8
10	11.3
15	10.2
20	9.2
25	8.4
30	7.6
35	7.1

Source: USEPA (1985).

immediately surrounding the outfall, neither the acute nor the chronic criterion is met. The acute criterion has to be met at the edge of the inner mixing zone. The chronic criterion has to be met at the edge of the second mixing zone. Commonly adopted mixing zones contain no more than 25% of the cross-sectional area and should not extend over more than 50% of the river width.

Example 3-15: BOD decay rate coefficients for a constituent are reported to be 0.221/day at 20°C and 0.026/day at 3°C. The re-aeration coefficient for the stream is 3.53/day at 20°C. Estimate the values at 2°C with 81% of ice cover in the stream.

Solution: The variation of BOD decay rate coefficient with temperature is given by Eq. (3-126):

$$K_L \text{ (at 3°C)} = K_L \text{ (at 20°C)} (A)^{3-20}$$

So, $0.026 = 0.221 (A)^{-17}$. This gives $A = 1.134$.

So, BOD decay rate coefficient at 2°C = K_L (at 2°C) = $0.221 \times (1.134)^{2-20} = 0.02298$/day.

The re-aeration coefficient at 2°C (Eq. 3-128) = K_a (at 2°C) = $3.53 \times (1.024)^{2-20} = 2.30$/day.

The recommended multiplying factors to adjust K_a for 0% and 100% ice cover conditions are 1.0 and 0.05, respectively (USEPA 1987a). By linear interpolation, the multiplying factor for 81% ice cover is estimated to be 0.23. Thus,

$$K_a \text{ (at 2°C with 81\% ice cover)} = 0.23 \times 2.30 = 0.53/\text{day}.$$

Example 3-16: In a stream, 33.85 mg/l of BOD is introduced at a point where the initial DO deficit is 2.0 mg/l. Due to downstream controls, the stream has a relatively slow flow velocity of 2,592 m/day. Estimate DO deficit at a distance of 1,200 m downstream from the point where the BOD was introduced. The re-aeration coefficient (K_a) and BOD decay coefficient (K_L) are 0.0933 day^{-1} and 0.01787 day^{-1}, respectively. Also, estimate the critical time and critical DO deficit. Assume that the effect of transverse dispersion on BOD dilution can be neglected.

Solution: Use Eq. (3-122), with $K_a = 0.0933$ day^{-1}, $K_L = 0.01787$ day^{-1}, $D_0 = 2.0$ mg/l, $u = 2,592$ m/day, $x = 1,200$ m, and $L_0 = 33.85$ mg/l:

$$K_a x/u = 0.0933 \times 1,200/2,592 = 0.0432$$

$$K_L x/u = 0.01787 \times 1,200/2,592 = 0.00827$$

DO deficit (at $x = 1,200$ m) $= D = 2.0 \times [\exp(-0.0432)] + \{33.85 \times 0.01787/(0.0933 - 0.01787)\}\{\exp(-0.00827) - \exp(-0.0432)\} = 2 \times 0.9577 + 8.019 [0.9918 - 0.9577] = 2.19$ mg/l

Using Eq. (3-123b), $t_c = \{1/(0.0933 - 0.01787)\} \ln [(0.0933/0.01787) \{1 - 2 \times (0.0933 - 0.01787)/(0.01787 \times 33.85)\}] = 13.257 \ln (3.9189) = 18.1$ days.

Using Eq. (3-123a), $D_{max} = 2.0 \times \exp(-0.0933 \times 18.1) + \{33.85 \times 0.01787/(0.0933 - 0.01787)\}[\exp(-0.01787 \times 18.1) - \exp(-0.0933 \times 18.1)] = 0.369 + 8.019 (0.7236 - 0.1846) = 4.69$ mg/l.

Pipe Flow

Flow through pipes or tunnels may be pressure flow or open-channel flow. Open-channel flow may be analyzed using the methods described previously. Pressure flow through pipes may be laminar if the Reynolds number, R_{ed}, (using pipe diameter) $\leq 2,000$; transitional if $2,000 < R_{ed} < 4,000$; and turbulent if $R_{ed} \geq 4,000$. Using pipe diameter,

$$R_{ed} = V d/\nu \qquad (3\text{-}131)$$

where d = diameter of pipe. For a pipe flowing full, $R = d/4$. So, $R_{ed} = 4 R_e$, where R_e is the Reynolds number computed using the hydraulic radius, R, in place of d.

Laminar Flow in Pipes

Usually, laminar flow occurs when viscous fluids (e.g., crude oil, glycerin, and certain lubricating oils) are transported through pipes at moderately low velocities. Velocity distribution along the cross section of the pipe for laminar flow is given by the Hagen-Poiseuille equation:

$$u = [(p_1 - p_2)/(4 \mu L)] [(d^2/4) - y^2] \qquad (3\text{-}132)$$

where

y = distance from center of pipe (m)

u = velocity (m/s) at y

p_1, p_2 = pressure head at the upstream and downstream points (kg/m^2)

L = distance between the two points (m)

μ = dynamic viscosity (kg-sec/m^2)

The maximum velocity in the cross section occurs at the center and is given by

$$u_{max} = [(p_1 - p_2)\, d^2/(16\, \mu\, L)] \tag{3-133}$$

and the average velocity over the cross section is

$$V = [(p_1 - p_2)\, d^2/(32\, \mu\, L)] \tag{3-134}$$

Example 3-17: Estimate the head and power required to transport crude oil over a distance of 1 km at the rate of 1,500 l/h through a 10-cm-diameter pipeline, $\mu = 0.12$ poise and density, $\rho = 860$ kg/m^3. The minimum expected temperature in the site vicinity is about 7°C. Also, compute R_{ed}.

Solution: 1,500 l/h = 0.0004167 m^3/s. $V = 1.5 \times 4/(3{,}600 \times \pi \times 0.10^2) = 0.053$ m/s.

$\mu = 0.12$ poise = 0.12 dyne-s/cm^2 = 0.12 gm/cm-s = $0.12 \times 1.020 \times 10^{-3}$ gm-s/cm^2 = 0.1224×10^{-2} kg-s/m^2 = 0.012 kg/m-s.

ν (m^2/s) = μ (kg/m-s)/ρ (kg/m^3) = 0.012/860 = 0.000014 m^2/s.

Using Eq. (3-134), $p_1 - p_2 = 32\, \mu$ (kg-s/m^2) $VL/d^2 = 32 \times 0.001224 \times 0.053 \times 1{,}000/(0.10^2) = 207.6$ kg/m^2.

Required head = 207.6/860 = 0.24 m.

Required power = $Q\,(p_1 - p_2) = 0.00041673 \times 207.6 = 0.0865$ kg-m/s.

$R_{ed} = V d\rho/\mu = 0.053 \times 0.10 \times 860/0.012 = 380$. This is less than 2,000. So, the flow is laminar.

Turbulent Flow in Pipes

Steady flow of incompressible fluids in pipes is governed by the Bernoulli equation:

$$p_1/\gamma + V_1^2/2g + z_1 = p_2/\gamma + V_2^2/2g + z_2 + h_L \tag{3-135}$$

where

p_1 = pressure (kg/m^2) at point 1

p_2 = pressure at point 2

V_1, V_2 = velocities (m/s) at points 1 and 2, respectively

z_1, z_1 = elevation at points 1 and 2, respectively

h_L = head loss due to friction and other minor losses between points 1 and 2

If the pressure in Eq. (3-134) is expressed in height of fluid (m), then

$$h = (p_1 - p_2)/\gamma = 32\, \mu\, VL/(\rho\, g\, d^2) = (64/R_{ed})\, [L\, V^2/(2\, g\, d)] \tag{3-136}$$

Setting $64/R_{ed} = f =$ roughness factor for the pipe, Eq. (3-136) becomes the Darcy-Weisbach equation for friction loss through pipes:

$$h = fL V^2/(2 g d) \tag{3-137}$$

The values of f may be obtained from Moody's diagram, which is available in most texts (e.g., Streeter 1971; Potter and Wiggert 1991; Brater et al. 1996). Alternatively, Manning's, Chezy's, or Hazen-Williams equations [Eqs. (3-3), (3-4), and (3-6)] may be used if the respective coefficients can be estimated. The roughness factor, f, can be estimated without knowing the pipe diameter if Chezy's, C, is known:

$$f = 8 g/C^2 \tag{3-138}$$

To estimate f from n, the pipe diameter must be known; and from C_H, both pipe diameter and gradient, S, must be known:

$$f = 124.58 \, n^2/(d^{1/3}) \tag{3-139}$$

$$f = 156.06/(C_H^2 \, d^{0.26} \, S^{0.08}) \tag{3-140}$$

Typical values of n and C_H for commonly used pipes are given in Tables 3-32(a) and 3-32(b) (Brater et al. 1996; ASCE 1976).

As an alternative to the use of Moody's diagram, friction loss, flow rate, or pipe diameter may be estimated explicitly if the pipe wall roughness height, e, kinematic viscosity of fluid, ν, and the other two of the three parameters are known (Swamee and Jain 1976):

$$h_L = [1.07 \, Q^2 \, L/(g D^5)] \, [\ln \{(e/(3.7D)) + 4.62 \, (\nu D/Q)^{0.9}\}]^{-2} \tag{3-141}$$

valid for $10^{-6} < e/D < 10^{-2}$ and $3{,}000 < R_{ed} < 3 \times 10^8$. Note that e and D must be in some units.

Table 3-32(a). Typical values of Manning's n for pipes

Type of pipe	Manning's n
Asbestos-cement	0.011–0.015
Clean uncoated cast iron	0.013–0.015
Clean coated cast iron	0.012–0.014
Galvanized iron	0.015–0.017
Corrugated metal	0.023–0.029
Corrugated metal, structural plate	0.030–0.033
Rough formed concrete	0.015–0.017
Steel formed concrete	0.012–0.014
Vitrified clay	0.011–0.017
Clay tiles	0.012–0.014
Unlined rock tunnel	0.038–0.041
Enameled steel	0.009–0.010
Brass or glass	0.009–0.013
Plastic (smooth)	0.011–0.015
PVC	0.009–0.010

Source: Brater et al. (1996); ASCE (1976).

Table 3-32(b). Typical values of Hazen-Williams, C_H

Type of pipe	Hazen-Williams, C_H
PVC, glass, or enameled steel pipe	130–150
Riveted steel pipe	100–110
Cast iron pipe	95–100
Smooth concrete pipe	120–140
Rough pipe (e.g., rough concrete pipe)	60–80

Source: Brater et al. (1996); ASCE (1976).

$$Q = -0.965 \, (g D^5 h_L/L)^{0.5} \ln [e/(3.7 D) + \{3.17 \nu^2 L/(g D^3 h_L)\}^{0.5}] \quad (3\text{-}142)$$

valid for $R_{ed} > 2{,}000$.

$$D = 0.66 \, [e^{1.25}(L Q^2/g h_L)^{4.75} + \nu Q^{9.4} \{L/(g h_L)\}^{5.2}]^{0.04} \quad (3\text{-}143)$$

valid for $10^{-6} < e/D < 10^{-2}$ and $5{,}000 < R_{ed} < 3 \times 10^8$.

Equations (3-141) to (3-143) are particularly useful for computer applications. Typical values of e are given in Table 3-33 (Brater et al. 1996; ASCE 1976; Chow 1959; Streeter 1971).

In addition to friction, there are minor losses in pipes at the entrance, exit, contractions and expansions, bends, and gates and valves. These losses are expressed as

$$h_L = k V^2/2g \quad (3\text{-}144)$$

where

h_L = loss of head

V = pipe flow velocity at or immediately upstream of the respective features

k = minor loss coefficient

Table 3-33. Typical values of pipe wall roughness, e

Pipe material	e (mm)
Drawn tubing	0.0015
Enameled steel	0.005
Wrought iron	0.06–2.4
Galvanized iron	0.15–4.6
Cast iron	0.12–3.0
Corrugated metal	0.30–60
Fully riveted steel	0.30–9.0
Concrete	0.30–3.0

Source: Brater et al. (1996); ASCE (1976); Chow (1959); Streeter (1971).

134 WATER RESOURCES ENGINEERING

Table 3-34. Saturated vapor pressure of water

Temperature °C	Saturated vapor pressure (m)
0	0.06
10	0.125
21	0.256
32	0.49
60	2.03
100	10.33

Source: Streeter (1971).

Values of k for different features may be found in standard texts (e.g., Streeter 1971; Potter and Wiggert 1991; Brater et al. 1996). At pipe enlargements or contractions, V pertains to the smaller pipe.

Sometimes, flow through a contraction is limited by cavitation, which occurs when absolute pressure at the contraction reaches the saturated vapor pressure of the liquid. Cavitation also may occur at the high point in a siphon. Saturated vapor pressure for water at selected temperatures is given in Table 3-34 (Streeter 1971).

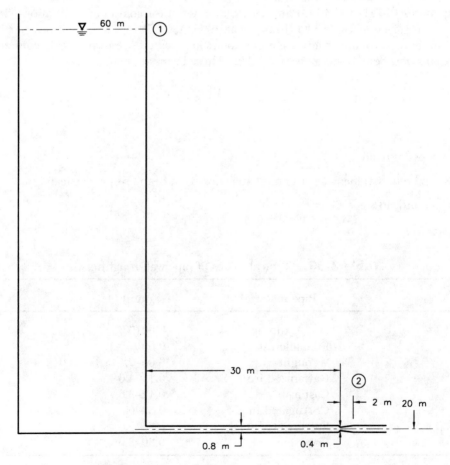

Figure 3-9. Schematic of low-level outlet

Hydraulic Analysis 135

Example 3-18: Estimate critical safe discharge of water at 16°C at which cavitation may begin at the location of the valve in the low-level outlet of a dam shown in Figure 3-9. Assume $f = 0.015$; k_c = loss coefficient at contraction = 0.30; k_e = loss coefficient at expansion = 0.35; p_a = atmospheric pressure = 10.33 m. Compare safe discharge of the system with the condition when a 0.8-m valve is installed without the contraction-expansion. Assume vapor pressure of water at 16°C = $p_v/\gamma = 0.18$ m. Note that both k_c and k_e are reported with respect to the velocity in the smaller pipe.

Solution: To avoid cavitation, pressure at the contraction where the valve is located should remain above vapor pressure. Point 1 is located at the reservoir water surface, and point 2 is at the center of the outlet pipe.

By continuity equation between points 1 and 2,

$$V_c \times (\pi/4)(0.4^2) = V \times (\pi/4)(0.8^2) \text{ and } V_c = 4V$$

where

V_c = velocity through contracted section

V = velocity through regular pipe section

By Bernoulli's equation between points 1 and 2,

$$z_1 + p_a/\gamma = z_2 + p_v/\gamma + fL\,V^2/2g\,d + 0.30\,V_c^2/2g + V_c^2/2g$$

$$60 + 10.33 = 20 + 0.18 + (0.015 \times 30 \times V^2)/(2g \times 0.8) + 0.30 \times 16\,V^2/2g + 16\,V^2/2g$$

So, $V = 6.79$ m/s and $Q = 6.79 \times (\pi/4)(0.64) = 3.411$ m³/s or 3,411 l/s.

If there was no contraction-expansion, point 2 may be taken at the pipe outlet. Now applying Bernoulli's equation between points 1 and 2,

$$z_1 + p_a/\gamma = z_2 + p_a/\gamma + fL\,V^2/2g\,d + V^2/2g$$

$$60 + 10.33 = 20 + 10.33 + 0.015 \times 32\,V^2/(2g \times 0.8) + V^2/2g$$

Or, $40 = 1.60\,V^2/2g$, $V = 22.15$ m/s; and $Q = 22.15 \times (\pi/4) \times 0.64 = 11.13$ m³/s or 11,130 l/s.

Example 3-19: Estimate the sizes of new subdivisions (assuming four persons per house) that can be supported by the water supply lines shown in Figure 3-10. The water is to be supplied at heads of 222 and 220 m, respectively. The quantity of available water is 1178.1 l/s. The average water requirement in the area is 650 l per capita per day. Also, estimate the pumping capacity required at the connection to the water source where the available head, without pumping, is 200 m. Although some valves will be installed for control of flow to either subdivision, assume no valves in the system for this preliminary analysis.

Solution: $H_1 = 222$ m; $d_1 = 0.6$ m; $L_1 = 1,200$ m; $f_1 = 0.018$; $H_2 = 220$ m; $d_2 = 0.8$ m; $L_2 = 1,500$ m; $f_2 = 0.015$; $H = 200$ m; $d = 1.0$ m; $L = 1,000$ m; and $f = 0.012$. Applying the continuity equation,

$$V_1\,(\pi/4)\,d_1^2 + V_2\,(\pi/4)\,d_2^2 = V\,(\pi/4)\,d^2 = Q = 1.1781 \quad \text{(i)}$$

$$V = 1.1781/[(\pi/4)\,1.0^2] = 1.5 \text{ m/s.}$$

Assume that, with pumping, the head at the connection to the source at O is H_o. Applying Bernoulli's equation between points O and 1 and points O and 2,

$$H_o = fL\,V^2/(2\,g\,d) + f_1\,L_1\,V_1^2/(2\,g\,d) + H_1 \tag{ii}$$

$$H_o = fL\,V^2/(2\,g\,d) + f_2\,L_2\,V_2^2/(2\,g\,d) + H_2 \tag{iii}$$

Equating (ii) and (iii),

$$0.018 \times 1200 \times V_1^2/(2\,g \times 0.6) + 222 = 0.015 \times 1500 \times V_2^2/(2\,g \times 0.8) + 220$$

Or, $1.83486 \times V_1^2 + 2 = 1.43349\,V_2^2$ \hfill (iv)

And (i) gives,

$$0.2827\,V_1 + 0.50265\,V_2 = 1.1781 \tag{v}$$

From (v), $V_1 = 4.1673 - 1.7780\,V_2$. Substituting in (iv) gives

$1.43349\,V_2^2 - 2 - 1.83486\,[3.1613\,V_2^2 - 14.8189\,V_2 + 17.36639] = 0$

Or, $4.3671\,V_2^2 - 27.1906\,V_2 + 33.86489 = 0$

This gives $V_2 = 4.50$ or 1.7214 m/s. It may be seen from (v) that $V_2 = 4.50$ m/s would result in a negative value for V_1. So, $V_2 = 1.7214$ m/s and $V_1 = 4.1673 - 1.7780 \times 1.7214 = 1.1067$ m/s.

$Q_1 = 1.1067 \times (\pi/4) \times 0.36 = 0.3129$ m^3/s and $Q_2 = 1.7214 \times (\pi/4) \times 0.64 = 0.8652$ m^3/s.

Water supply required per house = $650 \times 4 = 2,600$ l/day = 0.00003 m^3/s.

Max. number of houses in Subdivision 1 = $0.3129/0.00003 = 10,430$

Max. number of houses in Subdivision 2 = $0.8652/0.00003 = 28,840$

Required head at connection to source at point $O = H_o = 0.012 \times 1,000 \times 1.5^2/(2\,g \times 1.0) + 0.018 \times 1,200 \times (1.1067)^2/(2\,g \times 0.6) + 222 = 1.376 + 2.247 + 222 = 225.623$ m

Required pumping head at $O = 225.623 - 200 = 25.623$ m.

Required power = $\gamma\,Q\,H = 1,000 \times 1.1781 \times 25.623 = 30,186$ kg-m/s. Using an overall system efficiency of about 0.8, a pumping system with a capacity of 38,000 kg-m/s may be required.

For analyzing pipe networks involving several pipes and junctions, it is preferable to use available computer programs (e.g., Streeter 1971; Potter and Wiggert 1991; Brater et al. 1996; KYPIPE2 and KYPIPE3 1992).

Hydraulics of Diffusers

Wastewater is sometimes discharged into a receiving body of water through multiport diffusers. The diffuser consists of a pipe that extends nearly perpendicular to the shore and is connected to another pipe (diffuser) that may be parallel or normal to the current. The diffuser pipe has several closely spaced ports (nozzles) attached to it. If the receiving body of

Hydraulic Analysis   137

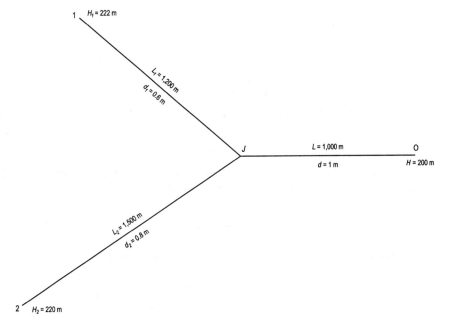

Figure 3-10. Schematic of water supply lines

water is uniform (not stratified) and static (e.g., a lake or an ocean), a preliminary estimate of dilution, S_a, in deep waters may be obtained by (Fischer et al. 1979)

$$S_a = C_0/C = 0.38 \, (g')^{1/3} \, y/q^{2/3} \tag{3-145}$$

where

- y = depth of water above source
- S_a and C are centerline dilution and concentration, respectively, at depth, y
- C_0 = initial concentration at diffuser port (i.e., at $y = 0$)
- q = wastewater discharge per unit length of diffuser (m²/s)
- $g' = g \, \Delta \rho/\rho$
- ρ = density of wastewater
- $\Delta \rho = \rho_0 - \rho$
- ρ_0 = density of receiving water

Eq. (3-145) may not provide conservative estimates because of the blocking effect of the waste field that may exist near the top of the rising plume. An appropriate factor of safety may be used in designs.

An approximate equation to estimate average dilution of a diffuser oriented normal to a deep, uniform current including the effect of blocking is (Fischer et al. 1979)

$$S_n = S_a/[1 + \{(Q_0 \, S_a)/(u \, L \, y_m)\}] \tag{3-146}$$

where

> S_n = average dilution in a perpendicular current including effect of blocking
> L = length of diffuser pipe
> u = velocity of ambient current normal to the diffuser
> y_m = maximum height of rising wastewater
> Q_0 = total wastewater discharge through the diffuser

Average dilution, S_p, for a diffuser oriented parallel to a deep, uniform current including effect of blockage may be approximated by (Fischer et al. 1979)

$$S_p = 0.38 \, (g')^{1/3} \, y_m / q^{2/3} \qquad (3\text{-}147)$$

Hydraulic analysis for a diffuser involves the use of the Darcy-Weisbach equation (Eq. (3-137)) and the orifice flow equation

$$Q_0 = C_d \, (\pi \, d^2/4) \, \sqrt{(2 \, g \, H)} \qquad (3\text{-}148)$$

where

> C_d = orifice discharge coefficient
> H = head above the center of the orifice
> d = diameter of orifice

The discharge coefficient varies from about 0.59 to 0.98, depending on the shape of the entrance.

The steps of computations are as follows:

1. Determine required dilution, S_n or S_p, at specified depth, y, and orientation of diffuser with respect to ambient current.
2. If the diffuser is oriented normal to the ambient current, assume $S_a = 1.2$ to $1.5 \, S_n$ and estimate q using Eq. (3-146). If the diffuser is oriented parallel to the ambient current, estimate q using Eq. (3-147) with required value of S_p.
3. Knowing the wastewater discharge, Q_0, estimate the preliminary diffuser length, $L = Q_0/q$.
4. Knowing u, L, y, S_a, and Q_0, use Eq. (3-146) to check if the required S_n is achieved. Otherwise, modify L. This step is not required if the diffuser is oriented parallel to ambient current.
5. By judgment, estimate reasonable port spacings ($L_1, L_2, L_3, \ldots, L_n$) between 1 to 10 m, approximately, depending on the total length of the diffuser pipe and number of ports, n.
6. Assuming average flow velocity of about 1.25 to 1.8 m/s, estimate diffuser pipe diameter, D.

Hydraulic Analysis

7. Estimate preliminary port diameters ($d_1, d_2, d_3, \ldots, d_n$, etc.) between 5 and 24 cm, approximately; the sum of the areas of all ports should be between $\frac{1}{3}$ and $\frac{2}{3}$ of the cross-sectional area of the diffuser pipe.

8. Number the last port adjacent to the dead end of the diffuser pipe as port No. 1, and estimate initial values of C_{d1}, d_1, and q_1 for this port.

9. Use Eq. (3-148) to estimate the required head, H_1, at this port.

10. Based on the bathymetry of the receiving water body, estimate the slope at which the diffuser pipe will be laid, and find the elevation change ($\Delta z_1, \Delta z_2, \Delta z_3, \ldots, \Delta z_n$) of the pipe invert from port to port.

11. Find the velocity in the farthest segment (adjacent to port No. 1) of the diffuser pipe, $V_1 = q_1/(\pi D^2/4)$.

12. Compute the velocity head $V_1^2/2g$.

13. Determine the specific gravity, s_o, of the receiving water, s, of the wastewater, and $\Delta s = s_o - s$.

14. Estimate the head, H_2, at the second port:

$$H_2 = H_1 + fL_1 V_1^2/(2gD) + \Delta s (\Delta z_1)/s.$$

15. Estimate C_{d2} for the second port using

$$C_{d2} = 0.975 \, [1 - V_1^2/(2gH_2)]^{0.375} \text{ for rounded port entrance} \quad (3\text{-}149)$$

$$C_{d2} = 0.63 - [0.58 \, V_1^2/(2gH_2)] \text{ for sharp-edged ports} \quad (3\text{-}150)$$

16. Estimate $q_2 = C_{d2} (\pi d_2^2/4) \sqrt{(2gH_2)}$.

17. Estimate $V_2 = V_1 + [q_2/(\pi D^2/4)]$.

18. Repeat steps (12) through (17) for all ports up to n in succession.

After these preliminary computations, appropriate changes may be made in port diameters, diffuser pipe diameter, and port spacings, etc., to obtain an acceptable diffuser design.

Example 3-20: It is desired to achieve a dilution factor of 85 at an average depth of about 45 m above the diffuser pipe for wastewater discharged at a rate of 6.5 m³/s through a diffuser into a receiving water body with $\Delta\rho/\rho = \Delta s/s = 0.025$, and $f = 0.012$. The diffuser pipe is to be laid at a uniform slope of 0.08 m/m. Assume that the diffuser is oriented normal to the ambient current and $u = 0.25$ m/s. Estimate preliminary diffuser dimensions. The wastewater storage is located 2 m above the water level in the receiving water body and will be connected to the diffuser pipe with a 1,000-m-long pipe of the same size and type. This pipe has two elbows ($k_e = 0.20$) and one tee ($k_T = 1.1$). Water surface in the receiving water body is 62 m above the first port (i.e., near the dead end of diffuser pipe). Determine the pumping capacity, if required.

Solution: For a preliminary estimate, assume $S_a = 1.3 \times S_n = 1.3 \times 85 = 110.5$. From Eq. (3-145), $110.5 = 0.38 \times (9.81 \times 0.025)^{1/3} \times 45/q^{2/3}$, or $q = 0.030$ m³/s.

$L = Q_0/q = 6.5/0.030 = 216.7$ m. Adopt $L = 200$ m.

Use uniform port spacing of 1.0 m; d for ports 1 to 30 = 15 cm; d for ports 31 to 60 = 14 cm; d for ports 61 to 130 = 13 cm; and d for ports 131 to 200 = 12 cm. Note that the sum of areas of all ports is about 60% of the cross-sectional area of the pipe.

Assume average velocity through diffuser pipe = 1.4 m/s. Then, $A = 6.5/1.4 = \pi D^2/4$. Or, $D = 2.4$ m.

Assume $C_{d1} = 0.975$ and $q_1 = 0.030$ m³/s. So, from Eq. (3-148), $0.030 = 0.975 \times (\pi d_1^2/4) \times \sqrt{(2 g H_1)}$. Or, $H_1 = 0.15$ m.

$V_1 = q_1/(\pi D^2/4) = 0.030/(\pi \times 2.4^2/4) = 0.00663$ m/s. If this velocity is judged to be too low, then the diffuser pipe diameter near the dead end may be reduced.

Using step (16), $H_2 = 0.15 + 0.012 \times 1.0 \times (0.00663)^2/(2 \times 9.81 \times 2.4) + 0.025 \times 0.08 = 0.152$ m.

Assume rounded port entrances. Then, using Eq. (3-149), $C_{d2} = 0.975 [1 - \{(0.00663)^2/(2 \times 9.81 \times 0.152)\}]^{0.375} = 0.975$.

$q_2 = 0.975 \times (\pi d_2^2/4) \times \sqrt{(2 g H_2)} = 0.975 \times (\pi \times 0.15^2/4) \times \sqrt{(2 g \times 0.152)} = 0.030$ m³/s.

The computations up to the 200th port may be carried out using a simple Fortran program or spreadsheet. Computed port discharges, velocity through diffuser pipe segments upstream of the port, C_d for the port, and head at the port at selected ports are shown in Table 3-35.

Computed head (above ambient water level) required for the diffuser = 0.582 m. The computations may be refined to achieve an acceptable diffuser design.

For $Q_0 = 6.5$ m³/s and $D = 2.4$ m for the 1,000-m-long pipe connecting the diffuser with the wastewater source, $V = 6.5/(\pi \times 2.4^2/4) = 1.44$ m/s.

Using Eq. (3-137), head loss through this pipe = $0.012 \times 1,000 \times (1.44)^2/(2 \times 9.81 \times 2.4) + (2 \times 0.2 + 1.1) \times 1.44^2/(2 \times 9.81) = 0.528 + 0.158 = 0.686$ m.

Total head required to force wastewater through diffuser ports = $0.686 + 0.582 = 1.268$ m.

Available head at wastewater storage point (measured above the farthest port) = $(62 + 2) s = 64 s$ kg/m² (s = unit weight of effluent).

Depth of ambient water at first port = 62 m; depth of ambient water at last port = $62 - 0.08 \times 200 = 46$ m.

Average head of ambient water against which wastewater is discharged = $\{(62 + 46)/2\} s_0 = 54 s_0$ kg/m² (s_0 = unit weight of ambient water).

Total head required to discharge wastewater through the ports = $54 s_0 + 1.268 s$ kg/m².

Since $\Delta s/s = 0.025$, $s_0 = 1.025 s$. So, $54 s_0 + 1.268 s = 56.618 s$ kg/m².

Required head is about $(56.618/64) \times 100 = 88\%$ of available head. So, pumping may not be required unless pipe corrosion and other factors increase the required head.

Table 3-35. Computed hydraulic parameters at selected diffuser ports

No. of port	Port discharge (m³/s)	Velocity in diffuser pipe (m/s)	C_d	Head at port (m)
1	0.030	0.00663	0.975	0.150
2	0.030	0.013	0.975	0.152
5	0.030	0.033	0.975	0.158
10	0.031	0.067	0.975	0.168
50	0.033	0.353	0.966	0.249
100	0.033	0.704	0.949	0.352
150	0.032	1.069	0.928	0.462
200	0.035	1.436	0.906	0.582

For shallow streams, vertical mixing is relatively rapid, and initial dilution, S_i, within approximately one diffuser length may be estimated by (Tchobanoglous and Burton 1991; Adams 1982)

$$S_i = [u D L/(2 Q_0)] [1 + \sqrt{\{1 + (2 Q_0 V_0 \cos \alpha)/(u^2 L D)\}}] \qquad (3\text{-}151)$$

where

u = velocity of flow

Q_0 = rate of wastewater discharge

L = length of diffuser

S_i = initial dilution

D = water depth in river

V_0 = velocity through diffuser ports

α = angle of inclination of port to the horizontal measured in a vertical plane parallel to the direction of river flow

Sometimes, dilution at the end of the initial mixing region has to be estimated for single-port vertical discharges (e.g., discharge through a single pipe discharging vertically upward) in a lake, an ocean, or a river. In such cases, initial dilution can be estimated by (Tchobanoglous and Burton 1991; Muellenhoff et al. 1985):

1. For a stagnant receiving water body (e.g., a lake or an ocean with low currents),

$$S_i = 0.13 \, (g'/D^5/Q_0^2)^{1/3} \qquad (3\text{-}152)$$

2. For a flowing receiving water (e.g., a river),

$$S_i = 0.29 \, (u D^2/Q_0) \qquad (3\text{-}153)$$

where

S_i = average dilution at end of initial mixing zone

g' = buoyancy of wastewater = $g(\rho_s - \rho)/\rho$

ρ_s = density of wastewater

ρ = density of receiving water

Minimum dilution at the centerline of the plume $S_m \cong 0.55\ S_i$.

Example 3-21: Estimate preliminary dimensions of a multiport diffuser in a stream to achieve near-field dilution of about 20 for a wastewater discharge of 1.5 m³/s when the 7-day average 10-yr low flow is 50 m³/s. The diffuser ports are oriented horizontally and parallel to the direction of river flow. The average velocity and depth of flow in the river are 1 m/s and 1.8 m, respectively.

Solution: From practical considerations and for expected initial dilution, $S_i = 20$, assume a port velocity, $V_0 = 3$ m/s. Using Eq. (3-151), with $u = 1$ m/s, $D = 1.8$ m, $Q_0 = 1.5$ m³/s, and $\alpha = 0$, by trial and error, $L = 16$ m.

Assume port spacing $\cong$ river water depth (for shallow streams) = 1.8 m. Therefore, the number of ports = $16/1.8 \cong 9$.

Also, assuming average diffuser pipe velocity of 1.4 m/s, average diameter of diffuser pipe for $Q_0 = 1.5$ m³/s is $D = \sqrt{[(1.5/1.4) \times 4/\pi]} = 1.17$ m.

Assuming nearly equal discharge through each port with $V_0 = 3$ m/s, diameter of port = $\sqrt{[\{1.5/(9 \times 3)\} \times 4/\pi]} = 0.266$ m.

The dilution, S_i, estimated by Eqs. (3-151), (3-152), and (3-153), increases as the plume rises up to the water surface. For preliminary estimates, this increase may be neglected, and the flow rate, Q_s, within the plume can be estimated by (Tchobanoglous and Burton 1991)

$$Q_s = S_i Q_0 \qquad (3\text{-}154)$$

The maximum (centerline) concentration in this plume, C_s, is estimated by

$$C_s = C_0/S_m \qquad (3\text{-}155)$$

The length of the plume may be assumed to be equal to the length of the diffuser, L. Thus, the height of the plume, h_s, is given by

$$h_s = Q_s/(u L) \qquad (3\text{-}156)$$

Computation for far-field dispersion can be made using Eqs. (3-105(a)) and (3-106) with Q_s = source discharge, C_s = source concentration, L = length of plane vertical source, and h_s = height of plane vertical source.

Water Hammer

Estimation of water hammer pressure is required to design pipelines where valves or gates located at the end of the pipeline are gradually or suddenly closed. A commonly encountered field situation is development of water hammer pressure in a penstock due to closure of turbine gates. A conservative value of water hammer pressure, ΔH (m), may be obtained assuming sudden valve or gate closure:

$$\Delta H = a\, V_0/g \qquad (3\text{-}157)$$

where

V_0 = initial flow velocity in pipe

a = velocity of pressure wave in pipe

If the valve is located at a distance, L, from the reservoir, then sudden valve closure may be assumed if $T \leq 2L/a$, where T = time of valve closure. The velocity of pressure wave is given by

$$a = \sqrt{[E_w/\{(\gamma/g)(1 + d\,E_w/(E_p\,t))\}]} \qquad (3\text{-}158)$$

where

E_w = modulus of elasticity of water (kg/m²)

E_p = modulus of elasticity of pipe walls (kg/m²)

t = thickness of pipe walls (m)

d = inside diameter of pipe (m)

For slow valve or gate closure, it may be preferable to use available computer models (e.g., Streeter 1971; Watters 1984; SURGE5 1996).

Example 3-22: A reservoir is connected to the point of distribution by a 1,000-m-long, 15-cm steel pipeline. The pipe carried 44 l/s when the valve located at the point of distribution was rapidly closed. Estimate water hammer pressure at the valve. After 0.5 s of the valve closure, a pressure gauge located at the valve exhibited a sudden pressure drop indicating a leak in the pipeline. Estimate the location of the leak. E_w = 21,100 kg/cm²; E_s = 2.04 × 10⁶ kg/cm²; t = 12 mm; and γ = 1,000 kg/m³.

Solution:

$\{(\gamma/g)(1 + d\,E_w/(E_p\,t))\} = (1{,}000/9.81)\,[1 + \{15 \times 21{,}100/(2.04 \times 10^6 \times 1.2)\}] =$

115.116

$a = \sqrt{[21{,}100 \times 10^4/115.116]} = 1353.86$ m/s.

$V = 0.044/(\pi \times 0.15 \times 0.15/4) = 2.49$ m/s.

Water hammer pressure = $a\,V/g$ = 1353.86 × 2.49/9.81 = 343.6 m. The pipe must be able to withstand this extra pressure.

It appears the pressure wave reaches the point of leak and the response travels back to the valve in 0.5 s, i.e., $2L'/a = 0.5$, where L' = distance of leak from the valve. This gives $L' = 0.5 \times 1353.86/2 = 338.46$ m.

Hydraulic Models

Some public-domain numerical computer models commonly used for various types of hydraulic analyses are listed and described below.

- **Water Surface Profiles Model, HEC-2** (USACE 1991c): This model computes water surface profiles for one-dimensional, gradually varied, steady-state flows in natural or man-made channels. It can be used to determine the effects of obstructions or structures in the channel. Hydraulic computations are made using the standard step method. Both supercritical and subcritical flows can be modeled. In subcritical flows, stream cross sections are arranged from downstream to upstream. In supercritical flows, they are arranged from upstream to downstream. The geometry of cross sections is defined from left to right looking downstream. The model has the capability to perform hydraulic computations through bridges, culverts, side flow weirs, floodplain encroachments, and split flows. In addition to water surface profiles, the model computes channel and overbank velocities, flow areas, and discharges and energy grade elevation, losses, and top width at every cross section. It also can provide velocity distribution in selected portions of the cross sections.

- **River Analysis System, HEC-RAS** (USACE 1998): This model is a Windows-based water surface profile program, which replaces the HEC-2 model described previously. It can use the data developed for the HEC-2 model, or the user can input data specifically for this model. The user interacts with the model through a graphical user interface for file management, data entry and editing, hydraulic analyses, graphical displays of input and output, and report preparation. It can perform subcritical, supercritical, and mixed flow computations and can simulate floodway encroachments, bridges and culverts, scour at bridges, and split flows. The input and output can be viewed and printed graphically. Also, the computed results can be exported to GIS and CADD files.

- **Water Surface Profiles, WSP-2** (USDA 1976): This model simulates flow characteristics for a given set of stream and floodplain conditions. It computes water surface profiles in open channels and estimates head losses at restrictive sections, including roadways with bridge or culvert openings. Generally, the model performs the same computations as the HEC-2 model (USACE 1991c).

- **Dam-Break Model, DAMBRK** (Fread 1988): DAMBRK is a relatively complex model to simulate water surface elevations resulting from dam failures. It includes a breach component to estimate the temporal and geometrical description of the breach for a given set of parameters. The model computes outflow hydrograph for the breach, including effects of reservoir storage depletion and reservoir inflows, and uses a dynamic routing technique to estimate water surface elevations and times of arrival of the dam-break flood wave at different cross sections downstream of the dam. It can perform computations for subcritical, supercritical, and mixed flows and can incorporate the effects of bridge embankments, other dams, debris flows, landslide-generated reservoir waves, and floodplains. Usually, dam-break analysis requiring the use of this model is undertaken as a special study.

- **Scour and Deposition in Rivers and Reservoirs, HEC-6** (USACE 1991d): HEC-6 is a one-dimensional model that computes the depths of scour or deposition at various cross sections of a channel or reservoir. Major input data include geometry of cross sections similar to the HEC-2 model (USACE 1991c), tables of suspended sediment loads versus discharges, and gradation of suspended sediment and bed material at various cross sections. The hydrograph of inflows is input as a step function of time durations. The model can perform bed load transport computations using one of the 12 transport functions included in the model. It can simulate deposition and erosion of clay and fine-silt–sized particles, and it adjusts sediment transport estimates for wash loads.

- **Culvert Analysis Program** (CAP 2001): This model uses standardized procedures of the USGS for computing flow through culverts. It can develop stage-discharge relationships for rectangular, circular, pipe arch, elliptical, and other nonstandard shaped culverts. The model solves the one-dimensional steady-state energy and continuity equations for upstream water surface elevation given a discharge and a downstream water surface elevation. It can be downloaded from http://www.waterengr.com/freeprog.htm.

- **Enhanced Stream Water Quality Models, QUAL2E and QUAL2E-UNCAS** (USEPA 1987a): These are water quality planning tools that can be operated as steady-state or dynamic models. They are used to study the impact of waste loads on instream water quality or to identify the magnitude and quality characteristics of nonpoint waste loads as part of a field sampling program. They also can be used to model the effects of diurnal variations in meteorological parameters on water quality (primarily dissolved oxygen and temperature) or to examine diurnal dissolved variations caused by algal growth and respiration. They have the capability for uncertainty analysis, an option to input reach-variable climatology for steady-state temperature simulation, and an option to plot observed and predicted dissolved oxygen concentrations.

- **Hydrodynamic Mixing Zone Model and Decision Support System for Pollutant Discharges into Surface Waters, CORMIX** (USEPA 1996a): This model predicts the dilution and trajectory of submerged single-port discharges, submerged multiport diffuser discharges, and surface discharges of arbitrary density (positive, neutral, or negative) into a stratified or uniform density environment with or without cross-flow. The model checks for data consistency, assembles and executes the appropriate hydrodynamic models, and interprets the results in terms of regulatory discharge criteria. It emphasizes rapid initial mixing and assumes a conservative pollutant discharge, neglecting any reaction or decay process.

- **Dilution Models for Effluent Discharges, PLUMES** (USEPA 1994): This model is useful for designing outfall diffusers where the ambient flow is unstratified and the receiving water body is relatively deep. It includes two relatively sophisticated initial dilution models and two relatively simple far-field algorithms. The required input includes port geometry, spacing, and total discharge; plume diameter and depth; effluent salinity and temperature; ambient conditions in receiving water; and far-field distance for computations. Compared to the CORMIX (USEPA 1996a) model, this model is simpler to use.

- **Water Quality for River-Reservoir Systems, WQRRS** (USACE 1978): This model consists of three separate modules: reservoir, stream hydraulics, and stream quality module. The three modules may be integrated for a complete river basin water quality analysis. The reservoir and stream hydraulics modules also can be executed as independent models. The stream quality module functions in conjunction with the stream hydraulics module. The reservoir module is applicable to relatively deep impoundments with long residence times. Shallow impoundments with rapid flow-through times may be treated as slow-moving streams. The stream hydraulics module computes hydraulic parameters for gradually varied flow in steady and unsteady flow regimes. The stream quality module simulates peak pollutant loads in a steady or unsteady hydraulic environment. This model is fairly complex, and projects involving the use of this model should be undertaken as special studies.

CHAPTER 4

GROUNDWATER

Occurrence of Groundwater and Types of Porous Media

Groundwater occurs in soil pores or rock fractures in the form of hygroscopic or extractable water. Hygroscopic water is tightly held by soil particles such that it cannot be extracted except by oven drying. Extractable water can be extracted by plants, by gravity drainage, or by pumping. The rates of extraction depend on permeability of the porous medium and the quantity of water held in soil pores or rock fractures. A soil mass or rock that can yield reasonable quantities of water is called an aquifer. From the standpoint of water transmissibility, aquifers are classified as follows:

1. **Confined Aquifer:** This type of aquifer is bounded by impervious formations at its top and bottom.

2. **Artesian Aquifer:** Artesian aquifer is a confined aquifer in which piezometric head is higher than the top of the aquifer. Piezometric head is the elevation up to which groundwater will rise in a tube screened in the aquifer.

3. **Phreatic, Water Table, or Unconfined Aquifer:** This type of aquifer has water table (at atmospheric pressure) as its upper boundary.

4. **Confining Unit:** A confined unit includes an aquitard, which is a semi-permeable formation that retards movement of water through it, and an aquiclude, which is an impermeable formation that prevents movement of water through it.

5. **Semi-Confined or Leaky Confined Aquifer:** An aquifer that may gain or lose water through underlying or overlying aquitards.

6. **Leaky Phreatic Aquifer:** An aquifer that rests on a semi-pervious layer through which it may gain or lose water.

7. **Perched Aquifer:** A phreatic aquifer where an impervious or semi-pervious layer of limited areal extent occurs between the water table of a phreatic aquifer and the ground surface. Groundwater is retained in the pervious medium overlying the impervious or semi-pervious layer.

Properties of Porous Media

Physical properties of porous media and water, which are commonly used in analyzing groundwater flow and transport, are described in this section.

Unit weight is the weight of material per unit volume. Density is unit weight divided by gravitational acceleration. Specific gravity is the ratio of unit weight of the material to unit weight of water.

If V_s = volume occupied by soil grains in a total soil volume V; V_v = volume occupied by voids; V_w = volume occupied by water; W = total weight of soil of volume V; γ_w or γ = unit weight of water; W_s = weight of soil grains in volume V; and W_w = weight of water in soil of volume V, then:

Particle unit weight or unit weight of soil grains = $\gamma_s = W_s/V_s$;
Dry bulk unit weight of soil = $\gamma_d = W_s/V$;
Moist unit weight of soil = $\gamma_m = (W_s + W_w)/V$;
Total porosity = $\varphi = V_v/V$;
Void ratio = $e = V_v/V_s$;
Water content = $w = W_w/W_s$;
Volumetric water content = $\varphi_w = V_w/V_s$;
Volumetric air or gas content in vadose (unsaturated) soil zone = $\varphi - \varphi_w$;
Degree of saturation or saturation = $S_r = V_w/V_v$;
Submerged unit weight of soil grains = $\gamma_s - \gamma_w$;
Total volume of soil = $V = V_s + V_v$; and
Volume occupied by air or gas = $V_g = V_v - V_w$.

It may be seen that

$\varphi = e/(1 + e)$ or $e = \varphi/(1 - \varphi)$;
$\gamma_d = (1 - \varphi)\gamma_s$; and
γ_m (saturated soil) $= \gamma_d + \varphi\,\gamma_w$; and $\varphi_w = (w\,\gamma_s)/\gamma_w$;

The porosity, particle unit weight, and dry bulk unit weight of selected porous media are given in Table 4-1 (USEPA 1985).

Physical properties of pure water at atmospheric pressure are summarized in Table 4-2 (USEPA 1985).

Permeability and Hydraulic Conductivity

The water-transmitting property of a porous medium is defined by its permeability or hydraulic conductivity. Permeability or intrinsic permeability (cm^2), k, is a property of the porous medium only and is defined as

$$k = C\,d^2 \qquad (4\text{-}1)$$

where

d = mean grain size of porous medium (cm)

C is a coefficient

Table 4-1. Porosity, particle unit weight, and dry bulk unit weight of porous media

Material	Porosity(%)	Particle unit weight (gm/cm^3)	Dry bulk unit weight (gm/cm^3)
Clay	34.2–56.9	2.51–2.77	1.18–1.72
Silt	33.9–61.1	2.47–2.79	1.01–1.79
Fine sand	26.0–53.3	2.54–2.77	1.13–1.99
Medium sand	28.5–48.9	2.60–2.77	1.27–1.93
Coarse sand	30.9–46.4	2.52–2.73	1.42–1.94
Fine gravel	25.1–38.5	2.63–2.76	1.60–1.99
Medium gravel	23.7–44.1	2.65–2.79	1.47–2.09
Coarse gravel	23.8–36.5	2.64–2.76	1.69–2.08
Loess	44.0–57.2	2.64–2.74	1.25–1.62
Dune sand	39.9–50.7	2.63–2.70	1.33–1.70
Till, predominantly silt	29.5–40.6	2.64–2.77	1.61–1.91
Till, predominantly sand	22.1–36.7	2.63–2.73	1.69–2.12
Till, predominantly gravel	22.1–30.3	2.67–2.78	1.72–2.12
Glacial drift, predominantly silt	38.4–59.3	2.70–2.73	1.11–1.66
Glacial drift, predominantly sand	36.2–47.6	2.65–2.75	1.36–1.83
Glacial drift, predominantly gravel	34.6–41.5	2.65–2.75	1.47–1.78
Sandstone (fine-grained)	13.7–49.3	2.56–2.72	1.34–2.32
Sandstone (medium-grained)	29.7–43.6	2.64–2.69	1.50–1.86
Siltstone	21.2–41.0	2.52–2.89	1.35–2.12
Claystone	41.2–45.2	2.50–2.76	1.37–1.60
Shale	1.4–9.7	2.47–2.83	2.20–2.72
Limestone	6.6–55.7	2.68–2.88	1.21–2.69
Dolomite	19.1–32.7	2.64–2.72	1.83–2.20
Granite (weathered)	34.3–56.6	2.70–2.84	1.21–1.78
Basalt	3.0–35.0	2.95–3.15	1.99–2.89
Schist	4.4–49.3	2.70–2.84	1.42–2.69

Source: USEPA (1985).

Table 4-2. Physical properties of pure water

Temperature (°C)	Unit weight (gm/cm^3)	Dynamic viscosity (gm/cm-s)	Kinematic viscosity (cm^2/s)	Compressibility (cm-s^2/gm)
4	1.00000	0.01567	0.0157	4.959×10^{-11}
10	0.99973	0.01307	0.0131	4.789×10^{-11}
15	0.99913	0.01139	0.0114	4.678×10^{-11}
20	0.99823	0.01002	0.01004	4.591×10^{-11}
25	0.99708	0.00890	0.00893	4.524×10^{-11}

Source: USEPA (1985).

The coefficient, C, depends on porosity, shape of pores, and tortuosity of the porous pathways. The values of C for sand vary from about 10^{-6} to 10^{-4}. Typical values of mean grain size and k are included in Table 4-3 (USEPA 1985).

Depending on spatial variation of water-transmitting characteristics, porous media are classified as homogeneous or heterogeneous. A homogeneous porous medium has the same permeability at all points. Otherwise, it is classified as inhomogeneous or heterogeneous. The porous medium is isotropic if its permeability is the same in all directions (i.e., lateral, vertical, etc.). Otherwise, it is said to be anisotropic. A porous medium is anistropic and homogeneous if hydraulic conductivities in different directions (e.g., K_x and K_y in x and y directions, respectively) are different from one another but do not change from point to point. It is isotropic and inhomogeneous if $K_1 = K_{x1} = K_{y1}$ at point 1 and $K_2 = K_{x2} = K_{y2}$ at point 2 (i.e., hydraulic conductivities in different directions remain the same but values change from point to point). It is anisotropic and inhomogeneous if $K_{x1} \neq K_{y1} \neq K_{x2} \neq K_{y2}$.

Equivalent hydraulic conductivity, $\overline{K}$, for flow perpendicular to n layers of different thicknesses ($H_1, H_2, H_3, \ldots,$ and H_n) and hydraulic conductivities ($K_1, K_2, K_3, \ldots,$ and K_n) is given by

$$\overline{K} = (H_1 + H_2 + H_3 + \ldots + H_n)/[(H_1/K_1 + H_2/K_2 + H_3/K_3 + \ldots + H_n/K_n)] \tag{4-2}$$

In some groundwater flow models (e.g., MODFLOW, USEPA 2000), conductance, C, for flow through a unit area of a layer of thickness, H, and hydraulic conductivity, K, is defined as

$$C = K/H \tag{4-3}$$

Table 4-3. Typical values of permeability (k) and hydraulic conductivity (K)

Material	Mean grain size (mm)	k (cm^2)	K (cm/s)
Unweathered marine clay	0.001–0.009	10^{-16}–10^{-12}	10^{-11}–10^{-7}
Silt, loess	0.009–0.074	10^{-12}–10^{-8}	10^{-7}–10^{-3}
Silty sand	0.074–0.297	10^{-10}–10^{-6}	10^{-5}–10^{-1}
Clean sand	0.297–4.76	10^{-9}–10^{-5}	10^{-4}–1.0
Clean gravel	4.76–76.2	10^{-6}–10^{-3}	0.10–100
Sandstone	—	10^{-13}–10^{-9}	10^{-8}–10^{-4}
Limestone	—	10^{-12}–10^{-9}	10^{-7}–10^{-4}
Karst limestone	—	10^{-9}–10^{-5}	10^{-4}–1
Shale	—	10^{-16}–10^{-12}	10^{-11}–10^{-7}
Fractured igneous and metamorphic rocks	—	10^{-11}–10^{-7}	10^{-6}–10^{-2}
Peat[a]	—	—	1.2×10^{-4} 2.3×10^{-3}

Source: USEPA (1985).
[a] Source: Bradley and Gilvear (2000).

Thus, in Eq. (4-2), $\overline{C} = \overline{K}/(H_1 + H_2 + H_3 + \ldots + H_n)$, and

$$(1/\overline{C}) = [1/C_1 + 1/C_2 + 1/C_3 + \ldots + 1/C_n] \tag{4-4}$$

Equivalent hydraulic conductivity for flow parallel to n layers of different thicknesses and hydraulic conductivities is given by

$$\overline{K} = (K_1 H_1 + K_2 H_2 + K_3 H_3 + \ldots + K_n H_n)/[(H_1 + H_2 + H_3 + \ldots + H_n)] \tag{4-5}$$

Example 4-1: In a groundwater flow model, layer 1 is 16 m and layer 2 is 23.5 m thick. The vertical hydraulic conductivities of layer 1 and 2 are 2.0 and 5.0 m/day, respectively. Determine conductance between these two layers.

Solution:
$C_1 = K_1/H_1 = 2.0/16.0 = 0.125$ day^{-1}, and $C_2 = K_2/H_2 = 5.0/23.5 = 0.2128$ day^{-1}.
So, $1/C = 1/C_1 + 1/C_2$, or $C = (C_1 C_2)/(C_1 + C_2) = (0.125 \times 0.2128)/(0.125 + 0.2128) = 0.0787$ day^{-1}.

One-Dimensional Steady-State Groundwater Flow

Darcian Flow

Groundwater flow is governed by Darcy's law:

$$V = K i, \text{ if } R_e = \text{Reynolds number} = V d/\nu < \text{about 1 to 10} \tag{4-6}$$

where

V = velocity of flow per unit area of soil mass including area of soil pores

K = hydraulic conductivity

i = hydraulic gradient represented by sine of the slope angle of the hydraulic grade line

d = mean grain size of porous medium

ν = kinematic viscosity of water

Hydraulic conductivity depends on permeability of the porous medium and kinematic viscosity of water and is defined as

$$K = k g/\nu \tag{4-7}$$

Typical values of K are given in Table 4-3. The velocity, v, through voids of the porous media at which contaminants dissolved in water may travel is known as seepage velocity, linear velocity, or pore velocity and is given by

$$v = V/\varphi \tag{4-8}$$

where ϕ = porosity. Steady groundwater flow from a reservoir to a stream per unit width (perpendicular to flow) of a confined aquifer is given by

$$q = KiA = K[(H_1 - H_2)/L](B \times 1.0) \tag{4-9}$$

where

q = flow (m^3/s)

i = hydraulic gradient = $[(H_1 - H_2)/L]$

A = area of porous medium normal to the direction of the hydraulic gradient

H_1 = head in the reservoir above an arbitrary datum

H_2 = head in the stream above the same datum

L = distance from reservoir to stream

B = thickness of confined aquifer

If the aquifer is unconfined, then it is assumed that the slope of the water table is relatively small and is given by tan θ instead of sin θ, where θ = angle of slope of hydraulic grade line and the groundwater flow is assumed to be uniform and horizontal. These are known as Dupuit's assumptions (Todd 1980). With these assumptions,

$$q = -K(h \times 1.0)\, dh/dx = -(K/2)\, dh^2/dx \tag{4-10}$$

where h = head above datum at a distance x from the reservoir. The minus sign indicates that head decreases in the positive x-direction (i.e., as x increases).

Integrating Eq. (4-10) with heads H_1 and H_2 in the reservoir and stream, respectively, gives

$$q = (K/2L)[H_1^2 - H_2^2] \tag{4-11}$$

Once q is estimated, head, h, at any distance, x, and shape of the phreatic surface can be estimated by

$$q = (K/2x)[H_1^2 - h^2] \tag{4-12}$$

Eqs. (4-11) and (4-12) do not include the seepage face that forms along the slope of the stream bank above the water surface in the river. Because of this, and because of Dupuit's assumptions, the phreatic surface computed by these equations is not accurate. However, for relatively flat slopes and away from the stream bank, the predicted and actual phreatic surfaces are reasonably close. Eq. (4-10) may be used for preliminary estimates of tile drain spacing in agricultural areas where continuous infiltration from irrigated fields may be assumed at a constant rate. For situations where infiltration changes with irrigation season, a more complex approach may be used (see the section in this chapter entitled "Flow toward Drains and Drain Spacing").

Figure 4-1 is a vertical cross section of the aquifer, showing two parallel drains located at a spacing, L, between them. It may be seen that $Q_x = fx$ = flow passing through a cross

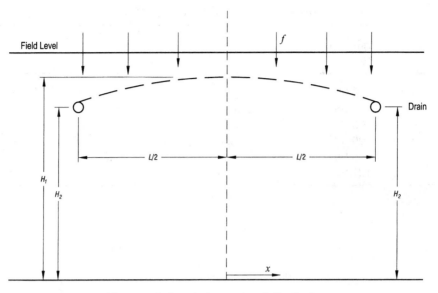

Figure 4-1. Schematic of tile drains

section of unit length parallel to the drain located at distance, x, from the midpoint between the two drains, where f = infiltration rate. Also, from Eq. (4-10), $Q_x = -(K/2) \, dh^2/dx$. Equating the two values of Q_x and integrating gives $h^2 = -fx^2/K + C$. At $x = 0$, $h = H_1$ and at $x = L/2$, $h = H_2$. So,

$$H_1^2 - H_2^2 = fL^2/4K, \quad \text{or} \quad L = 2\sqrt{[(H_1^2 - H_2^2)K/f]} \qquad (4\text{-}13)$$

Example 4-2: In an irrigated area, infiltration may be assumed to be uniform at an average rate of 0.98 m/yr. An impervious formation is located 9 m below the field level. Tile drains are to be located at a depth of 2.3 m below the field level. The drainage system is to be designed so that the water table below the fields is maintained at least 1.08 m below the field level. Estimate preliminary drain spacing. The hydraulic conductivity of soils below the fields is estimated to be 3.05 m/day.

Solution: $H_1 = 9.0 - 1.08 = 7.92$ m; $H_2 = 9.0 - 2.3 = 6.7$ m; and $f = 0.98/365 = 0.002685$ m/day. So, $L = 2\sqrt{[(7.92^2 - 6.7^2) \times 3.05/0.002685]} = 285$ m.

Non-Darcian Flow

Darcy's law is not valid for flows with $R_e \gg 10$ where R_e = Reynolds number = qd/ν, q = specific discharge or Darcy velocity, d = representative pore diameter (mean or d_{10}), and ν = kinematic viscosity of groundwater. This may occur for flows through coarse porous media, such as rockfill, riprap, and coarse gravel or boulders. The following is an equation for flow through mono-sized (uniformly sized) rocks with a specific gravity of 2.87 (Leps 1973):

$$V_V = W \, m^{0.50} \, i^{0.54} \qquad (4\text{-}14)$$

Table 4-4. Typical values of $Wm^{0.50}$

Grain size (cm)	$m^{0.50}$ (cm$^{0.5}$)	$Wm^{0.50}$ (cm/s)
1.90	0.48	25.40
5.08	0.78	40.64
15.24	1.38	71.12
20.32	1.56	81.28
60.96	2.81	147.32
121.92	4.04	213.36

Source: Leps (1973).

where

V_V = average velocity in voids of the coarse porous media (cm/s)

W = empirical constant for given material

m = hydraulic mean radius (cm)

i = hydraulic gradient

Typical values for the term ($Wm^{0.50}$) are shown in Table 4-4.

An empirical equation based on experiments for flow through riprap with d_{10} ranging from 15 to 97 mm is as follows (Abt et al. 1991):

$$V_V = 0.79 \, (g \times d_{10} \times i)^{0.50} \qquad (4\text{-}15)$$

where

V_V = velocity through voids (m/s)

g = 9.81 m/s^2

d_{10} = grain size (m) than which 10% of the grains are finer

---

Example 4-3: A mountain rivulet has a bed slope of 0.025 and a 1-m-thick boulder bed underlain by bedrock. During low flow periods, a water depth of about 10 cm is observed to flow above the boulders. Estimate the total quantity of flow this rivulet contributes to the low flows of the stream into which it discharges. Average bed width of the rivulet is 5 m. For the boulder bed, average porosity = 0.30; average stone size is 60 cm; and d_{10} = 25 cm.

Solution: From Eq. (4-14) and Table 4-4, V_V = 147 × (0.025)$^{0.54}$ = 20 cm/s. Alternatively, using Eq. (4-15), V_V = 0.79 $\sqrt{(9.81 \times 0.25 \times 0.025)}$ = 0.196 m/s, which is nearly the same as obtained from Eq. (4-14).

Using a kinematic viscosity of 0.13 × 10^{-5} m^2/s for water, R_e = Vd/ν = 0.20 × 0.60 × 10^5/0.13 = 9.23 × 10^4, i.e., flow through the boulder bed is far beyond the range of validity of Darcy's law.

The area of flow through boulders = $5.0 \times 1.0 \times 0.30 = 1.5$ m^2.

Q (through boulder bed) = $1.5 \times 0.20 = 0.30$ m^3/s.

Open channel flow above the boulders may be estimated using Manning's equation (Eq. 3-3). Manning's n for boulder bed = 0.05; water depth = y = 0.10 m $\cong R_h$ (hydraulic radius); and area of flow = $A = 5.0 \times 0.10 = 0.50$ m^2.

Q (above boulder bed) = $(1/n) R_h^{2/3} (\sqrt{S}) A = (1/0.05)(0.10^{2/3})(\sqrt{0.025})(0.50) = 0.34$ m^3/s.

Total flow = $0.30 + 0.34 = 0.64$ m^3/s.

Steady-State Radial Flow

Flow to a Single Well

The head, h, at a radial distance, r, from a well pumping at a constant rate, Q, in a confined aquifer is given by the following (Bear 1979; Todd 1980):

$$h - h_w = [Q/(2\pi K B)] \ln(r/r_w) \qquad (4\text{-}16)$$

where

h_w = head above the base of the confined aquifer at the well face

r_w = radius of well

K = hydraulic conductivity

B = aquifer thickness

Farther from the well, the influence of pumping gradually vanishes. This occurs at a distance, R (called radius of influence), from the well. The drawdown at the well face, s_w, is the difference between the head, h_0, at R and h_w at the well face:

$$s_w = [Q/(2\pi KB)] \ln(R/r_w) \qquad (4\text{-}17)$$

Also, drawdown at distance, r, from the well:

$$s_r = h_0 - h = [Q/(2\pi K B)] \ln(R/r) \qquad (4\text{-}18)$$

The specific capacity, S_p, of the well is given by

$$S_p = (Q/s_w) = (2\pi K B)/\ln(R/r_w) \qquad (4\text{-}19)$$

Hydraulic conductivity of a confined aquifer can be estimated if heads at two observation wells are known; for example,

$$K = [Q/\{2\pi B(h_2 - h_1)\}] \ln(r_2/r_1) \qquad (4\text{-}20)$$

where h_1 and h_2 are heads in observation wells located at distances r_1 and r_2 from the extraction well.

Using Dupuit's assumptions, radial flow in an unconfined aquifer is given by

$$h^2 - h_w^2 = (Q/\pi K) \ln (r/r_w) \tag{4-21}$$

and

$$h_0^2 - h_w^2 = (Q/\pi K) \ln (R/r_w) \tag{4-22}$$

Setting $h_0 + h_w = 2H$, drawdown in an unconfined aquifer may be approximated by

$$s_w = [Q/(2\pi K H)] \ln (R/r_w) \tag{4-23}$$

With this approximation, confined flow equations may be used for preliminary analysis of flow in unconfined aquifers as well.

Groundwater Mound

Infiltration from a recharge basin, surface impoundment, or open landfill results in groundwater mounding. To estimate the extent of groundwater mounding in such cases, the recharge area may be approximated by an equivalent circular area of radius, R. Using Dupuit's assumptions, the groundwater table within the mound can be approximated by two parabolic curves. The first curve extends from the center to the perimeter of the recharge area. The second curve extends from the perimeter to the constant water table in the region, which is relatively far from the recharge area and is not appreciably affected by mounding (Bouwer et al. 1999). The equation to the first curve is

$$H_1^2 - h^2 = i\, r^2/(2K),\, 0 \le r \le R \tag{4-24}$$

where

H_1 = groundwater elevation at the center of the recharge area above the impervious base

h = groundwater elevation at radial distance, r

i = recharge rate (L/T)

K = hydraulic conductivity

The equation of the second curve is

$$h^2 - H_2^2 = (i\, R^2/K) \ln (R_n/r),\, R \le r \le R_n \tag{4-25}$$

where

h = groundwater elevation at radial distance, r, from the center of the recharge area

H_2 = elevation of constant water table in the area above impervious base

R_n = radial distance of constant water table from center of recharge area

The radial distance, R_n, up to which groundwater mounding extends can be estimated by

$$H_1^2 - H_2^2 = (i\, R^2/2\, K)\, [1 + 2 \ln (R_n/R)] \tag{4-26}$$

Example 4-4: The rate of infiltration from a 500 m × 480 m lagoon at an industrial site is estimated to be 0.0015 m/day. Average water depth in the lagoon is 3 m. The bottom of the lagoon is at El. 180 m and the average water table elevation in the area is 179 m. Bedrock is encountered at El. 160 m. The hydraulic conductivity of the aquifer is 2 m/day. Estimate the extent of groundwater mounding and the shape of the altered water table around the lagoon.

Solution:

Equivalent radius of lagoon = $R = \sqrt{(500 \times 480)/\pi} = 276$ m.

Hydraulic head at the center of the lagoon above bedrock = $H_1 = 180 + 3 - 160 = 23$ m.

Hydraulic head at constant water table above bedrock = $H_2 = 179 - 160 = 19$ m.

From Eq. (4-26), $R_n = R \exp[\{(H_1^2 - H_2^2) K/(iR^2)\} - 0.5] = 276 \times \exp[\{(23^2 - 19^2) \times 2.0/(0.0015 \times 276^2)\} - 0.5] = 3{,}168$ m.

From Eq. (4-24), shape of mound from $r = 0$ to $r = R = 276$ m is given by $(23^2 - h^2) = 0.0015 \times r^2/(2 \times 2)$ or $h^2 = 529 - 0.000375 r^2$.

From Eq. (4-25), shape of mound from $r = R = 276$ m to $r = R_n = 3{,}168$ m is given by $h^2 = 361 + (0.0015 \times 276 \times 276/2) \ln(3{,}168/r) = 361 + 57.132 \ln(3{,}168/r)$.

Height of mound above bedrock at the perimeter of the lagoon (i.e, at $r = R = 276$ m) = $\sqrt{[529 - (0.000375 \times 276 \times 276)]} = \sqrt{500.43} = 22.37$ m.

Elevation of mound at the perimeter of the lagoon = $160 + 22.37 = 182.37$ m.

Capture Zone

Capture zone is the area from which groundwater is extracted by a pumping well. If there is no ambient groundwater flow and no river or impermeable boundary within the cone of influence of the well, the capture zone may be defined by the radius of influence. Usually, there is ambient groundwater flow in aquifers where extraction or recharge wells are installed, and capture zones have to be estimated for specified rates of pumping to estimate the area from which groundwater and dissolved substances in groundwater can be extracted by a system of wells. For multiple aquifer systems and variable rates of extraction or injection, through multiple wells, numerical models (e.g., USGS 2000b) must be used. For multiple wells in an isotropic and homogeneous aquifer, relatively simpler models may be used (e.g., Bair et al. 1992). For one, two, or three wells pumping at the same constant rate in an isotropic and homogeneous aquifer with uniform groundwater flow, the capture zone may be estimated using analytical equations (Javandel and Tsang 1986; Prakash 1995). For a single well located at the origin, with uniform groundwater flow velocity, v, in the negative x-direction,

$$x_s = -Q/(2\pi v B) \qquad (4\text{-}27)$$

$$y/x = \pm \tan(2\pi v B y/Q) \qquad (4\text{-}28)$$

where

x_s = location of stagnation point on the x-axis

Q = well discharge

B = aquifer thickness

x and y are coordinates of the line defining the capture zone

Stagnation point gives the location along the x-axis up to which the capture zone extends downgradient from the well, and Eq. (4-28) gives the equation of the capture zone. The right-hand side of Eq. (4-28) is given a positive sign to plot capture zone on the positive side of the y-axis and a negative sign for capture zone on the negative side of the y-axis. For two wells located along the x-axis at $(0, b)$ and $(0, -b)$, the corresponding equations are

$$x_s = [-a \pm \sqrt{\{a^2 - 4v^2b^2\}}]/(2v) \quad (4\text{-}29)$$

and

$$2xy/(x^2 - y^2 + b^2) = -\tan(2\pi vBy/Q) \quad (4\text{-}30)$$

where

$$a = Q/(\pi B), \, a \geq 2vb$$

For $a < 2vb$, each well would have a separate capture zone. On the y-axis, as $x \to \infty$, the capture zone approaches asymptotically the lines $y = \pm Q/2vB$.

Example 4-5: An extraction well discharging at a constant rate of 6.68 l/s penetrates a 30-m-thick confined aquifer where normal groundwater flow velocity is 0.20 m/day. Estimate the capture zone of the well. Also, estimate the capture zone if two wells, each discharging at a rate of 6.68 l/s, are located at 20 m from each other.

Solution: For a single well, $x_s = -Q/(2\pi vB) = -(0.00668 \times 86{,}400)/(2\pi \times 0.20 \times 30) = -15.3$ m.

The capture zone is estimated using Eq. (4-28). Selected coordinates of the capture zone boundary are shown in Table 4-5(a). For two wells,

$$a = Q/(\pi B) = (0.00668 \times 86{,}400)/(\pi \times 30) = 6.124 \text{ m}^2/\text{day}$$

and

$$2vb = 2 \times 0.20 \times 10 = 4.0.$$

Since $a \geq 2vb$, $x_s = [-a \pm \sqrt{\{a^2 - 4v^2b^2\}}]/(2v) = [-6.124 \pm \sqrt{\{6.124^2 - 16.0\}}]/(2 \times 0.20) = -26.90$ or -3.7 m. The capture zone for two wells has to be larger than a single well, so select $x_s = -26.90$ m.

The capture zone is estimated using Eq. (4-30). Selected coordinates of the capture zone boundary are shown in Table 4-5(b). Since there are two values of x for each y, the correct value has to be selected based on judgment so as to obtain a smooth capture zone.

Table 4-5(a). Coordinates of capture zone for a single well in uniform flow

y (m)	x (m)
0	−15.3
10	−13.1
15	−10.1
20	−5.4
30	12.2
35	30.3
40	68.2
45	217.3

Table 4-5(b). Coordinates of capture zone for two wells in uniform flow

y (m)	x (m)
0	−26.9
5	−26.7
10	−26.1
15	−25.1
25	−21.4
35	−14.9
45	−4.4
48	−0.04
50	2.9
60	23.9
70	59.9
80	135.8

Partially Penetrating Well

If a well does not penetrate the entire depth of a confined aquifer, then drawdown, s_{wp}, at the well face due to a partially penetrating well may be estimated by the following (Bear 1979):

$$s_{wp} = s_w / [(L/B)\{1 + 7 \cos(\pi L / 2B) \sqrt{(r_w/2L)}\}] \quad (4\text{-}31)$$

where

$s_w = (Q/2\pi KB) \ln(R/r_w)$

L = length of well screen of partially penetrating well

For the same drawdown, s_w,

$$Q_p/Q = (L/B)[1 + 7 \cos\{\pi L/(2B)\} \sqrt{(r_w/2L)}] \quad (4\text{-}32)$$

where

Q_p = discharge of partially penetrating well

Q = discharge of fully penetrating well

An approximate equation for discharge of a well partially penetrating an unconfined aquifer is as follows (USACE 1971b):

$$Q_p = [\pi K\{(H-s)^2 - t^2\}/\ln(R/r_w)][1 + \{0.30 + (10r_w/H)\}\sin(1.8\,s/H)] \quad (4\text{-}33)$$

where

s = height of well bottom above impervious layer

H = height of initial phreatic surface above impermeable base at $r = R$

t = depth of water in well, such that $(s + t) \cong h_w$

An alternative equation to estimate drawdown in a partially penetrating well is as follows (Bear 1979):

$$s_{wp} = [Q/(2\pi KB)][(1-p)/p]\ln\{(1.2-p)L/(\beta r_w)\} + (Q/2\pi KB)\ln(R/r_w) \quad (4\text{-}34)$$

where

$p = L/B$

$\beta = 1$ if well screen starts from the top or bottom of the confined aquifer

$ = 2$ if the screen is located in the middle of the confined aquifer

Eq. (4-34) is valid for $10r_w \leq L \leq 0.8B$.

Radius of Influence

Some semi-empirical and empirical equations to estimate radius of influence, R (m), are as follows (Bear 1979):

$$R = 1.9 \text{ to } 2.45\sqrt{(HKt/S)} \quad (4\text{-}35)$$

$$R = 3{,}000\, s_w \sqrt{K} \quad (4\text{-}36)$$

$$R = 575\, s_w \sqrt{(HK)} \quad (4\text{-}37)$$

where

H = initial saturated thickness for unconfined aquifer or thickness of confined aquifer (m)

K = hydraulic conductivity (m/s)

S = specific yield for unconfined aquifer or storativity for confined aquifer

t = time (s)

s_w = drawdown at well face (m)

Generally, R is larger for coarser materials than for finer soils and greater for confined than for unconfined aquifers.

Well Interference

The effect of interference between closely spaced wells discharging at a given rate may be approximated by comparing the discharge, Q, of a single well with drawdown, s_w, at its well face with the discharge of two, three, and four wells with the same drawdown at their well faces (Todd 1980); that is,

1. For two wells with a distance, b, between them,

$$Q_1/Q = \ln(R/r_w)/[\ln\{R^2/(r_w b)\}] \quad (4\text{-}38)$$

2. For three wells located on the vertices of an equilateral triangle of side b,

$$Q_1/Q = \ln(R/r_w)/[\ln\{R^3/(r_w b^2)\}] \quad (4\text{-}39)$$

3. For four wells located on the vertices of a square of side b,

$$Q_1/Q = \ln(R/r_w)/[\ln\{R^4/(\sqrt{2}\, r_w b^3)\}] \quad (4\text{-}40)$$

where Q_1 = discharge of each well in the group of two, three, or four wells. These equations assume a confined aquifer with $R \gg b$. They also may provide approximate results for unconfined aquifers with relatively small drawdown.

Example 4-6: Four wells are located in a confined aquifer on the four corners of a 152.4 m × 152.4 m square, with a fifth well at the center of the square. If the wells are pumped so as to create the same drawdown at the well faces, estimate the reduction in discharge due to interference. Assume $R = 304.8$ m and $r_w = 0.075$ m.

Solution: If the wells were located far apart so that there was no interference, then, from Eq. (4-17),

$$Q = (2\pi K B s_w)/\ln(R/r_w) \quad (i)$$

It will be seen later that, if drawdown at the faces of the five wells has to be the same, then $Q_1 \neq Q_c$, where Q_1 = discharge of each corner well and Q_c = discharge of the central well. Let the side of the square be b.

From Eq. (4-18), drawdown at the outer wells, s_{w1}, is given by

$$s_{w1} = [Q_1/(2\pi KB)]\,[\ln(R/r_w) + 2\ln(R/b) + \ln(R/\{\sqrt{2}\,b\})] + [Q_c/(2\pi KB)]\,[\ln\{R\sqrt{2}/b\}] \quad (ii)$$

Drawdown at the central well, s_{wc}, is given by

$$s_{wc} = [Q_1/(2\pi KB)]\,[4\ln\{\sqrt{2}\,R/b\}] + [Q_c/(2\pi KB)]\,[\ln(R/r_w)] \quad (iii)$$

162 WATER RESOURCES ENGINEERING

Equating (ii) and (iii),

$$Q_1 \ln [b/\{4\sqrt{(2)}r_w\}] = Q_c \ln [b/\{\sqrt{(2)}r_w\}] \qquad \text{(iv)}$$

Also, substituting (iv) in (iii),

$$s_{w1} = s_{wc} = [Q_1/(2\pi KB)] [\ln (4R^4/b^4) +$$
$$\ln (R/r_w) \cdot \ln (b/\{4\sqrt{(2)}r_w\})/\ln (b/\{\sqrt{(2)}r_w\})] \qquad \text{(v)}$$

From (i) and (v), for $s_{w1} = s_{wc} = s_w$,

$$Q_1/Q = \ln (R/r_w) \cdot \ln [b/\{\sqrt{(2)}r_w\}]/[\{\ln (4R^4/b^4)\} \cdot \ln (b/\{\sqrt{(2)}r_w\}) +$$
$$\ln (R/r_w) \cdot \ln (b/\{4\sqrt{(2)}r_w\})]$$

$$Q_c/Q = \ln (R/r_w) \cdot \ln [b/\{4\sqrt{(2)}r_w\}]/[\{\ln (4R^4/b^4)\} \cdot \ln (b/\{\sqrt{(2)}r_w\}) +$$
$$\ln (R/r_w) \cdot \ln (b/\{4\sqrt{(2)}r_w\})]$$

Using the given data,

$$Q_1/Q = [\ln (304.8/0.075)] \ln [152.4/\{0.075\sqrt{(2)}\}]/$$
$$[\ln \{4 \times 304.8^4/152.4^4\} \cdot \ln \{152.4/0.075\sqrt{(2)}\} +$$
$$\ln \{304.8/0.075\} \cdot \ln \{152.4/(4 \times 0.075\sqrt{(2)})\}] =$$
$$[\{8.31 \times 7.270\}/[(4.159 \times 7.270) + (8.31 \times 5.884)] = 0.76$$

$$Q_c/Q = [\{8.31 \times 5.884\}/[(4.159 \times 7.270) + (8.31 \times 5.884)] = 0.62.$$

Well interference reduces the discharge of the corner wells to 76% and that of the central well to 62% of that which would be available if the wells were located far apart.

Induced Recharge

Induced recharge from a nearby perennial stream to a pumping well located in a confined aquifer with uniform groundwater flow toward the stream may be estimated by the following (Bear 1979):

$$Q_r/Q = (2/\pi)[\tan^{-1}\sqrt{\{(Q/\pi\, d\, v\, B) - 1\}} -$$
$$(\pi\, d\, v\, B/Q) \sqrt{\{(Q/\pi\, d\, v\, B) - 1\}}] \qquad (4\text{-}41)$$

where

Q = total well discharge

v = velocity of uniform groundwater flow

Q_r = discharge induced from the stream

d = distance between the well and stream

If $Q \leq \pi dvB$, there is no contribution from the stream. Eq. (4-41) also may provide approximate results for unconfined aquifers with relatively small drawdowns where the average saturated thickness can be used for B.

Example 4-7: A fully penetrating well is located 60 m away from a stream in a 25-m-thick confined aquifer where normal ground flow velocity toward the stream is 0.15 m/day. The well is pumped at 26 l/s. Estimate the portion of the total discharge that is extracted from the stream.

Solution: $B = 25$ m; $d = 60$ m; $v = 0.15$ m/day; and $Q = 26 \times 86{,}400/1{,}000 = 2{,}246.4$ m^3/day; $Q/\pi dvB = 3.1780$; and $\pi dvB/Q = 0.31466$.

Using Eq. (4-41), $Q_r/Q = (2/\pi)[\tan^{-1}\sqrt{(3.1780-1)} - 0.31466\sqrt{(3.1780-1)}] = (2/\pi)[0.97527 - 0.46438] = 0.325$. Note that $\tan^{-1}(2.1780)$ has to be expressed in radians.

Approximately 32.5% of the well discharge is induced from the stream.

Recharge by Precipitation

For many field situations, recharge to an aquifer due to precipitation has to be estimated without adequate site-specific data. Often, a reasonable value is assumed and a modified value is obtained by using recharge as a calibration parameter in groundwater flow models. A reasonable practice is to estimate recharge by several alternative methods and select a plausible value by judgment. Some approximate methods include the following:

1. Precipitation recharge may be estimated based on available data for annual or seasonal recharge in the region or in regions of similar climatic and hydrogeologic conditions.

2. The Soil Conservation Service (SCS) Curve Number (CN) method may be used to estimate runoff, and groundwater recharge may be assumed to be nearly equal to the difference between precipitation and computed runoff (see "Soil Losses" in Chapter 2). Estimated CN for a storm may be higher than that for annual precipitation and may result in lower annual groundwater recharge. On the other hand, the difference in precipitation and computed runoff may include infiltration plus evaporation and interception losses, resulting in higher annual recharge. The positive and negative contributions of these factors may have to be accounted for by judgment. The following are equations used in this method:

$$S_a = (2{,}540/CN) - 25.4 \tag{2-6}$$

$$QR = (P - 0.2\,S_a)^2/(P + 0.8\,S_a) \tag{2-40}$$

where

QR = runoff (cm)

P = precipitation (cm)

3. Preliminary estimates of recharge may be made based on empirical ratios (Maxey-Eakin coefficients) shown in Table 4-6. These coefficients were developed for the arid climate in the state of Nevada (Avon and Durbin 1994).

The following is an alternative regression equation for arid watersheds (20 cm $< P <$ 51 cm) (Donovan and Katzer 2000):

$$r = 4.15 \times 10^{-6} (P)^{3.75} \qquad (4\text{-}42)$$

where

r = annual recharge (cm)

P = annual precipitation (cm)

For $P < 20$ cm, $r = 0$, and for $P > 51$ cm, $r = 0.25P$.

Seawater Intrusion

Under steady-state conditions, seawater and freshwater interface in a coastal aquifer can be approximated using the Ghyben-Herzberg relation (Figure 4-2):

$$h_s = [\rho_f/(\rho_s - \rho_f)] h_f \qquad (4\text{-}43)$$

where

h_s = elevation of interface below seawater surface elevation

h_f = elevation of water table above seawater surface elevation

ρ_f, ρ_s = density of freshwater and seawater, respectively

If $\rho_s = 1.025$ gm/cm^3 and $\rho_f = 1.0$ gm/cm^3, $h_s = 40\, h_f$. Thus, if the water table elevation at a certain distance from the shoreline is 0.25 m above the seawater surface elevation, the saltwater–freshwater interface at that location may be about 10 m below the seawater surface elevation. Using Eq. (4-43), upcoming, s_u, under a well pumping above the interface can be approximated by

$$s_u = [\rho_f/(\rho_s - \rho_f)] s_w \qquad (4\text{-}44)$$

Table 4-6. Maxey-Eakin coefficients

Annual precipitation (cm)	Recharge/precipitation
>51	0.25
38–51	0.15
30.5–51	0.07
20–30.5	0.03
<20	0.0

Source: Avon and Durbin (1994).

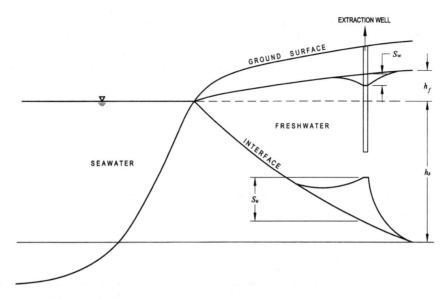

Figure 4-2. Schematic of saltwater interface and upconing

It must be recognized that seawater is miscible with freshwater and a sharp interface may not exist. Eqs. (4-43) and (4-44) provide preliminary estimates, which must be refined based on more sophisticated analysis and field data.

Example 4-8: In a coastal aquifer, saltwater concentrations were observed up to a distance of 300 m from the shoreline and a depth of 30 m below seawater surface elevation, which is also the elevation of the impervious layer at the bottom of the aquifer. Observed water table elevations are shown in Table 4-7(a).

$K = 30$ m/day, $\rho_f = 1.0$ g/cm³, and $\rho_s = 1.025\,0$ g/cm³. Estimate the location of the saltwater–freshwater interface. To be conservative, neglect seepage face at the intersection of the water table with the shoreline. Also, estimate freshwater flow per unit width of the shoreline.

Solution: Using Eq. (4-43), $h_s = [1.0/(1.025 - 1.0)]h_f = 40h_f$. The approximate position of the interface is shown in Table 4-7(b).

Table 4-7(a). Water table elevations near shoreline

Distance from shoreline (m)	Water table elevation above seawater elevation (m)
300	0.75
275	0.70
200	0.60
150	0.40
100	0.30
50	0.15
0	0

Table 4-7(b). Position of saltwater–freshwater interface

x (m)	h_f (m)	h_s (m)
300	0.75	30
275	0.7	28
200	0.6	24
150	0.4	16
100	0.3	12
50	0.15	6
0	0	0

x = distance from shoreline
h_f = elevation of water table above seawater elevation
h_s = elevation of interface below seawater elevation

Example 4-9: A 15-cm-diameter well is located at a distance of 275 m from the shoreline of Example 4-8. To avoid withdrawal of saline water, the minimum required vertical distance between the saltwater interface and the water table is 5 m. Estimate safe well discharge. Assume $R = 250$ m and $K = 30$ m/day.

Solution: From Example 4-8, the vertical distance between the water table and interface at $x = 275$ m is $28 + 0.70 = 28.70$. The lowest water table will be at the well face. So, $28.70 - s_u - s_w = 5$. Also, from Eq. (4-44), $s_u = [1.0/(1.025 - 1.0)] s_w = 40 s_w$. So, $s_w = (28.70 - 5)/41 = 0.578$ m.

For freshwater withdrawal, average saturated thickness may be assumed to be $H = 30.75/2 = 15.375$ m. Thus, $Q = 2\pi K H s_w / \ln(R/r_w) = 2\pi \times 30 \times 15.375 \times 0.578 / \ln(250/0.075) = 207$ m³/day or 2.4 l/s.

Because of variations in the saturated thickness and partial penetration of the well in the freshwater aquifer, this should be treated as a preliminary estimate. A more sophisticated analysis may be required for refined estimates.

Effect of Barometric Pressure Fluctuations

An approximate relationship between fluctuations of atmospheric pressure and water levels in a well in a confined aquifer is given by the following (Rouse 1950; Todd 1980):

$$dh/(dp_a/\gamma) = -1/[1 + \{(\alpha_s/(\theta \beta)\}] \qquad (4\text{-}45)$$

where

dh = change in water level in well
dp_a = change in barometric pressure
α_s = vertical compressibility of aquifer material
θ = aquifer porosity
β = compressibility of water

The negative sign in Eq. (4-45) indicates that water level in an observation well falls when barometric pressure increases, and vice versa.

For an unconfined aquifer, compressibility of aquifer material and water is relatively less significant compared to changes in water volume that result from water table fluctuations. The changes in atmospheric pressure are transmitted directly and simultaneously to the water table and observation well. Therefore, there is little or no effect of barometric pressure fluctuations on water levels observed in the well. The change in water level in the well is almost the same as in the unconfined aquifer. It has been observed that fluctuations in barometric pressure may result in small fluctuations in the water table in unconfined aquifers.

Example 4-10: Estimate the change in water level in a well fully penetrating a confined aquifer when barometric pressure changes by 7.72 cm of mercury. Assume $\alpha_s = 11.8 \times 10^{-6}$ cm²/kg; $\phi = 0.35$; and $\beta = 47 \times 10^{-6}$ cm²/kg.

Solution: $(dp_a/\gamma) = 0.0772 \times 13.6 = 1.05$ m of water. Using Eq. (4-45), $dh/(dp_a/\gamma) = dh/(1.05) = -1/[1 + \{11.8/(0.35 \times 47)\}] = -0.582$. The negative sign indicates that increase in barometric pressure results in depressing the water level in the well. Thus, $dh = -1.05 \times 0.582 = -0.61$ m.

Subsidence

The stress caused by total weight of soil and water above a point in an aquifer is balanced by effective (compressive) stress on the aquifer material and fluid (hydrostatic) pressure (Delleur 1999):

$$p_T = \sigma + p_w \tag{4-46}$$

where

p_T = total pressure due to weight of soil and water

p_w = fluid (hydrostatic) pressure

σ = effective or compressive stress on aquifer material

The hydrostatic pressure can be measured by a piezometer. An increase in the compressive stress on the aquifer material causes reduction in its volume (or in its thickness in one-dimensional compression), which may result in subsidence. Excessive groundwater extraction may result in some reduction in the total pressure, but it could result in relatively greater reduction in the hydrostatic pressure in the aquifer. This may result in an increase in the compressive stress on the aquifer material and cause land subsidence. Groundwater pumping has to be controlled to minimize potential for subsidence. Computational steps for preliminary estimates of subsidence due to lowering of groundwater levels in an aquifer are listed below.

1. Estimate total load (pressure) at the position of lowered groundwater level before pumping due to the weight of overlying soils and water held in pores.

2. Estimate hydrostatic pressure at that level due to head of groundwater above that level.
3. Estimate compressive stress on aquifer material at that level as the difference of steps (1) and (2) before pumping.
4. Following the same procedure, estimate compressive stress at the same level after lowering of the groundwater level.
5. Find the difference, $\Delta\sigma$, in the initial and final compressive stresses at that level [(4) − (3)].
6. The change in compressive stress at the level of the initial groundwater level is zero. Thus, the average change in compressive stress in the aquifer column between these two levels is $\Delta\sigma/2$.
7. Estimate subsidence, δ, in this aquifer column between the initial and lowered groundwater levels,

$$\delta = \alpha_s(\Delta\sigma/2)\Delta h \qquad (4\text{-}47a)$$

where

α_s = compressibility of aquifer material

Δh = change in groundwater levels

8. Change in compressive stress in aquifer material below the lowered groundwater level will be $\Delta\sigma$.
9. If there are two or more layers of soils below the lowered groundwater level, estimate subsidence in each:

$$\delta_1 = \alpha_1(\Delta\sigma)L_1, \quad \text{and} \quad \delta_2 = \alpha_2(\Delta\sigma)L_2, \text{ etc.} \qquad (4\text{-}47b)$$

where

δ_1, δ_2 = subsidence in layers 1 and 2

α_1, α_2 = compressibility of layers 1 and 2

L_1, L_2 = thickness of layers 1 and 2, respectively

10. Estimate total subsidence at the bottom of layer 2 = $\delta + \delta_1 + \delta_2$.

Example 4-11: Extensive groundwater pumping in an area is expected to lower groundwater levels by 25 m. Initial water level is 10 m below the ground surface. The aquifer material is sand up to a depth of 50 m below the ground surface. Below the sand is a 30-m-thick silty clay layer overlying the bedrock. Estimate potential subsidence in the soils above the bedrock. Assume degree of saturation in the unsaturated soil zone above groundwater level to be 0.10, unit weight of sand grains = 2,600 kg/m³; unit weight of water = 1,000 kg/m³; porosity of sand = 0.33; compressibility of sand = 12×10^{-8} m²/kg; and compressibility of silty clay = 100×10^{-8} m²/kg.

Solution: Let the positions of initial and lowered groundwater level be denoted by B and C, the bottom of the sand unit by D, and the bottom of clay by E.

Vertical distance between ground surface and C = 10 + 25 = 35 m.

After expected lowering of groundwater level, total load above point C at the elevation of lowered groundwater level = 35 × (1 − 0.33) × 2,600 + 35 × (0.33 × 0.10) × 1,000 = 60,970 + 1,155 = 62,125 kg/m².

Hydrostatic pressure at point C (after lowering of the water table to this level) = 0.

Compressive stress in soils at this elevation (at point C) = 62,125 − 0 = 62,125 kg/m².

Under initial groundwater conditions, total load above point C = 10 × (1 − 0.33) × 2,600 + 10 (0.33 × 0.10) × 1,000 + 25 × (1 − 0.33) × 2,600 + 25 × 0.33 × 1,000 = 17,420 + 330 + 43,550 + 8,250 = 69,550 kg/m².

Under initial groundwater conditions, hydrostatic pressure at C = 25 × 1,000 = 25,000 kg/m².

Under initial groundwater conditions, compressive stress in soils at C = 69,550 − 25,000 = 44,550 kg/m².

Increase in compressive stress at C due to lowering of water level = 62,125 − 44,550 = 17,575 kg/m².

Change in compressive stress in soils at initial groundwater level (point B) = 0.

Change in compressive stress at lowered groundwater level (point C) = 17,575 kg/m².

Average subsidence over a soil column of Δh = 25 m is given by $\delta = \alpha_s(\Delta\sigma/2)\Delta h$ = 12 × 10⁻⁸ × (17,575/2) × 25 = 0.026 m.

For sand below the lowered groundwater level (point C to D), α_1 = 12 × 10⁻⁸ m²/kg and L_1 = 15 m. So, $\delta_1 = \alpha_1(\Delta\sigma)L_1$ = 12 × 10⁻⁸ × 17,575 × 15 = 0.032 m.

For silty clay below the sand (point D to E), α_2 = 100 × 10⁻⁸ m²/kg, and L_2 = 30 m. So, $\delta_1 = \alpha_1(\Delta\sigma)L_1$ = 100 × 10⁻⁸ × 17,575 × 30 = 0.527 m.

Total subsidence above bedrock = 0.026 + 0.032 + 0.527 = 0.585 m.

Safe Yield, Specific Capacity, and Efficiency

There is no precise definition of the safe yield of an aquifer or well. Usually, it is the rate of groundwater withdrawal that can be maintained without creating potential for subsidence; saltwater intrusion; undue depletion of the groundwater table, which cannot be replenished by natural groundwater recharge; undue interference with the yield of existing groundwater wells; undue induced recharge from nearby surface water bodies; and encroachment on existing groundwater contaminant plumes. The limitations on drawdowns at specific locations may be specified by local, state, or federal regulatory agencies. Sometimes, the term "optimum yield" is used to identify safe yield, which can be economically obtained if various alternative groundwater resource management strategies are considered.

Specific capacity of a well is its yield per unit drawdown (Q/s), where s includes drawdown in the aquifer at the boundary of the well screen and well loss resulting from turbulent flow of groundwater through the well screen. The efficiency of a well is the ratio of its actual specific capacity to the theoretical specific capacity. It is also defined as the ratio of the aquifer head loss to the total head loss. The total head loss (or total drawdown at the well face) includes aquifer (i.e., theoretical) drawdown and well drawdown. Aquifer drawdown varies linearly with discharge and can be estimated by steady-state or non–steady-state equations for drawdown at the well face (e.g., Eqs. (4-17) and (4-81)). Well loss includes a linear component and a nonlinear component. The linear component includes drawdown in the gravel pack and screen entrance, and the nonlinear component includes losses due to turbulent flow in the well. A simple method to estimate well efficiency is as follows:

- Plot drawdown, s, on the y-axis on the natural scale, and plot the distance from the well, r, on the logarithmic scale on the x-axis.
- Draw the best-fitting straight line through these points by visual judgment.
- Extend the straight line to $r = r_w$ (well radius) and read the theoretical drawdown, s_0, at this location.
- Estimate well efficiency = s_0/s_w, where s_w is the actual drawdown measured in the well.

In the case of an unconfined aquifer, well operation may cause considerable reduction in the saturated thickness of the aquifer. As a result, extra drawdown is observed. This extra drawdown does not represent inefficiency in the well. If the reduction in saturated thickness is more than 20%, then the drawdown, s_0, may be corrected before well efficiency is computed (see the section in this chapter entitled "Unsteady Radial Flow to a Well Fully Penetrating an Unconfined Aquifer"):

$$s_0 \text{ (corrected)} = s_0 - \{s_0^2/(2H)\} \tag{4-48}$$

where H = initial saturated thickness of the aquifer. The corrected drawdown should be used in preparing the aforementioned plot. A well efficiency of about 70 to 80% is acceptable for a well-designed well.

Transient (Unsteady) Groundwater Flow

Unsteady One-Dimensional Flow

Continuity equation for unsteady one-dimensional groundwater flow in a confined aquifer of thickness, B, and hydraulic conductivity, K, is

$$\partial^2 h/\partial x^2 = S/T \, \partial h/\partial t \tag{4-49}$$

where

S = dimensionless storage coefficient or storativity

T = transmissivity or transmissibility of the aquifer

$S_s = S/B$ = specific storage, defined as the volume of water that is released by a unit volume of the aquifer per unit decline in hydraulic head:

$$S = \rho \, g(\alpha_s + \theta \, \beta)B \tag{4-50}$$

$$T = K B \tag{4-51}$$

Typical values of specific storage, S_s, are given in Table 4-8 (USEPA 1985).

Table 4-8. Typical values of specific storage

Material	Specific storage (cm^{-1})
Plastic clay	$2.0 \times 10^{-4} - 2.5 \times 10^{-5}$
Stiff clay	$2.5 \times 10^{-5} - 1.3 \times 10^{-5}$
Medium hard clay	$1.3 \times 10^{-5} - 6.9 \times 10^{-6}$
Loose sand	$9.8 \times 10^{-6} - 5.1 \times 10^{-6}$
Dense sand	$2.1 \times 10^{-6} - 1.3 \times 10^{-6}$
Dense sandy gravel	$9.8 \times 10^{-7} - 5.1 \times 10^{-7}$
Rock (fissured, jointed)	$6.9 \times 10^{-7} - 3.2 \times 10^{-8}$
Sound rock	Less than 3.2×10^{-8}

Source: USEPA (1985).

Linearized continuity equation for non–steady-state one-dimensional flow in an unconfined aquifer is

$$\partial^2 h/\partial x^2 = [S_y/(KH)] \, \partial h/\partial t \tag{4-52}$$

where

S_y = specific yield

H = average saturated thickness

Specific yield is that portion of water held in soil pores that can be extracted from the aquifer per unit area per unit drop in the water table. It is also called effective porosity and is less than total porosity. Typical values of specific yield are given in Table 4-9 (USEPA 1985).

Bank Storage

During flood stages, river water enters the porous material of the banks. As a result, the groundwater table on both sides of the river may rise. After the flood recedes, water surface elevation in the river returns to the normal condition in a relatively short time. The water stored in the bank material is slowly released into the stream under the head difference between the elevated water table and the normal (lowered) water surface elevation in the stream. In some cases, bank material may consist of contaminated sediments resulting from past industrial activities, and the water returning to the river by way of seepage from banks also may be contaminated. In certain situations, estimation of the rates of return flow and quantities of contaminated groundwater likely to enter the river are required. Analytical equations for preliminary analysis of this type of situations are given below (Carslaw and Jaeger 1984; Glover 1985):

1. $h = H \, \mathrm{erf}[x/\sqrt{(4\alpha t)}]$ (4-53)
2. $q(x, t) = \{HKD/\sqrt{(\pi \alpha t)}\} \exp\{-x^2/(4\alpha t)\}$ (4-54)
3. $q(0, t) = \{HKD/\sqrt{(\pi \alpha t)}\}$ (4-55)
4. $Q(t) = 2 HKD \sqrt{t/(\pi \alpha)}$ (4-56)

where

H = height of the water table resulting from the flood stage, measured above normal (lowered) water surface elevation in the river

h = height of the water table above normal (lowered) water surface elevation in the river at distance, x, and time, t

x = distance into the bank from the edge of the river

t = time after the flood event (i.e., after the river has receded to normal water surface elevation)

α = aquifer diffusivity = KD/S_y

D = initial saturated thickness of the bank material (i.e., depth of raised water surface in the bank above the impervious base)

$q(x, t)$ = rate of return flow at distance, x, and time, t, per unit length of bank (parallel to river flow)

$q(0, t)$ = rate of return flow at the edge of the river at time, t, per unit length of bank

$Q(t)$ = total volume of flow that has returned to the river up to time, t, per unit length of the bank

Table 4-9. Typical values of specific yield

Material	Range (%)	Mean (%)
Clay	1.1–17.6	6
Silt	1.1–38.6	20
Fine sand	1.0–45.9	33
Medium sand	16.2–46.2	32
Coarse sand	18.4–42.9	30
Fine gravel	12.6–39.9	28
Medium gravel	16.9–43.5	24
Coarse gravel	13.2–25.2	21
Loess	14.1–22.0	18
Dune sand	32.3–46.7	38
Till (predominantly silt)	0.5–13.0	6
Till (predominantly sand)	1.9–31.2	16
Till (predominantly gravel)	5.1–34.2	16
Glacial drift (predominantly silt)	33.2–48.1	40
Glacial drift (predominantly sand)	29.0–48.2	41
Sand stone (fine-grained)	2.1–39.6	21
Sandstone (medium-grained)	11.9–41.1	27
Siltstone	0.9–32.7	12
Shale	0.5–5.0	—
Limestone	0.2–35.8	14
Schist	21.9–33.2	26

Source: USEPA (1985).

The notation erf$[x/\sqrt{(4\alpha t)}]$ represents the error function of the quantity within the brackets. The values of error function are available in tabulated form (e.g., Abramowitz and Stegun 1972). A rational approximation suitable for computer use is

$$\text{erf}(x) = 1 - (a_1 t + a_2 t^2 + a_3 t^3 + a_4 t^4 + a_5 t^5) \exp(-x^2) \quad (4\text{-}57)$$

where

$t = 1/(1 + 0.3275911x)$

$a_1 = 0.254829592$

$a_2 = -0.284496736$

$a_3 = 1.421413741$

$a_4 = -1.453152027$

$a_5 = 1.061405429$

Eqs. (4-53) to (4-56) are based on the assumption that bank material is isotropic and homogeneous and the river returns to normal water surface elevation in a relatively short time following the flood event so that water table in the bank material is still at the elevated level.

Example 4-12: During flood season, the water table in the banks of a stream is found to have risen by 3 m above the normal water surface elevation of the river. Estimate return flow from one side of the bank after the river has returned to normal water surface elevation, which may be assumed to have occurred in a relatively short time. The bank material on one side of the stream contains 0.4 gm/kg of adsorbed lead. The distribution coefficient of lead is found to be 10,000 l/kg. Estimate the mass of lead that may have entered the river by leaching from the bank materials in a period of 90 days. Assume that adsorbed lead is in a soluble state and can be leached during this period. For the bank materials, use $KD = 1,766$ m^2/day and $S_y = 0.15$.

Solution: $\alpha = KD/S_y = 1,766/0.15 = 11,773.3$ m^2/day.

From Eq. (4-55), $q(0, t) = 3 \times 1,766/\sqrt{(\pi \times 11,773.3 \times 90)} = 2.9$ m^3/day per meter length of bank.

From Eq. (4-56), $Q(t) = 2 \times 3 \times 1,766 \sqrt{\{(90/(\pi \times 11,773.3)\}} = 522.7$ m^3 per meter length of bank.

Assuming equilibrium conditions, $S_d = K_d C$, where S_d = mass of lead adsorbed per unit dry mass of bank material = 0.0004 kg/kg; K_d = distribution coefficient for lead = 10 m^3/kg; and C = concentration of lead in water contained in bank materials. So, $C = 0.0004/10 = 0.00004$ kg/m^3.

Mass of lead entering the river in 90 days = $0.00004 \times 522.7 = 0.02$ kg per meter length of bank.

This estimate is preliminary because establishment of equilibrium conditions may take longer and leaching of lead may be slower.

Flow toward Drains and Drain Spacing

An approximate method for spacing of tile drains in agricultural areas was given in a previous section entitled "Darcian Flow." If an initial elevated water table is to be lowered to a certain level within a specified period, then drain spacing has to be estimated using the transient flow equation (Eq. (4-49)). Relevant analytical equations for this case are listed below (Carslaw and Jaeger 1984; Glover 1985):

1. $h(x, t) = (4H/\pi) \Sigma\{1/(2n+1)\} [\exp\{-(2n+1)^2 \pi^2 \alpha\, t/L^2\}]$
   $\sin\{(2n+1)\pi x/L\}, n = 0, 1, 2, \ldots, \infty$ (4-58)

2. $h(x = L/2) = (4H/\pi) \Sigma\{1/(2n+1)\} [\exp\{-(2n+1)^2 \pi^2 \alpha\, t/L^2\}]$
   $\sin\{(2n+1)\pi/2\}, n = 0, 1, 2, \ldots, \infty$ (4-59)

3. $q(x = 0, t) = (4KDH/L) \Sigma[\exp\{-(2n+1)^2 \pi^2 \alpha\, t/L^2\}]$,
   $n = 0, 1, 2, \ldots, \infty$ (4-60)

4. $p = (8/\pi^2) \Sigma[\exp\{-(2n+1)^2 \pi^2 \alpha\, t/L^2\}]/(2n+1)^2$,
   $n = 0, 1, 2, \ldots, \infty$ (4-61)

where

H = height of initial water table above drains

D = height of drains above impervious layer

L = drain spacing

$h(x, t)$ = height of water table at distance, x, and time, t, above drains

x = distance from drain

t = time since groundwater starts to drain from initial water table elevation

$q(x = 0, t)$ = flow to drain from one side per unit length of drain

p = fraction of drainable volume of water that remains to be drained at time, t

It is assumed that $H \ll D$. Otherwise, D may be taken to be the average saturated thickness of the aquifer. The minimum lowering of the water table will occur at the center of two parallel drains, i.e., at $x = L/2$. Thus, Eq. (4-59) can be used to estimate drain spacing for a minimum water table lowering to h at $x = L/2$ above the drains. The infinite series of Eqs. (4-58) to (4-61) converge fairly rapidly for $(\alpha t/L^2) \gg 0.01$. For such cases, the second term is $<2\%$ of the first, and the remaining terms are even smaller. Thus, these equations may be approximated by

1. $h(x, t) = (4H/\pi)[\exp\{-\pi^2 \alpha\, t/L^2\}] \sin\{\pi x/L\}$ (4-62)

2. $h(x = L/2) = (4H/\pi)[\exp\{-\pi^2 \alpha\, t/L^2\}]$ (4-63)

3. $q(x = 0, t) = (4KDH/L)[\exp\{-\pi^2 \alpha\, t/L^2\}]$ (4-64)

4. $p = (8/\pi^2)[\exp\{-\pi^2 \alpha\, t/L^2\}]$ (4-65)

From Eq. (4-63), for $(\alpha t/L^2) \gg 0.01$, $L = \pi \sqrt{[\alpha t/\ln\{4H/(\pi h)\}]}$ (4-66)

For other cases where $(\alpha t/L^2) \leq 0.01$, $n = 0, 1, 2,$ and 3 may have to be used in Eqs. (4-58) to (4-61). Usually, terms involving $n > 3$ may be too small to consider.

Example 4-13: In an irrigated area, the impervious soil layer is about 12 m below the field level. Tile drains are to be installed about 3 m below the field level. During the first irrigation season, the water table rises to within 0.75 m below the field level. The next irrigation period is 30 days after the first. Before the second irrigation, the water table has to be lowered to a minimum of 1.5 m below the field level. Estimate drain spacing for this situation. Use $K = 3.05$ m/day and $S_y = 0.18$.

Solution: Height of maximum water table above drains = $H = 3.0 - 0.75 = 2.25$ m.
Height of maximum water table above impervious layer = $12 - 0.75 = 11.25$ m.
Height of drains above impervious layer = $12 - 3 = 9$ m, and $h(x = L/2) = 3 - 1.5 = 1.5$ m.
Average saturated thickness = $D \cong (11.25 + 9)/2 = 10.125$ m, and $\alpha = KD/S_y = 3.05 \times 10.125/0.18 = 171.56$ m²/day.
Using Eq. (4-66), $L = \pi\sqrt{[(171.56 \times 30)/\ln\{(4 \times 2.25)/(\pi \times 1.5)\}]} = 280.2$ m.
Check the validity of Eq. (4-66), $\alpha t/L^2 = 171.56 \times 30/(280.2)^2 = 0.066$. So, the approximation of Eq. (4-66) is valid.

The following is an approximate equation to estimate steady-state groundwater flow toward a single circular drain or tunnel (Freeze and Cherry 1979):

$$q_T = 2\pi K H/\ln(2H/r) \qquad (4\text{-}67)$$

where

q_T = flow into the drain or tunnel per unit length
H = head above tunnel centerline
r = radius of drain or tunnel

An approximate equation for the transient case is as follows (Freeze and Cherry 1979):

$$q_T(t) = \sqrt{(C K H^3 S_y\, t)} \qquad (4\text{-}68)$$

where

$q_T(t)$ = flow into the drain or tunnel per unit length at time, t, after the breakdown of steady flow

C is a constant

The values of C may vary from 4/3 to 2. Eqs. (4-67) and (4-68) may be useful for preliminary analyses. Numerical models must be used for more refined analyses.

At some industrial sites, trenches or underground drains are provided to collect or intercept contaminated groundwater from the site area, which may be pumped out through sumps located at suitable locations on the trench or drain. The pumps and trenches are designed for groundwater flows that may be expected during high groundwater table conditions following storm events. If, initially, the water table is approximately horizontal, the following equations estimate the lowering of groundwater levels and flows entering the trench or drain (Carslaw and Jaeger 1984):

1. $h(x, t) = h_0 + [4(H - h_0)/\pi] \Sigma \{(-1)^n/(2n + 1)\} [\exp\{-(2n + 1)^2 \pi^2 \alpha t/(4L^2)\}] \cos\{(2n + 1) \pi x/(2L)\}, n = 0, 1, 2, \ldots, \infty$ (4-69)

2. $q(x = 0, t) = \{2 KD(H - h_0)\}/L \Sigma [\exp\{-(2n + 1)^2 \pi^2 \alpha t/(4L^2)\}], n = 0, 1, 2, \ldots, \infty$ (4-70)

3. $Q(t) = \{8 KDL(H - h_0)\}/(\pi^2 \alpha) \Sigma [\exp\{-(2n + 1)^2 \pi^2 \alpha t/(4L^2)\}]/(2n + 1)^2, n = 0, 1, 2, \ldots, \infty$ (4-71)

As in the case of Eqs. (4-55) to (4-58), for $\alpha t/(4L^2) > 0.01$, simplified approximate equations are as follows:

1. $h(x, t) = h_0 + [4(H - h_0)/\pi)][\exp\{-\pi^2 \alpha t/(4L^2)\}] \cos\{\pi x/(2L)\}$ (4-72)

2. $q(x = 0, t) = [\{2 KD(H - h_0)\}/L][\exp\{-\pi^2 \alpha t/(4L^2)\}]$ (4-73)

3. $Q(t) = [\{8 KDL(H - h_0)\}/(\pi^2 \alpha)][\exp\{-\pi^2 \alpha t/(4L^2)\}]$ (4-74)

where

H = height of initial water table above impervious layer

D = average saturated thickness

L = distance from groundwater divide to trench or drain

$h(x, t)$ = height of water table above impervious layer at distance, x, and time, t

x = distance from groundwater divide

t = time since groundwater starts to drain from initial water table elevation

h_0 = height of drain above impervious layer

$q(x = 0, t)$ = flow to trench from one side per unit length of trench

$Q(t)$ = total flow that entered the trench from one side up to time, t

If initial water table can be approximated by a sloping straight line, then groundwater lowering due to the trench can be estimated by

$$h(x, t) = h_0 + [8(H - h_0)/\pi^2] \Sigma \{1/(2n + 1)^2\} [\exp\{-(2n + 1)^2 \pi^2 \alpha t/(4L^2)\}] \cos\{(2n + 1) \pi x/(2L)\}, n = 0, 1, 2, \ldots, \infty \quad (4\text{-}75)$$

or

$$h(x, t) = h_0 + (H - h_0)\{(L - x)/L\} - [2(H - h_0)\sqrt{(\alpha t)}/L] \Sigma (-1)^n \{\text{ierfc}(U_1) - \text{ierfc}(U_2)\}, n = 0, 1, 2, \ldots, \infty \quad (4\text{-}76)$$

where

$U_1 = (2nL + x)/\sqrt{(4\alpha t)}$

$U_2 = \{(2n + 2)L - x\}/\sqrt{(4\alpha t)}$

initial water table is given by the straight line $h(x) = H - (H - h_0)x/L$

ierfc(U) = integral of erfc(U) (from $u = 0$ to ∞) = $(1/\sqrt{\pi})\exp(-U^2) - U\,\mathrm{erfc}(U)$

erfc$(x) = 1 - \mathrm{erf}(x)$

Eq. (4-75) converges rapidly when $\{\alpha t/(4L^2)\} \gg 0.01$ and Eq. (4-76) when $\{\alpha t/(4L^2)\} \ll 0.01$.

Response of Groundwater Levels to River Stage Fluctuations

If groundwater monitoring wells are located in an isotropic and homogeneous aquifer, which is in hydraulic communication with a river, water levels in the wells fluctuate with changes in river stages. If initially the water table is approximately horizontal at the same level as the initial river stage, then response of groundwater levels in a well to river stage fluctuations may be approximated by the following (Pinder et al. 1969; Prakash 1997):

$$h(x, t) = \Sigma\, e(m)\,\mathrm{erfc}\,\{(u)/\sqrt{(n - m + 1)}\},\ m = 1, 2, 3, \ldots, n \qquad (4\text{-}77)$$

where

$u = [x/\{2\sqrt{(\alpha \Delta t)}\}]$

$h(x, t)$ = height of groundwater level above initial steady-state stage in the river at distance, x, and time, t

x = distance from river

t = time since the start of rise or fall in river stage = $n\Delta t$

n = number of equal time intervals selected to divide rise or fall in river stages into different, smaller depth increments

Δt = time interval during which river stage changes by $e(1)$, $e(2)$, or $e(3)$, etc.

$e(m)$ = rise or fall in river stage in the mth time increment

Division of the time, t, into several time steps is required if the rate of changes in river stages is not uniform (i.e., differs from one time interval, Δt, to another). Eq. (4-77) may be used to estimate aquifer diffusivity, α, if groundwater level and river stage fluctuations are known from field observations. If the rise or fall in river stages lasts for a finite time, t_0, then,

$$h(x, t, t_0) = h(x, t) - h(x, t - t_0) \qquad (4\text{-}78)$$

where $h(x, t, t_0)$ = change in groundwater level at distance, x, and time, t, due to river stage rise or fall occurring during time period t_0.

For uniformly rising river stages and approximately horizontal water table at the beginning of river stage fluctuations, changes in groundwater levels may be approximated by the following (Carslaw and Jaeger 1984):

$$h(x, t) = c\,t[\{(1 + 2U^2)\,\mathrm{erfc}(U)\} - \{(2/\sqrt{\pi})\,U\exp(-U^2)\}] \qquad (4\text{-}79)$$

where

c = rise or fall in river stages per unit time

$U = x/\sqrt{(4\alpha t)}$

Table 4-10. Computations of water level fluctuations

(1) Time (h)	(2) $5.303/\sqrt{t}$	(3) $h(x, t)$ (m)	(4) $h(x, t)$ lagged by $t_0 = 48$ h	(5) $h(x, t, t_0)$ (m)
24	1.083	0.051	0	0.051
48	0.765	0.301	0	0.301
72	0.625	0.698	0.051	0.647
96	0.541	1.193	0.301	0.892
120	0.484	1.757	0.698	1.059
144	0.442	2.373	1.193	1.180
168	0.409	3.030	1.757	1.273
192	0.383	3.722	2.373	1.349
216	0.361	4.441	3.030	1.411
240	0.342	5.184	3.722	1.462

Example 4-14: Estimate groundwater levels in an observation well located 45 m from a river when river stage rises at 0.05 m/h for 48 h and thereafter stays at that level for a sufficiently long time. Use $\alpha = 18$ m^2/h.

Solution: $U = x/\sqrt{(4\alpha t)} = 5.303/\sqrt{t}$. Computations using Eqs. (4-79) and (4-78) are shown in Table 4-10.

Computations in Eq. (4-78) are performed by lagging $h(x,t)$ by t_0 (column (4)) and subtracting column (4) from (3). The results are included in column (5).

Unsteady Radial Flow to a Well Fully Penetrating a Confined Aquifer

The differential equation governing unsteady radial flow to a fully penetrating well in a confined homogeneous and isotropic aquifer of infinite extent is

$$\partial^2 h/\partial r^2 + 1/r \, \partial h/\partial r = (S/T) \, \partial h/\partial t \qquad (4\text{-}80)$$

The boundary conditions for Eq. (4-80) are listed below:

1. Limit $r = r_w \to 0$, $(r\partial h/\partial r) = (Q/2\pi T)$, $t > 0$ (indicating constant well discharge)
2. $h(r, 0) = H$, $r_w \leq r \leq \infty$ (indicating no drawdown before pumping)
3. $h(\infty, t) = H$, $t \geq 0$ (indicating no drawdown at infinity)

With these boundary conditions, the solution of Eq. (4-80) is,

$$H - h(r, t) = s(r, t) = -(Q/4\pi T) \, \text{Ei}\,(-u) = (Q/4\pi T) W(u) \qquad (4\text{-}81)$$

where

H = initial hydraulic head in the aquifer

$h(r, t)$ = head at distance, r, from the well at time, t

$s(r, t)$ = drawdown at distance, r, from the well at time, t

$u = Sr^2/(4\,Tt)$

$-Ei\,(-u)$ = exponential integral

$W(u)$ = well function

Eq. (4-81) is known as the Theis equation (Bear 1979). The well function is given by the infinite series

$$W(u) = -0.5772 - \ln(u) + u - u^2/(2.\,!2) + u^3/(3.\,!3) - u^4/(4.\,!4) + \ldots \quad (4\text{-}82)$$

For computations on a computer, polynomial and rational approximations are available to estimate $W(u)$ with an accuracy of up to 2×10^{-7} for values of u between 0 and infinity (Abramowitz and Stegan 1965). Well functions for selected values of u are given in Table 4-11 (Bear 1979; Freeze and Cherry 1979; Wenzel 1942). An expanded table of $W(u)/2$ for various values of $\sqrt{u}$ is given by Glover (1985).

For $u < 0.01$, $W(u)$ may be estimated by the first two terms on the right side of Eq. (4-82) and Eq. (4-81) may be approximated by

$$s(r, t) = \{Q/(4\,\pi\,T)\}\ln\{2.25\,Tt/(Sr^2)\} \quad (4\text{-}83)$$

If drawdown for a given r is known at two times, t_1 and t_2,

$$s_1 - s_2 = \{Q/(4\,\pi\,T)\}\ln(t_1/t_2) = \{2.3\,Q/(4\,\pi\,T)\}\log(t_1/t_2) \quad (4\text{-}84)$$

Eq. (4-83) may be used up to $u < 0.10$ with errors of about 5% or less in estimating drawdown (Delleur 1999). Note that r tends to R as $s(r, t)$ tends to zero. Setting $s(r, t) = 0$ in Eq. (4-83), it is seen that the radius of influence, R, increases with duration of pumping and is given by the following (see the section in this chapter entitled "Radius of Influence"):

$$R = 1.5\,\sqrt{(T\,t/S)} \quad (4\text{-}85)$$

If the well is pumped for a duration, t_0, and then stopped, the resulting drawdown is given by

$$s(r, t, t_0) = \{Q/(4\,\pi\,T)\}\ln\{t/(t - t_0)\} \quad (4\text{-}86)$$

where $s(r, t, t_0)$ = drawdown at distance, r, from the well at time, t, due to pumping for duration, t_0, at a constant rate, Q.

If a constant head boundary (e.g., a perennial river) is located at distance, b, from the well face along the x-axis and the well is located at the origin, then, using an image well (i.e., a recharge well) located at distance, b, from the river on the other side,

$$s(x, y, t) = \{Q/(2\,\pi\,T)\}\ln(r/r') \quad (4\text{-}87)$$

where

$s(x, y, t)$ = drawdown at point (x, y) at time, t

$r = \sqrt{(x^2 + y^2)}$

$r' = \sqrt{\{(2b - x)^2 + y^2\}}$

Table 4-11. Well function $W(u)$ for confined aquifers

n	$u = n$	$u = n \times 10^{-1}$	$u = n \times 10^{-2}$	$u = n \times 10^{-3}$	$u = n \times 10^{-4}$	$u = n \times 10^{-5}$	$u = n \times 10^{-6}$	$u = n \times 10^{-7}$
1.0	0.2194	1.8229	4.0379	6.3315	8.6332	10.9357	13.2383	15.5409
1.5	0.1000	1.4645	3.6374	5.9266	8.2278	10.5303	12.8328	15.1354
2.0	0.04890	1.2227	3.3547	5.6394	7.9402	10.2426	12.5451	14.8477
2.5	0.02491	1.0443	3.1365	5.4167	7.7172	10.0194	12.3220	14.6246
3.0	0.01305	0.9057	2.9591	5.2349	7.5348	9.8371	12.1397	14.4423
3.5	0.00697	0.7942	2.8099	5.0813	7.3807	9.6830	11.9855	14.2881
4.0	3.779×10^{-3}	0.7024	2.6813	4.9482	7.2472	9.5495	11.8520	14.1546
4.5	2.073×10^{-3}	0.6253	2.5684	4.8310	7.1295	9.4317	11.7342	14.0368
5.0	1.148×10^{-3}	0.5598	2.4679	4.7261	7.0242	9.3263	11.6280	13.9314
5.5	6.409×10^{-4}	0.5034	2.3775	4.6313	6.9289	9.2310	11.5330	13.8361
6.0	3.601×10^{-4}	0.4544	2.2953	4.5448	6.8420	9.1440	11.4465	13.7491
6.5	2.034×10^{-4}	0.4115	2.2201	4.4652	6.7620	9.0640	11.3665	13.6691
7.0	1.155×10^{-4}	0.3738	2.1508	4.3916	6.6879	8.9899	11.2924	13.5950
7.5	6.580×10^{-5}	0.3403	2.0867	4.3231	6.6190	8.9209	11.2234	13.5260
8.0	3.760×10^{-5}	0.3106	2.0269	4.2591	6.5545	8.8563	11.1589	13.4614
8.5	2.160×10^{-5}	0.2840	1.9711	4.1990	6.4939	8.7957	11.0982	13.4008
9.0	1.240×10^{-5}	0.2602	1.9187	4.1423	6.4368	8.7386	11.0411	13.3437
9.5	0.71×10^{-5}	0.2387	1.8695	4.0887	6.3828	8.6845	10.9870	13.2896

Table 4-11. Well function $W(u)$ for confined aquifers *(Continued)*

n	$u = n \times 10^{-8}$	$u = n \times 10^{-9}$	$u = n \times 10^{-10}$	$u = n \times 10^{-11}$	$u = n \times 10^{-12}$	$u = n \times 10^{-13}$	$u = n \times 10^{-14}$	$u = n \times 10^{-15}$
1.0	17.8435	20.1460	22.4486	24.7512	27.0538	29.3564	31.6590	33.9616
1.5	17.4380	19.7406	22.0432	24.3458	26.6483	28.9509	31.2535	33.5561
2.0	17.1503	19.4529	21.7555	24.0581	26.3607	28.6632	30.9658	33.2684
2.5	16.9272	19.2298	21.5323	23.8349	26.1375	28.4401	30.7427	33.0453
3.0	16.7449	19.0474	21.3500	23.6526	25.9552	28.2578	30.5604	32.8629
3.5	16.5907	18.8933	21.1959	23.4985	25.8010	28.1036	30.4062	32.7088
4.0	16.4572	18.7598	21.0623	23.3649	25.6675	27.9701	30.2727	32.5753
4.5	16.3394	18.6420	20.9446	23.2471	25.5497	27.8523	30.1549	32.4575
5.0	16.2340	18.5366	20.8392	23.1418	25.4444	27.7470	30.0495	32.3521
5.5	16.1387	18.4413	20.7439	23.0465	25.3491	27.6516	29.9542	32.2568
6.0	16.0517	18.3543	20.6569	22.9595	25.2620	27.5646	29.8672	32.1698
6.5	15.9717	18.2742	20.5768	22.8794	25.1820	27.4846	29.7872	32.0898
7.0	15.8976	18.2001	20.5027	22.8053	25.1079	27.4105	29.7131	32.0156
7.5	15.8280	18.1311	20.4337	22.7363	25.0389	27.3415	29.6441	31.9467
8.0	15.7640	18.0666	20.3692	22.6718	24.9744	27.2769	29.5795	31.8821
8.5	15.7034	18.0060	20.3086	22.6112	24.9137	27.2163	29.5189	31.8215
9.0	15.6462	17.9488	20.2514	22.5540	24.8566	27.1592	29.4618	31.7643
9.5	15.5922	17.8948	20.1973	22.4999	24.8025	27.1051	29.4077	31.7103

Source: Bear (1979); Freeze and Cherry (1979); Wenzel (1942).

Table 4-12. Distances of additional wells from original well

Well number	Distance from well no. 1 (m)
2	$r_2 = 914.36$
3	$r_3 = 1,219.14$
4	$r_4 = 1,523.93$
5	$r_5 = 3,047.85$

Similarly, if an impervious or no-flow boundary (e.g., groundwater divide) is located at distance, b, from the well face along the x-axis, then, using a pumping well as the image well,

$$s(x, y, t) = \{Q/(2\pi T)\} \ln \{2.25\, T t/(S r r')\} \quad (4\text{-}88)$$

Also, for a number of wells pumping in the same aquifer,

$$s(x, y, t) = (Q_1/4\pi T) W(u_1) + (Q_2/4\pi T) W(u_2) + \ldots + (Q_n/4\pi T) W(u_n) \quad (4\text{-}89)$$

where

$Q_1, Q_2, \ldots, Q_n$ = pumping rates of well no. 1, 2, $\ldots$, n

$u_i = S r_i^2/(4 T t_i)$

r_i = distance of well, i, from point (x, y)

t_i = time since pumping started at well, i, up to time, t

$s(x, y, t)$ = drawdown at point (x, y) at time, t

Note that $t_i = t - t_{ia}$, where t_{ia} = time when pumping started at well, i. Eq. (4-89) can be used to estimate reduction in well yield for the same drawdown due to interference of nearby wells.

Example 4-15: A well is pumping in a homogeneous, isotropic, confined aquifer. The permissible drawdown at the well face after 180 days of pumping is not to exceed a given value. Determine the reduction in well yield due to interference if four additional wells of the same capacity are installed in the aquifer. Distances of the additional wells from the original well are shown in Table 4-12.

Use $S = 0.00066$, $T = 1,207.6$ m^2/day, and r_w = well radius = 7.62 cm.

Solution: For well no. 1, $s = \{Q_1/(4\pi T)\} W(u_1)$, where s = permissible drawdown at well face, Q_1 = rate of pumping without any well interference, and $u_1 = S r_w^2/(4 T t)$.

With well no. 2 installed, $s = \{Q_2/(4\pi T)\}[W(u_1) + W(u_2)]$, where Q_2 = rate of pumping when two wells are operated so that drawdown at face of well no. 1 is the same (i.e., s) and $u_2 = S r_2^2/(4 T t)$.

So, $Q_2/Q_1 = 1/[1 + \{W(u_2)/W(u_1)\}]$

For five wells, $Q_5/Q_1 = 1/[1 + \{W(u_2)/W(u_1)\} + \ldots + \{W(u_5)/W(u_1)\}]$

$u_1 = 0.00066 \times (0.0762)^2/(4 \times 1{,}207.6 \times 180) = 4.4 \times 10^{-12}$

$W(u_1) = 25.5706$

$u_2 = 0.00066 \times (914.6)^2/(4 \times 1{,}207.6 \times 180) = 0.000635$

$W(u_2) = 6.7859$

$u_3 = 0.00066 \times (1{,}219.14)^2/(4 \times 1{,}207.6 \times 180) = 0.0011282$

$W(u_3) = 6.2110$

$u_4 = 0.00066 \times (1{,}523.93)^2/(4 \times 1{,}207.6 \times 180) = 0.001763$

$W(u_4) = 5.7654$

$u_5 = 0.00066 \times (3{,}047.85)^2/(4 \times 1{,}207.6 \times 180) = 0.007051$

$W(u_5) = 4.3844$

Therefore, $Q_2/Q_1 = 1/[1 + 6.7859/25.5706] = 0.79$.

$$Q_3/Q_1 = 1/[1 + \{(6.7859 + 6.2110)/25.5706\}] = 0.663.$$

Similarly, $Q_4/Q_1 = 0.577$, and $Q_5/Q_1 = 0.525$.

Since values of u_1, u_2, u_3, u_4, and u_5 are less than 0.01, the approximation of Eq. (4-83) is also valid. Thus,

$$s(r, t) = \{Q_1/(4\pi T)\} \ln \{2.25Tt/(Sr_w^2)\} = \{Q_2/(2\pi T)\} \ln \{2.25Tt/(Sr_w r_2)\}$$

and

$Q_2/Q_1 = \ln \{2.25Tt/(Sr_w^2)\}/[2 \cdot \ln \{2.25Tt/(Sr_w r_2)\}] = \ln [2.25 \times 1{,}207.6 \times 180/\{0.00066 \times (0.0762)^2\}]/(2 \times \ln [2.25 \times 1{,}207.6 \times 180/\{0.00066 \times 0.0762 \times 914.36\}]) = 25.5723/\{2 \times 16.1797\} = 0.79$.

The values of Q_3/Q_1, Q_4/Q_1, and Q_5/Q_1 may be estimated in the same way.

Unsteady Radial Flow to a Well Fully Penetrating an Unconfined Aquifer

The differential equation governing unsteady radial flow to a fully penetrating well in an unconfined homogeneous and isotropic aquifer of infinite extent is

$$1/r \, \partial/\partial r(r h \, \partial h/\partial r) = (S_y/K) \, \partial h/\partial t \quad (4\text{-}90)$$

where S_y = specific yield of the aquifer. There are several ways to linearize and solve Eq. (4-90):

1. Assume h (in the brackets on the left-hand side) $\cong H_0$ = average saturated thickness and $KH_0 = T_0$ = average aquifer transmissivity, then Eq. (4-90) becomes

$$1/r \, \partial/\partial r \, (r \, \partial h/\partial r) = (S_y/T_0) \, \partial h/\partial t \quad (4\text{-}91)$$

As for Eq. (4-80), the solution of Eq. (4-91) is

$$s(r, t) = (Q/4\pi T_0) W(u) \qquad (4\text{-}92)$$

where $u = S_y r^2/(4 T_0 t)$.

2. Set $s' = s - (s^2/2H)$ where H = initial saturated thickness (see Eq. (4-48)), then Eq. (4-90) becomes

$$1/r\, \partial/\partial r\, (r\, \partial s'/\partial r) = (S'_y/T)\, \partial s'/\partial t \qquad (4\text{-}93)$$

where $S'_y = S_y[H/(H - s)]$ and $T = KH$. As for Eq. (4-80), the solution of (Eq. (4-93)) is

$$s'(r, t) = (Q/4\pi T) W(u') \qquad (4\text{-}94)$$

where $u' = S'_y r^2/(4Tt)$. This approximation is reasonable as long as $s \ll H$. Since s is not known a priori, an initial estimate of S'_y is required, which may be modified after computing s. The initial value of S'_y may be taken to be slightly greater than the known value of specific yield, S_y, for the aquifer. The computations may then be refined with the modified value of S'_y.

Estimation of Aquifer Parameters

Aquifer parameters (e.g., T and S) are estimated using time-drawdown data from pumping tests on wells in the aquifer. Eq. (4-83) is useful in analyzing pumping test data to determine S and T for confined, leaky confined, and unconfined aquifers. This is because there is a linear (straight line) relationship between s (natural scale) on the y-axis and t (logarithmic scale) on the x-axis. The slope of the straight line with the x-axis is $2.3 Q/4\pi T$. Knowing Q, T can be calculated. Usually, linear regression analysis using various measured values of s and t and computation of T and S is accomplished through computer programs (e.g., AQTESOLV, Duffield and Rumbaugh 1989). These computer programs can process a large number of data values using Eq. (4-83) or modified equations for unconfined aquifers and leaky confined aquifers and using data from partially penetrating wells. Other methods to analyze pumping test data for different types of aquifers also are included in computer programs such as AQTESOLV. These include the Theis method (using the well function, $W(u)$), the Cooper-Jacob method (using Eq. (4-83)), and the Hantush method for leaky aquifers (Bear 1979).

If the maximum expected yield of a well cannot be estimated from the yields of other wells in the vicinity or from other hydrogeologic information, then it may sometimes be advisable to conduct a step-drawdown test to estimate the optimum pumping rate. This test also may be useful to assess the performance of wells where significant turbulent flow is expected (Driscoll 1989).

A step-drawdown test may include five to eight pumping steps, each lasting for 1 to 2 h. The pumping rate is increased in a stepwise manner during successive periods of time. In each time step, the well is pumped at a constant rate until the water level stabilizes. Then the pumping rate is increased to the next higher level. A constant pumping rate should be maintained within any one pumping step, but the rate should be different from one step to another. The duration of the entire test may be about 5 to 16 h. Since both laminar and turbulent flow may occur during pumping, the drawdown may be expressed as follows (Driscoll 1989; Bear 1979):

$$s = BQ + CQ^2 \qquad (4\text{-}95)$$

where B and C are dimensional constants. With five to eight sets of s and Q values obtained from the step-drawdown test, a polynomial regression may be used to obtain the values of B and C. An approximate and simpler method may be to estimate the values of B and C using linear regression or a graphical straight line fit between values of s/Q and Q to obtain values of B and C. These values of B and C and maximum permissible value of s for the aquifer may then be used to estimate the optimum value of Q.

If a step-drawdown test has been conducted, the constant-rate–pumping test may be started after the water level has recovered to pre-pumping static levels. The constant-rate–pumping test should be conducted with a constant discharge at least equal to 10% (but may be up to 100%) of the maximum anticipated yield of the well.

The duration of the constant-rate–pumping test may be about 24 h for a confined aquifer and 72 h for an unconfined aquifer. Time-drawdown data should be collected during both the pumping and recovery periods. Usually, data loggers are used for recording these data. It is advisable to record barometric pressure, rainfall, and water levels in surface water bodies within the anticipated cone of depression of the test well during the period of the test. It may be noted that a barometric pressure increase of 1 cm of mercury may result in a fall of about 8 cm in the water level in the observation well in a confined aquifer (see the section in this chapter entitled "Effect of Barometric Pressure Fluctuations").

The test well should be screened at least through one-third of the thickness of the aquifer except for thin aquifers where up to 75% of the entire thickness may be screened. The diameter of the test well may be sufficient to accommodate the pumping equipment. Typical well diameters are in the range of 10 to 30 cm and may be as large as 60 cm for larger wells. A preliminary estimate of the diameter of the well casing for a given well discharge may be made using a velocity of less than 1.5 m/s through the casing. Thus, a well discharge of 100 l/s may require a casing diameter of about 30 cm. This size may have to be modified to accommodate the required pump. A preliminary estimate of the open area of the well screen may be made using a velocity of about 0.03 m/s through the well screen. Thus, the open area of the screen required for a well discharge of 100 l/s is about 3.5 m^2. The required screen length can be calculated if open area per unit length of the screen for different sizes of screens is known from the manufacturer.

Due to pump operation, there is turbulence in the test well. As a result, time-drawdown data for the test well may not be accurate. Therefore, at least one observation well should be used to record time-drawdown data in addition to the test well. The diameter of the observation well may be as small as practicable to permit accurate measurement of water levels and assess aquifer response. Typical diameters are in the range of 8 to 12 cm. The screen depth of the test well may be about 1 to 2 m and should correspond to the central portion of the screen depth of the test well. Depending on available right-of-way and aquifer transmissivity, the location of the observation well may be about 50 to 200 m from the test well in confined aquifers, and about 30 to 100 m for unconfined aquifers where the cone of depression develops at a relatively slower rate. Generally, smaller distances should be used for pervious aquifers.

Time-drawdown data should be recorded during both the pumping and recovery periods. The recovery period begins immediately after the pump is shut down. If analyses of pumping test data are based on Eq. (4-83), then it may be advisable to record time-drawdown data so that there are a number of observations within each log cycle of time (i.e., 1 to 10 min, 11 to 100 min, 101 to 1,000 min, etc., after the start of pumping or recovery). Typical time intervals for recording drawdown during a constant-rate–pumping test are shown in Table 4-13 (Driscoll 1989; Delleur 1999). The same time intervals may be used for recording water level rise during the recovery period.

In some cases, particularly for large-diameter wells, early time-drawdown data may not fit a straight line on the s versus log t plot because these data may reflect removal of water from the well casing. Also, early time-drawdown data (less than about 5 min or so) for the observation well may not fit a straight line because u may be greater than 0.10 and Eq. (4-83) may not be applicable. Therefore, these data may have to be ignored. This may exclude data for the first 10 min or so. Time-drawdown data for much later times may be affected by recharge or by impervious boundaries located farther from but within the cone of depression of the well. Presence of a recharge boundary is indicated if the s versus log t plot flattens after a certain point. Conversely, an impervious boundary is indicated if the plot steepens after a certain time. In such cases, data for the earlier periods should be used to estimate aquifer parameters.

Designing a pumping test program involves the following steps:

1. Site selection and installation of the extraction (test) and observation wells with equipment for pumping, water level recording, and data logging.

2. Monitoring of ambient groundwater levels within the anticipated cone of depression of the test well, identification of aquifer boundaries (e.g., rivers, lakes, recharge zones, uniform groundwater flow, and groundwater divides or impervious boundaries), and extraction or injection wells with their rates of discharge or recharge.

3. Installation of barometric pressure gauge and rain gauge in the vicinity.

4. Conduct of pump test, including step-drawdown (if required) and constant rate tests.

5. Data analysis using the appropriate method for the aquifer (i.e., confined, unconfined, or leaky aquifer).

6. Report preparation.

Table 4-13. Time intervals to record time-drawdown data

Pumping well		Observation well	
Time since start of test (min)	Time interval between measurements (min)	Time since start of test (min)	Time interval between measurements (min)
0–3	0.5	0–10	1
4–15	1	11–120	5
16–60	5	121–240	10
61–360	30	241–360	30
361–1,440	60	361–1,440	60
After 1,440	480	After 1,440	480

Source: Driscoll (1989); Delleur (1999).

GROUNDWATER

Example 4-16: An extraction well is to be installed in an unconfined aquifer. The anticipated well discharge is 100 l/s. Manufacturer's data indicate a screen opening of 2,860 cm²/m for 20- and 30-cm-diameter screens. Estimate preliminary sizes of well casing and screen. Initial water table elevation is 20 m above base, and drawdown at the well face is not to exceed 2 m. Use $K = 0.003$ m/s and a radius of influence of 130 m.

Solution: Desired velocity through the casing pipe = 1.5 m/s and $Q = 0.1$ m³/s.

So, casing diameter = $\sqrt{[(0.1/1.5)(4/\pi)]}$ = 0.291 m. Use 30-cm-diameter casing pipe.

If screen length is L and maximum velocity through the screen slots is 0.03 m/s, then $L \times 0.2860 \times 0.03 = 0.1$, and $L = 11.67$ m. Use 12-m-long screens.

Assume $r_w = 10$ cm $= 0.1$ m. Then from Eq. (4-22), $Q = 0.1 = \pi K(H^2 - h_w^2)/\ln(R/r_w) = \pi \times 0.003(20^2 - h_w^2)/\ln(130/0.1)$. This gives $h_w = 18$ m and drawdown at well face = 2m, which is acceptable. So, use 20-cm-diameter, 12-m-long screens. Based on strata encountered during drilling, these dimensions may be adjusted.

Example 4-17: Time-drawdown data at an observation well located 60 m from a well fully penetrating a confined aquifer and pumping at 50 l/s are given in columns (1), (3), (4), (6), (7), and (9) of Table 4-14. Estimate the transmissivity and storage coefficient for the aquifer.

Solution: The data for this pumping test may be analyzed using computer models such as AQTESOLV (Duffield and Rumbaugh 1989). An alternative approach based on Eq. (4-83) is illustrated here using a spreadsheet. This may be useful for preliminary analyses when Eq. (4-83) is applicable and access to models is not readily available.

The logarithms of time, t, are entered in columns (2), (5), and (8). Linear regression is then performed with $\log(t)$ as the independent and drawdown, s, as the dependent variable. The regression equation using all 36 data points (ignoring $t = 0$) is $s = 0.201 + 0.431747 \log(t)$ with coefficient of determination $(r^2) = 0.9994$.

Computing values of s at $t = 10$ and 100 min gives s_1 ($t = 10$ min) = $0.201 + 0.431747 \log(10) = 0.632747$ m.

$$s_2 (t = 100 \text{ min}) = 0.201 + 0.431747 \log(100) = 1.0645 \text{ m}$$

$$s_1 - s_2 = 0.431747 \log(10/100) = -0.431747$$

From Eq. (4-84), $s_1 - s_2 = \{2.3 Q/(4\pi T)\} \log(t_1/t_2) = \{2.3 Q/(4\pi T)\} [\log(10)/\log(100)] = -\{2.3 Q/(4\pi T)\} = -0.431747$.

With $Q = 0.050$ m³/s = 4,320 m³/day, $T = 2.3 \times 4,320/(4\pi \times 0.431747) = 1,831$ m²/day = 1.27 m²/min.

From Eq. (4-83), $s_2 = 1.0645 = \{2.3 Q/(4\pi T)\} \log[2.25 Tt/(S r^2)] = 0.431747 \log[2.25 \times 1.27 \times 100/(S \times 60 \times 60)] = 0.431747 \log(0.079486/S)$.

$\log(0.079486/S) = 1.0645/0.431747 = 2.46556$

$S = 0.079486/(10^{2.46556}) = 0.000272$

Table 4-14. Time-drawdown data for observation well ($r = 60$ m)

(1) t (min)	(2) log (t)	(3) s (m)	(4) t (min)	(5) log (t)	(6) s (m)	(7) t (min)	(8) log (t)	(9) s (m)
0	—	0	11	1.041	0.644	60	1.778	0.966
1	0	0.221	12	1.079	0.660	90	1.954	1.033
1.5	0.176	0.292	13	1.114	0.681	120	2.079	1.01
2	0.301	0.332	14	1.146	0.697	150	2.176	1.147
2.5	0.398	0.372	15	1.176	0.731	210	2.322	1.210
3	0.477	0.406	20	1.301	0.764	240	2.380	1.230
4	0.602	0.456	25	1.398	0.801	270	2.431	1.251
5	0.699	0.499	30	1.477	0.835	300	2.477	1.271
6	0.778	0.533	35	1.544	0.862	360	2.556	1.304
7	0.845	0.56	40	1.602	0.888	420	2.623	1.338
8	0.903	0.587	45	1.653	0.912	480	2.681	1.375
9	0.954	0.607	50	1.699	0.932			
10	1	0.624	55	1.740	0.949			

Slug Tests

In some field situations, the hydraulic conductivity of the porous medium may be too small or the diameter, depth, or yield of the well may be too small (e.g., in the range of 5 to 200 m³/day) to conduct a pumping test, or the scope of the investigations may not warrant a pumping test. In such cases, a slug test may be useful as a relatively quick and cost-effective method to estimate aquifer hydraulic conductivity. This test is applicable to completely or partially penetrating wells in unconfined aquifers and fully penetrating wells in confined aquifers (Bouwer and Rice 1976; Bouwer 1989; Cooper et al. 1967). The test consists of quickly lowering or raising the water level in the well from its equilibrium position and measuring subsequent rise or fall with time.

Referring to Figure 4-3, Eq. (4-17) gives

$$Q = 2 \pi K L_e y / [\ln (R_e/r_w)] \tag{4-96}$$

where

Q = discharge entering the well (if water level in the well is instantaneously lowered) or leaving the well (if water level in the well is instantaneously raised)

K = hydraulic conductivity of the aquifer within a radius of R_e around the well and depth slightly greater than L_e

L_e = length of screened, perforated, or otherwise open section of the well

y = rise or fall in water level in the well

r_w = radial distance of undisturbed portion of aquifer from well centerline

R_e = effective radial distance over which y is dissipated

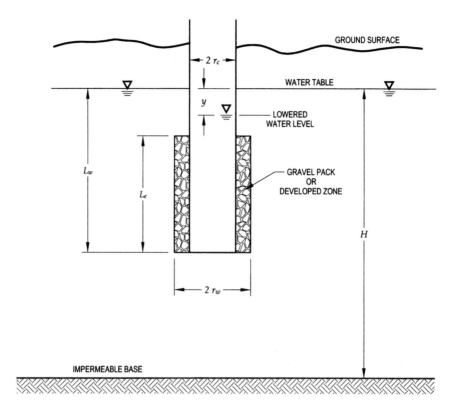

Figure 4-3. Schematic of slug test well

The rate of rise of water level in the well after the water level has been quickly lowered is

$$dy/dt = -Q/\pi r_c^2 \tag{4-97}$$

where

r_c = radius of well
πr_c^2 = cross-sectional area of the well where the water level is rising
t = time

Thus,

$$dy/dt = -2\,K\,L_e\,y/[r_c^2 \ln(R_e/r_w)] \tag{4-98}$$

If $y = y_0$ at $t = 0$, and $y = y_t$ at time, t, then

$$K = [\{r_c^2 \ln(R_e/r_w)\}/(2L_e)][(1/t)\ln(y_0/y_t)] \tag{4-99}$$

or

$$\log(y_0/y_t) = [0.8686KL_e/\{r_c^2 \ln(R_e/r_w)\}]t \tag{4-100}$$

For $L_w < H$,

$$\ln(R_e/r_w) = [\{1.1/\ln(L_w/r_w)\} + \{A + B\ln(H - L_w)/r_w\}/(L_e/r_w)]^{-1} \qquad (4\text{-}101)$$

For $L_w = H$,

$$\ln(R_e/r_w) = [\{1.1/\ln(L_w/r_w)\} + \{C/(L_e/r_w)\}]^{-1} \qquad (4\text{-}102)$$

where

L_w = depth of well penetration below water table

H = height of water table above impermeable base (Figure 4-3)

A, B, and C are dimensionless coefficients

Approximate values of these coefficients are given in Table 4-15.

Slug tests can be used for production wells, observation wells, or monitoring wells. The test can be carried out by immersing a section of pipe filled with sand or ballast and closed with caps on both ends or other similar object in the borehole. The water level in the borehole is allowed to return to equilibrium, and then the submerged object is quickly removed. The subsequent rise in water level in the well is recorded with time. Usually, a data logger is used to record and store the data for analysis. A plot of the data may be prepared with y (logarithmic scale) on the y-axis and t (natural scale) on the x-axis. The slope, θ (with the x-axis), of the straight line fitted to the data points on this plot is given by

$$\tan\theta = 0.8686\, K L_e / [r_c^2 \ln(R_e/r_w)] \qquad (4\text{-}103)$$

Knowing $\tan\theta$, L_e, r_c, R_e, and r_w, K can be estimated from Eq. (4-103). Usually, the plotting and computations are performed using a computer program (e.g., AQTESOLV, Duffield and Rumbaugh 1989).

Table 4-15. Approximate values of dimensionless coefficients A, B, and C

L_e/r_w	A	B	C
4	1.7	0.3	0.8
5	1.8	0.3	0.9
10	1.9	0.3	1.2
25	2.3	0.4	1.8
50	3.0	0.5	2.7
100	4.2	0.8	4.2
250	6.5	1.4	7.8
500	8.0	2.2	10.5
750	8.6	2.5	11.6
1,000	9.1	2.8	12.3
2,000	9.6	3.3	13.0

Source: Bouwer and Rice (1976); Bouwer (1989); Cooper et al. (1967).

In some cases, data may fit two straight lines, the one at smaller values of t being steeper than that at larger values of t. In such cases, the first (steeper) straight line may reflect drainage from the gravel pack. The second straight line may be representative of the hydraulic conductivity of the undisturbed aquifer. If the gravel pack is surrounded by a less permeable zone, the data may fit three straight lines, one at very small values of t, the second at intermediate values of t, and the third at larger values of t. Again, the last straight line may be representative of the hydraulic conductivity of the undisturbed aquifer.

In slug tests, the head difference between the static water table and water level in the well is dissipated mostly in the vicinity of the well around the screened or perforated section. Therefore, if the top of the screened or perforated section is sufficiently below the bottom of an upper confining layer, the test may provide reasonable values of the hydraulic conductivity even for confined aquifers. Slug tests may not be useful in estimating hydraulic conductivity of larger aquifers (i.e., deep or larger in areal extent). Comparison of hydraulic conductivities estimated by slug tests with values estimated by pumping tests indicate that the slug test results are generally low.

Example 4-18: Slug test data (water level rise with time) for a well with radius of casing = 2.54 cm, radius of gravel pack = 12.7 cm, and screen length = 152.4 cm are given in Table 4-16. The saturated thickness of the shallow aquifer is 175.3 cm, and the well penetrates the full depth of the aquifer. Initially, the water level in the well was lowered by 37.19 cm. Estimate the hydraulic conductivity of the aquifer.

Solution: The slug test data can be analyzed using a computer model like AQTESOLV (Duffield and Rumbaugh 1989). When access to such computer models is not readily available, a simpler and approximate analysis can be done using linear regression using a spreadsheet based on Eq. (4-100).

In this case, $y_0 = 37.19$ cm; $r_c = 2.54$ cm; $r_w = 12.7$ cm; $L_e = 152.4$ cm; $H = L_w = 175.3$ cm; and $L_e/r_w = 12$. For $L_w = H$ and $L_e/r_w = 12$, $C = 1.28$ from Table 4-15. Computed values of y_0/y and log (y_0/y) are shown in columns (3), (4), (7), and (8).

Data for the first 0.6 s may be ignored because of the effect of gravel pack, and those after 73.8 s may be ignored because the straight line drawn through the later points deviates significantly from the one through the points between 0.6 and 73.8 s.

Linear regression between the remaining 30 data points gives,

log $(y_0/y_t) = 0.0176\ t + 0.156$, with a correlation coefficient (r) of 0.99.

From Eq. (4-100), log $(y_0/y_t) = [0.8686\ KL_e/\{r_c^2 \ln (R_e/r_w)\}]\ t$.

Therefore, $[0.8686\ KL_e/\{r_c^2 \ln (R_e/r_w)\}] = 0.0176$.

From Eq. (4-102), $\ln (R_e/r_w) = [\{1.1/\ln (175.3/12.7)\} + \{1.28/(152.4/12.7)\}]^{-1} = [0.41906 + 0.10667]^{-1} = 1.90213$.

So, $K = [0.0176 \times (2.54^2) \times 1.90213]/(0.8686 \times 152.4) = 0.00163$ cm/s. The value estimated using the AQTESOLV model is 0.0016 cm/s.

It may be seen that linear regression with all the 50 data points (excluding the initial value $y_0 = 37.19$ cm) with non-zero intercept gives

log $(y_0/y_t) = 0.57 + 0.0038\ t$, with a relatively low correlation coefficient (r) of 0.57.

Table 4-16. Slug test data

(1) t (s)	(2) y (cm)	(3) y_0/y	(4) $\log(y_0/y)$	(5) t (s)	(6) y (cm)	(7) y_0/y	(8) $\log(y_0/y)$
0	37.19	1	0	38.8	4.88	7.6209	0.8820
0.2	33.53	1.1092	0.045	43.8	3.96	9.3914	0.9727
0.4	33.22	1.1195	0.049	48.8	3.35	11.1015	1.0454
0.6	32.00	1.1622	0.0653	53.8	2.74	13.573	1.1327
0.8	31.09	1.1962	0.0778	58.8	2.44	15.2418	1.1830
1.8	27.43	1.3558	0.1322	63.8	2.13	17.4601	1.2420
2.8	24.99	1.4882	0.1727	68.8	1.83	20.3224	1.308
3.8	23.16	1.6058	0.2057	73.8	1.52	24.4671	1.3886
4.8	21.64	1.7186	0.2352	78.8	1.52	24.4671	1.3886
5.8	20.42	1.8213	0.2604	83.8	1.22	30.4836	1.4841
6.8	19.51	1.9062	0.2802	88.8	1.22	30.4836	1.4841
7.8	18.89	1.9677	0.294	93.8	1.22	30.4836	1.4841
8.8	17.98	2.0684	0.3156	98.8	1.22	30.4836	1.4841
9.8	17.37	2.1410	0.3306	103.8	1.22	30.4836	1.4841
10.8	16.76	2.219	0.3462	108.8	1.22	30.4836	1.4841
11.8	16.15	2.3028	0.3623	113.8	1.22	30.4836	1.4841
12.8	15.54	2.3932	0.379	118.8	1.22	30.4836	1.4841
13.8	14.63	2.5420	0.4052	148.8	0.91	40.8681	1.6114
14.8	14.02	2.6526	0.4237	178.8	0.61	60.9672	1.7851
15.8	13.41	2.7733	0.443	208.8	0.61	60.9672	1.7851
16.8	12.80	2.9055	0.4632	238.8	0.61	60.9672	1.7851
17.8	11.89	3.1278	0.4952	268.8	0.61	60.9672	1.7851
18.8	11.58	3.2116	0.5067	388.8	0.61	60.9672	1.7851
23.8	9.14	4.0689	0.6095	508.8	0.61	60.9672	1.7851
28.8	7.32	5.0806	0.7059	598.8	0.30	123.9667	2.0933
33.8	5.79	6.4231	0.8077				

Contaminant Transport in Saturated Zone

Contaminants found in groundwater may be miscible (soluble) or immiscible in water. The property of contaminants that governs their miscibility in water is solubility. It is the maximum mass of a chemical that can dissolve in a specific amount of solvent (e.g., water) at a specific temperature and is expressed as mass of the chemical per unit volume of the solvent. Chemicals that have very low aqueous solubility (i.e., less than about 20,000 mg/l) can exist as a separate liquid phase in an aquifer. These chemicals are called nonaqueous phase liquids (NAPLs). If the density of an NAPL is less than water, it is called light nonaqueous phase liquid (LNAPL). Examples of LNAPLs include gasoline, fuel oil, benzene, toluene, ethyl benzene, and xylene (BTEX). If the density is greater than water, it is called dense nonaqueous phase liquid (DNAPL). Examples of DNAPLs include tetrachloroethylene (PCE), trichloroethylene (TCE), and chloroform. An organic liquid can exist as a stable separate phase in equilibrium with water only after its dissolved concentration in water has reached its saturation limit (Pankow and Cherry 1996). Different values of solubility, organic carbon partition coefficient (K_{oc}), decay coefficient (λ), density, and kinematic viscosity (ν)

Table 4-17. Typical values of solubility, K_{oc}, λ, density, and ν

Chemical	Solubility (mg/l)	K_{oc} (ml/gm)	λ (day^{-1})	Density (gm/ml)	ν (cm²/s) at 25°C
LNAPL					
Benzene	1,750	58.9	0.0009	0.88	0.0061
Toluene	526	182	0.011	0.87	0.0056
Ethylbenzene	169	363	0.003	0.87	0.0068
Xylene	186	260	0.0019	0.88 (o-xylene)	0.0076 (o-xylene)
Vinyl chloride	2,760	18.6; 57	0.00024	0.92	—
DNAPL					
Tetrachloroethylene (Perc, PCE)	200	155; 364	0.00096	1.63	0.0054
Trichloroethylene (TCE)	1,100	126; 166	0.00042	1.46	0.0039
Chloroform	7,920	39.8; 47	0.00039	1.49	0.0038
Cis-1,2dichloroethylene	3,500	35.5; 49	0.00024	1.28	0.0038
Trans-1,2dichloroethylene	6,300	52.5; 59	0.00024	1.26	0.0032
Chlorobenzene	472	219; 330	0.0023	1.11	0.0072

Source: Weast (1987); Maidment (1993); Pankow and Cherry (1996); IPCB (2001).

for the same organic compound are reported in the literature (Weast 1987; Maidment 1993; Pankow and Cherry 1996; IPCB 2001). Typical values are given in Table 4-17.

Miscible contaminants may include naturally occurring minerals in rock formations, domestic pollutants contributed by leakage from septic systems or sewer lines, various types of industrial chemicals, and pollutants carried by water infiltrating from agricultural areas. Industrial chemicals may include volatile or nonvolatile soluble organic compounds (VOCs or non-VOCs) or soluble inorganic compounds. The transport of soluble substances in groundwater involves advection, mechanical dispersion, molecular diffusion, adsorption/desorption, and decay or biodegradation. The mass conservation equation for the transport of soluble contaminants in groundwater is

$$R_d \, \partial C/\partial t = D_x \, \partial^2 C/\partial x^2 + D_y \, \partial^2 C/\partial y^2 + D_z \, \partial^2 C/\partial z^2 - u \, \partial C/\partial x - \lambda \, R_d \, C \qquad (4\text{-}104)$$

where

$D_x = \alpha_x u$ = longitudinal dispersion coefficient (x-direction)

$D_y = \alpha_y u$ = lateral dispersion coefficient (y-direction)

$D_z = \alpha_z u$ = vertical dispersion coefficient (z-direction)

$\alpha_x, \alpha_y, \alpha_z$ = longitudinal, lateral, and vertical dispersivities, respectively, of the porous medium

u = average uniform pore velocity in x-direction = u_d/φ

u_d = Darcy velocity in x-direction

φ = porosity

C = concentration of dissolved substance

t = time

x, y, z = Cartesian coordinates, such that the average uniform pore velocity is in x-direction

R_d = retardation factor

λ = decay or degradation coefficient of the chemical

Eq. (4-104) is called the advective-dispersion, convective-dispersion, or hydrodynamic dispersion equation and is valid only if $v = w = 0$, where v and w are pore velocities in the y- and z-directions, respectively. Theoretically, for the case with $v = w = 0$, $\alpha_y = \alpha_z$ (Bear 1979). However, in most practical cases, different values of α_y and α_z are used. Dispersivity has the dimension of length. The retardation factor is dimensionless and is defined as

$$R_d = 1 + \{(1 - \varphi)/\varphi\} \rho_s K_d = 1 + \rho_b K_d/\varphi \qquad (4\text{-}105)$$

where

ρ_s = density of soil grains

ρ_b = dry bulk density of porous medium

K_d = distribution coefficient of the chemical

The distribution coefficient is usually defined by the linear equilibrium isotherm, although other nonlinear isotherms also may be applicable in some cases:

$$S = K_d C \qquad (4\text{-}106)$$

where S = mass of the constituent adsorbed per unit mass of soil grains. Typical values of K_d for selected inorganic substances (metals) are given in Table 4-18 (Rai and Zachara 1984; Dragun 1988). Significantly different values of K_d are reported for the same substances in different types of soils and different types of environments. Reported values of K_d in streams with suspended solids concentrations of 1,000 mg/l are also shown in Table 4-18 (USEPA 1985).

Table 4-18. Values of K_d for selected inorganic substances (metals)

Substance	K_d(ml/gm) in saturated soils	K_d(ml/gm) in streams with suspended solids concentrations of 1,000 mg/l
Arsenic	1.9–18 (mean 6.7)	3,000
Cadmium	1.2–25 (river sediments)	2,000
Copper	2.2–43 (kaolinite)	6,000
Lead	4.5–7,640 (mean 99.5)	90,000
Selenium	1.2–8.6 (mean 2.7)	—
Mercury	30,000–224,000 (bentonite)	1,000

Source: Rai and Zachara (1984); USEPA (1985).

For organic chemicals,

$$K_d = f_{oc} K_{oc} \tag{4-107}$$

where

f_{oc} = fraction of organic carbon in soil by weight

K_{oc} = organic carbon partition coefficient of the organic constituent

Typical values of f_{oc} in different soils are given in Table 4-19 (Maidment 1993; Pankow and Cherry 1996). Suggested default values of f_{oc} are 0.006 for surface soils and 0.002 for subsurface soils (IPCB 2001). However, where possible, field measured values of f_{oc} should be used.

The decay or biodegradation process is defined by

$$C = C_0\, e^{-\lambda t} \tag{4-108}$$

where

C_0 = concentration at time, $t = 0$

C = concentration at time, t

If $C = C_0/2$,

$$\lambda = \ln(2)/t_L \tag{4-109}$$

where t_L = half-life of the constituent.

Organic chemicals are also classified as VOCs and semi-volatile organic chemicals (SVOCs). Chemicals with vapor pressures of less than 10^{-7} mm of mercury volatilize to a negligible degree in air, while chemicals with vapor pressure of greater than 10^{-2} mm of mercury will volatilize and be present in the atmosphere or soil air (Dragun 1988). Organic chemicals with relatively high diffusivity in air (e.g., greater than about 0.059 cm^2/s) and high vapor pressure are called volatile organic compounds. Volatile organic chemicals with

Table 4-19. Typical values of organic carbon content in soils

Type of soil	f_{oc}
Silty clay	0.01–0.16
Sandy loam	0.10
Silty loam	0.01–0.02
Unstratified silts, sands, gravels	0.001–0.006
Medium to fine sand	0.0002
Sand	0.0003–0.1
Sand and gravel	0.00008–0.0075
Coarse gravel	0.0011

Source: Maidment (1993); Pankow and Cherry (1996).

lower diffusivity and vapor pressure are called semi-volatile compounds. Examples of VOCs include chloroform, TCE, PCE, trans-1,2dichloroethylene, cis-1,2dichloroethylene, vinyl chloride, benzene, toluene, ethyl benzene, and xylene. Examples of SVOCs include phenol, hexachloroethane, hexachlorobenzene, and naphthalene.

For steady-state conditions, a commonly used solution of Eq. (4-104) for centerline (i.e., at $y = 0$, $z = 0$) concentrations is

$$C/C_0 = \exp[\{x/(2\,\alpha_x)\}\{1 - \sqrt{(1 + 4\lambda'\,\alpha_x/u)}\}] \cdot \mathrm{erf}[S_w/\{4\sqrt{(\alpha_y\,x)}\}] \cdot \mathrm{erf}[S_d/\{4\sqrt{(\alpha_z\,x)}\}] \quad (4\text{-}110\mathrm{a})$$

where

C_0 = concentration at the source

C = centerline concentration at x at $y = z = 0$

$\lambda' = \lambda R_d$

S_w = width of source in y-direction

S_d = depth of source in z-direction

For $R_d = 1$ (i.e., no adsorption), Eq. (4-110a) reduces to the Domenico equation (Domenico 1987; ASTM 1995; IPCB 2001). The corresponding non–steady-state solution for centerline concentrations at time, t, is

$$C/C_0 = (1/2)\exp[\{x/(2\alpha_x)\}\{1 - \sqrt{(1 + 4\lambda'\alpha_x/u)}\}] \cdot \mathrm{erfc}[\{x - ut\sqrt{(1 + 4\lambda'\alpha_x/u)}\}/\{2\sqrt{(\alpha_x ut)}\}] \cdot \mathrm{erf}[S_w/\{4\sqrt{(\alpha_y x)}\}] \cdot \mathrm{erf}[S_d/\{4\sqrt{(\alpha_z x)}\}] \quad (4\text{-}110\mathrm{b})$$

The steady-state concentration at any point (x, y, z) is given by

$$C/C_0 = (1/4)\exp[\{x/(2\alpha_x)\}\{1 - \sqrt{(1 + 4\lambda'\alpha_x/u)}\}] \cdot \{\mathrm{erf}[(y + S_w/2)/\{2\sqrt{(\alpha_y x)}\}] - \mathrm{erf}[(y - S_w/2)/\{2\sqrt{(\alpha_y x)}\}]\} \cdot \{\mathrm{erf}[(z + S_d/2)/\{2\sqrt{(\alpha_z x)}\}] - \mathrm{erf}[(z - S_d/2)/\{2\sqrt{(\alpha_z x)}\}]\} \quad (4\text{-}110\mathrm{c})$$

Finally, the non–steady-state concentration at any point (x, y, z) at time, t, is given by

$$C/C_0 = (1/8)\exp[\{x/(2\alpha_x)\}\{1 - \sqrt{(1 + 4\lambda'\alpha_x/u)}\}] \cdot \mathrm{erfc}[\{x - ut\sqrt{(1 + 4\lambda'\alpha_x/u)}\}/\{2\sqrt{(\alpha_x ut)}\}] \cdot \{\mathrm{erf}[(y + S_w/2)/\{2\sqrt{(\alpha_y x)}\}] - \mathrm{erf}[(y - S_w/2)/\{2\sqrt{(\alpha_y x)}\}]\} \cdot \{\mathrm{erf}[(z + S_d/2)/\{2\sqrt{(\alpha_z x)}\}] - \mathrm{erf}[(z - S_d/2)/\{2\sqrt{(\alpha_z x)}\}]\} \quad (4\text{-}110\mathrm{d})$$

If the upper surface of the contaminant plume coincides with the water table so that the plume spreads only in the downward direction, the quantities $S_d/2$ in Eqs. (4-110a) through (4-110d) are replaced by S_d (Domenico and Robbins 1985). For instance, the solution for steady-state centerline concentrations becomes

$$C/C_0 = \exp[\{x/(2\,\alpha_x)\}\{1 - \sqrt{(1 + 4\lambda'\,\alpha_x/u)}\}] \cdot \mathrm{erf}[S_w/\{4\sqrt{(\alpha_y\,x)}\}] \cdot \mathrm{erf}[S_d/\{2\sqrt{(\alpha_z\,x)}\}] \quad (4\text{-}111)$$

Commonly used empirical equations to estimate the longitudinal, lateral, and vertical dispersivities are as follows (USEPA 1985):

$$\alpha_x = 0.1\ L;\ \alpha_y = \alpha_x/3; \quad \text{and} \quad \alpha_z = \alpha_x/20 \qquad (4\text{-}112)$$

where L = length of flow path.

Computations using Eqs. (4-110a), (4-110b), (4-110c), (4-110d), and (4-111) may be performed using a spreadsheet or a Fortran program. Approximate analytical solutions for steady-state transports, accounting for no flux across the water table or upper confining layer and through the bottom confining layer, may be obtained using the method of images and recognizing that the contribution of D_x in steady-state transport is negligible (Prakash 1982). Similar equations may be used for non–steady-state transports (Prakash 1984). Computations using these equations require use of a computer program (e.g., a spreadsheet or a Fortran program).

Percolation of contaminated water from a source into the underlying saturated porous medium results in the development of a mixing zone below the source. The source depth, S_d, in Eq. (4-110) is the vertical depth of this mixing zone. If S_d is not known from field measurements, it can be estimated as the sum of transport distance due to vertical dispersion and advection (USEPA 1996b):

$$S_d = [\sqrt{\{2\ \alpha_z\ L_s\}}] + H[1 - \exp\{-(L_s\ I)/(u\ \varphi\ H)\}] \qquad (4\text{-}113)$$

where

L_s = length of source in the direction of groundwater flow in the saturated zone (m)

H = saturated thickness of aquifer (m)

I = rate of vertical leakage from the source (m/yr)

φ = total porosity

u = seepage (pore) velocity of groundwater flow in horizontal direction (m/yr)

α_z = vertical dispersivity of aquifer (m)

Because of mixing with ambient groundwater flow, there is dilution of contaminated water leaking from the source within the mixing zone under the source length, L_s. The dilution factor, DF, under the source is given by

$$DF = 1 + [(u\ \varphi\ H)/(L_s\ I)] \qquad (4\text{-}114)$$

Example 4-19: A leaking underground storage tank (LUST) has a length of 3 m along the direction of ambient groundwater flow in a shallow aquifer with saturated thickness of 5 m. The rate of leakage is estimated to be 25 cm/yr. The hydraulic conductivity, groundwater gradient, and porosity of the aquifer are 120 m/yr, 0.005, and 0.30, respectively. Estimate the thickness of the mixing zone at the downgradient edge of the LUST and the initial dilution factor under it. Assume $\alpha_z = 0.0056\ L_s$ (m).

Solution: $L_s = 3$ m; $H = 5$ m; $K = 120$ m/yr; $i = 0.005$; $\varphi = 0.30$; $I = 0.25$ m/yr; and $\alpha_z = 0.0056 \times 3 = 0.0168$ m. Seepage velocity $= u = Ki/\varphi = 120 \times 0.005/0.30$, and $u\varphi = Ki = 120 \times 0.005 = 0.60$ m/yr.

So, $S_d = [\sqrt{\{2 \times 0.0168 \times 3\}}] + 5[1 - \exp\{-(3 \times 0.25)/(0.60 \times 5)\}] = 0.3175 + 1.106 = 1.4235$ m.

$DF = 1 + [(0.60 \times 5)/(3 \times 0.25)] = 5.0$.

For complex groundwater flow and transport conditions, numerical models such as MODFLOW (USGS 2000b) should be used to establish the flow field, and models such as MT3D (USEPA 1992) to simulate contaminant transport. Several software packages are available for coupling such flow and transport models (Maidment 1993).

If it can be assumed that the LNAPLs or DNAPLs are not trapped in isolated pockets of the porous medium, then a preliminary approximation of the relative quantities of water and NAPL withdrawn by a pump-and-treat system may be made using Eqs. (4-7), (4-17), (4-23), (4-83), and (4-92). Thus, for steady-state conditions,

$$Q_1/Q_2 = B\nu_2/(b\nu_1) \tag{4-115}$$

where

Q_1 = volumetric rate of water withdrawal

Q_2 = volumetric rate of NAPL withdrawal

ν_1, ν_2 = kinematic viscosities of water and NAPL, respectively

B, b = water-saturated and NAPL-saturated thicknesses of the porous medium, respectively

For non–steady-state conditions,

$$Q_1/Q_2 = (T_1/T_2)[W(u_2)/W(u_1)] = B\nu_2/(b\nu_1)[W(u_2)/W(u_1)] \tag{4-116}$$

where

$u_1 = Sr_w^2/(4T_1 t)$

$u_2 = Sr_w^2/(4T_2 t)$

$T_1 = K_1 B$

$T_2 = K_2 b = K_1 \nu_1 b/\nu_2$

The estimates must be refined with sophisticated two-phase flow models.

Example 4-20: Make a preliminary estimate of the quantity of TCE that can be withdrawn through a pump-and-treat system designed to withdraw 20 l/s of groundwater under steady-state conditions and also within 6 mo (180 days) of pumping under non–steady-state conditions. Use $S = 0.01$; $B = 10$ m; $b = 1.0$ m; $\nu_1 = 0.011$ cm^2/s; $\nu_2 = 0.0039$ cm^2/s; $r_w = 6$ cm; and $K_1 = 0.1$ cm/s.

Solution: For steady-state conditions, using Eq. (4-115), $Q_1/Q_2 = 10 \times 0.0039/(1.0 \times 0.011) = 3.55$. So, $Q_2 = 20/3.55 = 5.6$ l/s. Allowing a safety factor of 2.0, assume that the system may withdraw about 2.5 l/s of TCE.

For non–steady-state conditions, $T_1 = 0.001 \times 10 = 0.01$ m²/s; $u_1 = 0.01 \times (0.06)^2 / (4 \times 0.01 \times 180 \times 24 \times 3,600) = 5.787 \times 10^{-11}$; $T_2 = 0.001 \times 0.011 \times 1.0/0.0039 = 0.00282$ m²/s; $u_2 = 0.01 \times (0.06)^2 / (4 \times 0.00282 \times 180 \times 24 \times 3,600) = 2.0518 \times 10^{-10}$; $W(u_1) = 23.0$; and $W(u_2) = 21.7$. Therefore, using Eq. (4-116), $Q_1/Q_2 = (0.01/0.00282) \cdot (21.7/23.0) = 3.346$, and $Q_2 = 20/3.346 = 5.98$ l/s. Allowing a safety factor of 2.0, assume that the system may withdraw about 3 l/s of TCE.

The thickness of immiscible LNAPLs (i.e., free products) measured in a well is greater than the actual product thickness in the aquifer. The following is a simple, approximate equation to estimate actual free product thickness (Hampton 1990):

$$H_T/h_T = G/(1 - G) \tag{4-117}$$

where

H_T = apparent thickness of free product measured in the monitoring well

h_T = actual thickness of free product in aquifer

G = specific gravity of free product

Flow and Contaminant Transport through Fractured Rock

There are several methods to analyze transport of miscible contaminants through fractured rocks. A relatively complex and sophisticated method is the dual-porosity approach, in which different hydraulic properties are assigned to the fractures and rock matrix and footprints of the actual or simulated fracture patterns are used (e.g., Valliappan and Naghadeh 1991; Sudicky and Therien 1999; Sudicky 1988). The second approach uses porous media equivalents of discontinuous fractures, particularly where density of fractures is high and distribution of fracture orientations is highly nonuniform (Freeze and Cherry 1979). In this case, an effective porosity (in the range of 10^{-2} to 10^{-5}) and effective hydraulic conductivity are assigned to the fractured rock mass and the transport is analyzed using methods applicable to granular porous media. The third approach estimates one-dimensional flow, dispersion, and retardation through individual fractures (apertures) and cumulates the results based on known or assumed fracture density. If there are n fractures (apertures) per unit width of the rock perpendicular to the direction of flow, then

$$\varphi = n\,a \tag{4-118}$$

$$K = \varphi\,a^2 g/(12\,\nu) = n\,a^3 g/(12\,\nu) \tag{4-119}$$

where a = width of a single aperture and n has the dimension $1/L$. Although quite uncommon, if the fractures can be approximated by n circular apertures of diameter, d, per unit area of the rock perpendicular to the direction of flow, then

$$\varphi = \pi\,d^2\,n/4 \tag{4-120}$$

$$K = \varphi\, d^2 g/(32\, \nu) \tag{4-121}$$

Knowing n, a, or d, φ and K can be estimated. Typical values of a or d are in the range of 0.001 to 0.005 cm, although smaller or larger values may be found at specific sites.

Example 4-21: Scanning of rock cores at a site indicated that the average fracture width is approximately 0.0035 cm and there are about two fractures per centimeter width of the rock perpendicular to the direction of flow. Estimate the equivalent porosity and hydraulic conductivity of the fractured rock. Use $\nu = 0.0131$ cm^2/s for groundwater.

Solution: Using Eqs. (4-118) and (4-119), $\varphi = na = 2 \times 0.0035 = 0.007$, and $K = \{2 \times (0.0035)^3 \times 981\}/(12 \times 0.0131) = 5.35 \times 10^{-4}$ cm/s.

Contaminant Transport through Unsaturated (Vadose) and Saturated Soil Zones

A field situation of interest is to analyze the transport of contaminants from a near-surface source to the water table through the vadose zone, followed by transport through the saturated zone to a potential receptor. The source may consist of soils of finite depth contaminated by leakage from storage tanks for liquid chemicals, seepage from waste management impoundments, or spills of liquid chemicals. The fate and transport of such chemicals is simulated using numerical groundwater flow and transport models. Often, screening level analyses are required for site remediation, evaluation of risk-based corrective actions, identification of potentially responsible parties, and apportionment of environmental liabilities.

The primary mode of transport of miscible contaminants through the vadose zone is in the dissolved form with infiltrating rainwater. Other modes of transport include diffusion, volatilization, and degradation. For screening-level analysis of the transport in dissolved form, infiltration is generally approximated by an average annual rate and it is assumed that advection is predominantly in the vertical direction, contribution of lateral (horizontal) dispersivities is negligible, the vadose zone is isotropic and homogeneous, and adsorption is linear and at equilibrium level (USEPA 1996b; USEPA 1985; IPCB 2001). The contaminant adsorbed to the soil grains at the source leaches (by desorption) with the infiltrating rainwater and may also undergo biodegradation with time. As a result, the mass of contaminant adsorbed to the soil grains decreases with time. The contaminant dissolved in the infiltrating rainwater travels nearly vertically through the partially saturated soil zone up to the water table. Upon arrival at the water table, it mixes with the ambient groundwater and travels along the dominant direction of groundwater flow.

With the aforementioned approximations and assumptions, contaminant transport with advection, dispersion, adsorption, and biodegradation can be described by the following (Bear 1979):

$$R_d\, \partial C/\partial t = D_z\, \partial^2 C/\partial z^2 - u\, \partial C/\partial z - \lambda\, R_d\, C \tag{4-122}$$

where

C = concentration

t = time

z = Cartesian coordinate in the vertical (positive downward) direction

D_z = dispersion coefficient in the vertical direction

u = seepage velocity in the vertical direction

λ = decay coefficient during transport through the vadose zone

R_d = retardation factor

For the sake of simplification, it is assumed that the solid and liquid phase concentrations at the source are in equilibrium, and partitioning between the two phases is governed by a linear equilibrium isotherm (Prakash 1996). Thus,

$$S = M/(\rho_b D) = K_d C \qquad (4\text{-}123)$$

$$dM/dt = -(q + \lambda_1) C \qquad (4\text{-}124)$$

$$C = C_0 e^{-\nu t} \qquad (4\text{-}125)$$

$$\nu = -(q + \lambda_1)/(\rho_b D K_d) \qquad (4\text{-}126)$$

where

S = mass of contaminant adsorbed to soil particles per unit dry (bulk) mass of soils

M = mass of contaminant adsorbed per unit surface area of the source

ρ_b = dry (bulk) density of contaminated soils

D = depth of contaminated soils at source

K_d = soil water partition coefficient

q = average rate of infiltration

λ_1 = degradation coefficient in soils at source

C_0 = aqueous phase concentration at source at time, $t = 0$

ν = overall decay coefficient at source

Assuming the water table to be sufficiently deep, the initial and boundary conditions for Eq. (4-122) are

$$C(0, t) = C_0 e^{-\nu t}, \; t \geq 0 \qquad (4\text{-}127a)$$

$$C(\infty, t) = 0, \; t \geq 0 \qquad (4\text{-}127b)$$

$$C(z, 0) = 0, \; z \geq 0 \qquad (4\text{-}127c)$$

Using Laplace transforms and their inversions, the solution of Eq. (4-122) is found to be the following (Carslaw and Jaeger 1984; USEPA 1985):

$$C(z, t) = C_0/2 \, \exp\{[u z/(2 D_z)] - \nu t\} \, [e^{-2AB} \text{erfc}\{-A\sqrt{t} + B/\sqrt{t}\} + e^{2AB} \text{erfc}\{A\sqrt{t} + B/\sqrt{t}\}] \qquad (4\text{-}128)$$

where

$C(z, t)$ = concentration at z, at time, t

$A = \sqrt{[\lambda + u^2/(4D_z R_d) - \nu]}$

$B = (z/2)\sqrt{(R_d/D_z)}$

Eq. (4-128) involves products of exponentials and complementary error functions of arguments, which may be real or complex. If the arguments are complex, the values may be obtained as described in (Prakash 2000a) or by table look-up of error functions of complex arguments (Abramowitz and Stegun 1972).

For a finite-time release of contaminants from the source, linear superposition may be used. Thus,

$$C(z, t, t_0) = P(z, t), \quad 0 \leq t \leq t_0 \qquad (4\text{-}129)$$

$$C(z, t, t_0) = P(z, t) - P(z, t - t_0)\exp(-\nu t_0), \quad -t \geq t_0 \qquad (4\text{-}130)$$

where

t_0 = time at which the source is remediated

$P(z, t)$ is the same as $C(z, t)$ given by Eq. (4-128)

Eqs. (4-128), (4-129), and (4-130) give pollutographs of contaminant concentrations reaching the water table, as shown in Figure 4-4. These pollutographs constitute the source for contaminant transport through the saturated zone. For computational convenience,

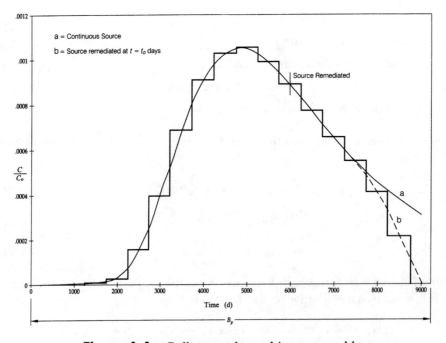

Figure 4-4. Pollutograph reaching water table

these pollutographs are approximated by successive rectangular pulses of constant concentrations and finite duration, Δt (Figure 4-4).

After entry into the saturated zone, transport of miscible contaminants to the receptor is governed by advection, dispersion, adsorption, degradation, and recharge of relatively fresh rainwater along the transport path. In a majority of field situations, dominant groundwater gradient is toward the receptor and advection can be treated as one-dimensional. For these cases, contaminant transport can be analyzed using the MULTIMED model (USEPA 1996b) or other similar models. For more complex groundwater flow conditions, simulation of three-dimensional flow and transport may be necessary. The MULTIMED model is used to generate a pollutograph of contaminant concentrations reaching the receptor due to a finite-duration (Δt) and unit-concentration pulse. Using this pollutograph as the kernel function, $C(L, \Delta t, t)$, and the constant concentration pulses as the input function, $V(\Delta t, t)$, the pollutographs of contaminant concentrations reaching the receptor due to continuous or finite-time releases from the source are computed by convolution. Here, L = horizontal distance of the receptor from the source. The following is a computationally convenient discretized form of the convolution process (Prakash 2000a):

$$Q(L, t) = \Sigma \, C[L, \Delta t, t - (i - 1)\, \Delta t] \cdot V(\Delta t, i\Delta t) \qquad (4\text{-}131)$$

where

Σ is from $i = 1$ to $n \leq j$

$Q(L, t)$ = ordinate of pollutograph reaching the receptor located at a distance, L, from the source at time, t

j (integer) $= t/\Delta t$

n (integer) $= B_P/\Delta t$

Δt = selected finite-time duration for rectangular pulses and for convolution

B_P = time base of the pollutograph reaching the water table (Figure 4-4)

Gas Phase Transport

The gas phase transport of VOCs from the source to the water table through the vadose zone may be analyzed assuming that gaseous diffusion is predominantly vertical. If the source is located from $x = -a$ to $x = a$, with its center at a distance, a_0, below the ground surface (which is paved and impervious to gaseous diffusion) and at b_0 above the water table (Figure 4-5), then, using the method of images (Carslaw and Jaeger 1984),

$$\begin{aligned} C(x, t) = (C_0/2)\, [&\text{erf}\{(a - x)/\sqrt{(4 D_d t)}\} + \text{erf}\{(a + x)/\sqrt{(4 D_d t)}\} + \\ &\text{erf}\{(a - (x - 2 a_0))/\sqrt{(4 D_d t)}\} + \text{erf}\{(a + (x - 2 a_0))/\sqrt{(4 D_d t)}\} + \\ &\text{erf}\{(a - (x + 2 b_0))/\sqrt{(4 D_d t)}\} + \text{erf}\{(a + (x + 2 b_0))/\sqrt{(4 D_d t)}\}] \end{aligned} \qquad (4\text{-}132)$$

where D_d = effective gas phase diffusion coefficient of the VOC in the vadose zone. The effective gas phase diffusion coefficient can be estimated by

$$D_d = D_b \, \phi_a \, T_a \qquad (4\text{-}133)$$

204 WATER RESOURCES ENGINEERING

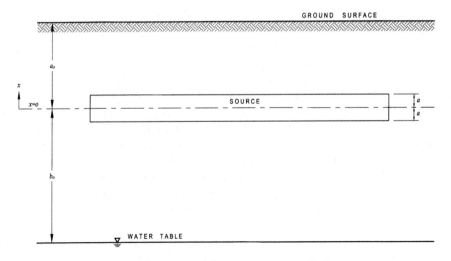

Figure 4-5. Schematic of gaseous diffusion

where

D_b = free-air diffusion coefficient

ϕ_a = gas-filled porosity

T_a = tortuosity

Values of free-air diffusion coefficient for different gases or VOCs may be obtained from tabulated values (e.g., Weast 1987; IPCB 2001; Lyman et al. 1984). Tortuosity may be estimated by the following (Baehr 1987; Pankow and Cherry 1996):

$$T_a = \phi_a^{7/3}/\phi^2 \qquad (4\text{-}134)$$

where ϕ = total porosity. The computations of Eq. (4-132) can be made using a spreadsheet (e.g., Excel).

Example 4-22: Estimate the gaseous concentration of TCE at a depth of 20 m below paved ground surface at $t = 180$ days due to a source located from 5 to 7 m below ground surface. The water table is 66 m below ground surface. Gaseous TCE concentration at source is 10,000 mg/l at the present time, $t = 0$. Free-air diffusion coefficient of TCE, D_b, is 7,100 cm²/day; total porosity of the porous medium = ϕ = 0.30; gas-filled porosity or volumetric content of TCE per unit volume of soil = ϕ_a = 0.25.

Solution: Using empirical equations to estimate tortuosity and effective diffusion coefficient (Baehr 1987; Pankow and Cherry 1996), T_a = tortuosity for the gaseous phase = $\phi_a^{2.333}/\phi^2 = 0.25^{2.333}/(0.3 \times 0.3) = 0.4375$. Effective diffusion coefficient of TCE = D_d = $D_b \phi_a T_a = 7{,}100 \times 0.25 \times 0.4375 = 777$ cm²/day = 0.0777 m²/day. Also, $x = -(20 - 6) = -14$ m; $a = (7 - 5)/2 = 1$ m; $a_0 = 5 + 1 = 6$ m; $b_0 = 66 - 6 = 60$ m. Then,

$a - x = 1 + 14 = 15$; $a + x = 1 - 14 = -13$; $a - x + 2a_0 = 1 + 14 + 12 = 27$; $a + x - 2a_0 = 1 - 14 - 12 = -25$; $a - x - 2b_0 = 1 + 14 - 120 = -105$; $a + x + 2b_0 = 1 - 14 + 120 = 107$.

$(a - x)/\sqrt{(4D_d t)} = 2.00;$

$(a + x)/\sqrt{(4D_d t)} = (1 - 14)/\sqrt{(4 \times 0.0777 \times 180)} = -13/7.4796 = -1.738;$

$(a - x + 2a_0)/\sqrt{(4D_d t)} = 3.6096;$

$(a + x - 2a_0)/\sqrt{(4D_d t)} = -3.342;$

$(a - x - 2b_0)/\sqrt{(4D_d t)} = -14.037;$ and

$(a + x + 2b_0)/\sqrt{(4D_d t)} = 14.305.$

Using Eq. (4-132) and values of erf(x) from tables of error functions (or Excel functions), $C(x, t) = (10,000/2) [0.9954 - 0.9860 + 1.0 - 0.999998 - 1 + 1] = 0.0094 \times 5,000 = 47.0$ mg/l.

Note that the last two terms are both nearly equal to 1 and cancel out. If the water table is too deep (as in this example), the last two terms of Eq. (4-132) may be neglected.

One measure for the remediation of soil gas contamination is using vapor extraction wells. Typical radii of soil vapor extraction wells range from about 1.27 to 5 cm. The radius of influence ranges from about 9 to 30 m depending on soil conditions. It is smaller for sandy soils and larger for silty and clayey formations. The vapor extraction rate for a well can be estimated by the following (Johnson et al. 1990):

$$Q = [\pi k B/\mu] [P_a^2 - (P_a - P_w)^2]/[(P_a - P_w) \ln (R/r_w)] \qquad (4\text{-}135)$$

where

Q = vapor extraction rate (cm^3/s)

k = soil permeability to gas flow (cm^2)

μ = dynamic viscosity of gas (gm-cm/s)

B = thickness of permeable zone or height of screens whichever is less (m)

P_a = atmospheric pressure (about 1.01×10^6 gm-cm/s^2)

P_w = pressure at the well created by the vacuum pump (gm-cm/s^2)

r_w = well radius

R = radius of influence

Example 4-23: Estimate the quantity of soil gas that can be extracted by a 5-cm diameter vapor extraction well with a 1-m-long screen. The vacuum pump is able to create an absolute pressure of 0.51×10^6 gm-cm/s^2 at the well face. Assume $\mu = 1.9 \times 10^{-4}$ gm-cm/s; $k = 10^{-6}$ cm^2; $P_a = 1.01 \times 10^6$ gm-cm/s^2; and $R = 9$ m.

Solution: $B = 100$ cm; $(P_a - P_w) = (1.01 - 0.51) \times 10^6 = 0.50 \times 10^6$ gm-cm/s^2 at the well face. Using Eq. (4-135), $Q = [\pi \times 10^{-6} \times 100 \,(1.01^2 - 0.5^2) \times 10^6]/[(1.9 \times 10^{-4}) \times 0.5 \times \ln(9/0.025)] = 43.27 \times 10^4$ cm^3/s = 433 l/s.

Monitoring Well Installation

Monitoring wells are required to record groundwater levels and collect samples of groundwater for water quality analysis. In addition, soil samples are collected during drilling. A monitoring well may be a single well or a pair of wells installed in the same borehole (nested monitoring wells), each screened in different formations or at different depths in the same formation. General specifications for monitoring well installation include the following:

1. **Drilling:** Monitoring well bores are drilled using a truck-mounted drill rig or other appropriate equipment that has hollow stem augers with internal diameters of 11 cm (4.25 in.) for single and 30.5 cm (12 in.) for nested wells. Soil samples are collected at 1.5-m (5-ft) intervals of depths using 5-cm (2-in.) diameter split-spoon samplers. The sizes and depths may vary depending on requirements at specific locations. Physical characteristics of soil samples are recorded on the boring logs. These physical characteristics include color, visual size classifications (e.g., fine, medium, coarse), wetness (e.g., dry, moist, very wet), blow counts, photo-ionization detector readings for VOCs, and smell. A portion of each soil sample is preserved and sent to the laboratory to analyze for the contaminants of concern and grain-size distribution.

2. **Monitoring Well Casing, Screen, Filter Pack, and Seal:** Monitoring well casing is constructed using a 5-cm (2-in.) diameter Schedule 40 Polyvinyl Chloride (PVC) pipe and extends from the bottom of the borehole to about 1 m above ground surface. The screen has the same diameter and material as the casing. The diameter and material of the casing and screen may vary depending on site conditions, economics, and purpose. The length is determined based on the thicknesses of water-bearing strata encountered during drilling. The annular space between the borehole and well screen is filled with clean sand and fine gravel, which is called filter pack. The filter pack extends up to about 1 m farther from the top and bottom of the well screen. Filter pack gradation is designed to retain about 70% of the soil particles in the formation, and the screen slots are sized to retain about 90% of the filter pack. Commonly used screen slot sizes for monitoring wells are 0.25 to 0.50 mm (0.01 to 0.02 in.). A seal of bentonite pellets is provided in a thickness of about 1 m above the filter pack. For nested wells, the bentonite pellet seal is provided in the entire length between the top of the filter pack of the deep well and bottom of the filter pack of the shallow well and in a length of about 1 m above the filter pack of the shallow well. The annular space from ground surface to the top of the bentonite pellet seal is filled with cement-bentonite mix with a cement-to-bentonite ratio of 19:1 by weight. The mix may change depending on specific site conditions.

3. **Well Development, Purging, and Sampling:** Monitoring wells are developed using submersible pumps until the pumped water is free of sediment. Water levels, pH, specific conductivity, and temperature are recorded during well development. The well should be purged before each sampling event. Purging should continue until water volume equivalent to a minimum of three well casings is removed. After purging, water level in the well should be allowed to recover. Water quality samples are collected using disposable polyethylene bailers. Specific conductance, pH, dissolved oxygen, and temperature of groundwater may be measured in the field. Retrieved water quality samples are stored, sealed, and transported to the designated laboratory for analysis.

4. **Decontamination and Waste Disposal:** The drilling augers, auger bits, split-spoon samplers, and other soiled equipment should be steam cleaned prior to use, double rinsed in tap water, and finally rinsed with distilled water. All drill cuttings and liquid wastes including well development water should be containerized for appropriate disposal.

5. **Well Survey:** The locations and elevations of top of casing of the monitoring wells should be surveyed and connected to appropriate state or other plane coordinates and referenced to the mean sea level. The survey information should be indicated on site maps and kept with other monitoring well records.

Groundwater Flow and Transport Models

The following are some commonly used public-domain groundwater flow and transport models:

- **A Modular Three-Dimensional Finite-Difference Groundwater Flow Model, MODFLOW** (USGS 2000b): This model simulates steady and transient groundwater flow in three dimensions. Its modular structure consists of a main program and a series of subroutines, called modules. The groundwater flow domain is divided into grids and blocks of grids that represent flow, no-flow, constant head, river, drain, pond, flow barrier, well, recharge, and evapotranspiration and layers representing different formations. Layers can be simulated as confined, unconfined, or a combination of the two. MODFLOW is a versatile model and can be used for various types of groundwater studies, including water supply evaluation and contaminant transport potential on a regional or local scale.

- **Well Head Protection Area Delineation Code, WHPA** (USEPA 1993): WHPA is a semi-analytical groundwater flow simulation model for delineating capture zones in a well head protection area. It is applicable to homogeneous, confined, unconfined, and leaky confined aquifers exhibiting two-dimensional, steady groundwater flow in an areal plane. It can simulate barrier or stream boundaries that penetrate the entire aquifer depth and can account for multiple pumping and injection wells.

- **The Hydrologic Evaluation of Landfill Performance, HELP** (USEPA 1995): This is a quasi-two-dimensional model for conducting water balance analysis for landfills, cover systems, and other solid waste containment facilities. It requires climatological, soil, and design data. A significant amount of climatological and soil information is built into the model itself. Water balance for landfill systems with different types of soil covers, vegetation covers, lateral drainage layers, low permeability soil barriers, and geomembranes can be simulated. The output includes amounts of runoff, evaporation, drainage, leachate collection, and liner leakage from different types of landfill designs.

- **Multimedia Exposure Assessment Model (MULTIMED) for Evaluating the Land Disposal of Wastes** (USEPA 1996b): This model simulates the movement of contaminants in one-dimensional groundwater flow with three-dimensional dispersion, retardation, first-order decay, and dilution due to direct infiltration into the groundwater plume. The model is user friendly and includes interactive pre- and postprocessing software. The output includes contaminant concentrations reaching a specified receptor under steady-state conditions or after specified time periods released from a rectangular patch or Gaussian source. The rectangular patch source is defined by given width and depth. The Gaussian source has an exponential distribution along the width and uniform over the depth. The concentration at the source may be constant (steady-state), a constant concentration pulse of finite duration, or an exponentially decaying pulse over time.

- **BIOSCREEN** (USEPA 1997): BIOSCREEN is a screening model that simulates remediation through natural attenuation of dissolved hydrocarbons at petroleum release sites. It is programmed on Excel and is based on Domenico's analytical solute transport model. It can simulate advection, dispersion, adsorption, aerobic decay, and anaerobic reactions. The package includes three models: solute transport without decay, with biodegradation modeled as a first-order decay process, and with biodegradation modeled as an instantaneous biodegradation reaction.

- **A Modular Three-Dimensional Transport Model for Simulation of Advection, Dispersion, and Chemical Reactions of Contaminants in Groundwater Systems, MT3D** (USEPA 1992): This model uses a modular structure similar to MODFLOW (USGS 2000b). It can be used with any block-centered, finite-difference groundwater flow model such as MODFLOW. It retrieves hydraulic heads and the various flow and source/sink terms saved by the groundwater flow model and automatically incorporates specified hydrologic boundary conditions. It can be used to simulate concentrations of single-species miscible contaminants in groundwater considering advection, dispersion, and some simple chemical reactions. The chemical reactions included in the model are linear or nonlinear sorption and first-order decay or biodegradation. This model is fairly complex to use, and projects involving the use of this model should be undertaken as special studies.

CHAPTER 5

Hydraulic Designs

Introduction

Water resources engineers are required to develop conceptual and preliminary designs of hydraulic structures. These designs form the basis for the preparation of detailed designs, construction plans, drawings, specifications, and cost estimates. Detailed designs may involve structural designs and geotechnical analysis of various components of the hydraulic structures. This chapter presents methods to prepare hydraulic designs of structures, which are commonly dealt with by a water resources engineer. Such hydraulic structures include channel transitions, flood and erosion protection measures, drop structures, dams and reservoirs, spillways, and hydraulic components of a hydropower plant.

Channel Transitions

Channel transitions are required to provide contractions or expansions of flow sections when the channel has to pass through constricted areas (e.g., reaches bounded by floodwalls or levees, bridge openings, etc.). General rules for the design of channel transitions for subcritical flow are listed below (USDA 1977):

1. The water surface should be smoothly transitioned to meet hydraulic conditions at the beginning and end of the transition.

2. The edge of the water surface on any one side of the bank should not converge at an angle greater than 14 degrees with the direction of flow or diverge at an angle greater than 12.5 degrees. This means the total channel section should not converge at an angle greater than 28 degrees or diverge at an angle greater than 25 degrees.

3. Losses through the transition should be minimized. Excluding friction losses, transition losses should not exceed 0.10 h_v through contraction and 0.20 h_v through expansion, where h_v is the velocity head based on average velocity through the transition.

4. As far as practicable, the bed slopes and side slopes should meet the end conditions tangentially.

Neglecting friction losses, conservation of energy through the transition gives

$$y_1 + z_1 + v_1^2/(2g) = y_2 + z_2 + v_2^2/(2g) + h_L \tag{5-1a}$$

or

$$WS_1 + v_1^2/(2g) = WS_2 + v_2^2/(2g) + h_L \tag{5-1b}$$

where

y_1, y_2 = water depths

z_1, z_2 = bed elevations

v_1, v_2 = velocities

WS_1, WS_2 = water surface elevation at the upstream and downstream end of the transition

g = acceleration due to gravity

h_L = losses through transition due to change in streamline patterns

In relatively short transitions, friction losses, which are usually small compared to losses due to contraction and expansion, can be neglected.
Thus,

$$WS_1 - WS_2 = \Delta WS = (v_2^2 - v_1^2)/(2g) + h_L \tag{5-2a}$$

Also, $h_L = C_c(v_2^2 - v_1^2)/(2g)$ for contraction and $h_L = C_e(v_2^2 - v_1^2)/(2g)$ for expansion, where $C_c \cong 0.15$ and $C_e \cong 0.25$. For contraction, $v_1 < v_2$ and ΔWS is positive or $WS_1 > WS_2$. For expansion, $v_1 > v_2$ and ΔWS is negative or $WS_1 < WS_2$. So,

$$WS_1 - WS_2 = \Delta WS = 1.15\,(v_2^2 - v_1^2)/(2g) \text{ for contraction} \tag{5-2b}$$

$$WS_1 - WS_2 = \Delta WS = -0.75\,(v_1^2 - v_2^2)/(2g) \text{ for expansion} \tag{5-2c}$$

The above-mentioned energy equations also are applicable to the design of flumed sections for concrete-lined channels except that sharper transitions may be used. The angle of convergence on any one side of the lined channel bank should not exceed 30 degrees, and that of divergence should not exceed 22.5 degrees.

Example 5-1: Design a contraction transition between two channel segments. The upstream channel is an earthen channel with $n = 0.025$, bed slope of 0.002, side slopes of $2H{:}1V$, and bottom width of 10 m. The downstream channel is riprap-protected with $n = 0.038$, side slopes of $2H{:}1V$, bed slope of 0.01, and bottom width of 6 m. The design discharge is 12.25 m³/s.

Solution: For the upstream channel, Manning's equation gives $12.25 = (1/0.025)\,[(10\,y_1 + 2\,y_1^2)/\{10 + 2\,(\sqrt{5})\,y_1\}]^{2/3}\,(10\,y_1 + 2\,y_1^2)\,\sqrt{(0.002)}$.

By trial and error, $y_1 = 0.777$ m; $A_1 = 8.977$ m^2; $P_1 = 13.475$ m; $R_1 = 0.666$ m; and $V_1 = 1.3645$ m/s.

For the downstream channel, $12.25 = (1/0.038) \, [(6 \, y_2 + 2 \, y_2^2)/\{6 + 2(\sqrt{5}) \, y_2\}]^{2/3} \, (6 \, y_2 + 2 \, y_2^2) \, \sqrt{(0.01)}$. By trial and error, $y_2 = 0.8165$ m; $A_2 = 6.2323$ m^2; $P_2 = 9.6515$ m; $R_2 = 0.6457$ m; and $V_2 = 1.966$ m/s.

Thus, $\Delta WS = 1.15 \, [(1.996^2 - 1.3645^2)/(2 \times 9.81)] = 0.1244$ m. The required change in bed elevation from section 1 to section 2 is given by $\Delta WS = (y_1 + z_1) - (y_2 + z_2)$ or $(z_1 - z_2) = \Delta WS - (y_1 - y_2) = 0.1244 - (0.777 - 0.8165) = 0.1639$ m.

The channel bed at section 2 needs to be 0.164 m lower than at section 1.

Top width (upstream) = $T_1 = 10 + 2 \times 0.777 = 11.554$ m.

Top width (downstream) = $T_2 = 6 + 2 \times 0.8165 = 7.633$ m.

Using a convergence of 4:1 or a convergence angle of 14 degrees on one side, the length of transition = $[(11.554 - 7.633)/2] \cot 14$ degrees = 7.84 m.

A slightly different method for the design of channel transitions is illustrated in Example 5-2.

Example 5-2: A trapezoidal channel is to be contracted to a rectangular concrete section to pass over a creek. The length of the rectangular section is 150 m. Thereafter, the channel has to be expanded back to the same trapezoidal section. The design discharge of the channel is 30 m^3/s. Channel bed elevation on the downstream side of the transition is 1,000 m. Other relevant dimensions are as follows: For the trapezoidal channel, $B_1 = 22$ m and side slopes are $2H{:}1V$ and for the flumed rectangular channel, $B_2 = 11$ m; contraction transition is 2.5:1 (contraction angle = 21.8°); and expansion transition is 3:1 (expansion angle = 18.4°). Design the flume so that water depth through the transition is maintained at 1.7 m. Assume contraction and expansion loss coefficients of 0.2 and 0.3, respectively, and Manning's coefficient of 0.015 for the concrete channel.

Solution: Referring to Figure 5-1:

Top width at section 1 = $T_1 = 22 + 2 \times 1.7 = 25.4$ m = Top width at section 4 = T_4.

Top width at section 2 = $T_2 =$ Top width at section 3 = $T_3 = 11$ m.

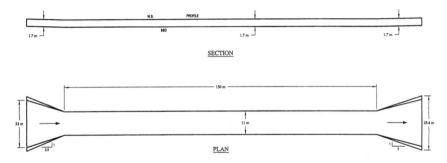

Figure 5-1. Channel contraction and expansion

Length of contraction from trapezoidal (section 1) to rectangular (section 2) = $\{(25.4 - 11)/2\} \times 2.5 = 18$ m (Figure 5-1).

Length of expansion transition from rectangular (section 3) to trapezoidal (section 4) = $\{(25.4 - 11)/2\} \times 3 = 21.6$ m. Neglecting friction loss, balance total energy in the expansion transition on the downstream side between section 4 and 3: $V_4 = 30/(22 \times 1.7 + 2 \times 1.7^2) = 0.695$ m/s; $y_4 = 1.7$ m and $V_3 = 30/(11 \times 1.7) = 1.604$ m/s; $y_3 = 1.7$ m.

$z_4 + y_4 + V_4^2/2g + 0.3\{(V_3^2 - V_4^2)/2g\} = z_3 + y_3 + V_3^2/2g$.

$1,000 + 1.7 + 0.695^2/(2 \times 9.81) + 0.3\{(1.604^2 - 0.695^2)/(2 \times 9.81)\} = z_3 + 1.7 + 1.604^2/(2 \times 9.81)$. So, $z_3 = 999.92544$ m.

For the flumed rectangular section, $y_2 = y_3$, $R_2 = R_3$, and $V_2 = V_3$. Using Manning's equation, estimate friction loss in the rectangular concrete channel with $L = 150$ m and $R_3 = 11 \times 1.7/(11 + 2 \times 1.7) = 1.2986$ m.

Friction loss = $V_3^2 n^2 L/R_3^{4/3} = 1.604^2 \times 0.015^2 \times 150/(1.2986)^{4/3} = 0.06129$ m, and friction slope = $S_f = 0.06129/150 = 0.0004086$. Flow through the rectangular section being uniform, $S_f = S_0$ = bed slope.

Bed elevation at section 2 = $z_2 = 999.92544 + 0.06129 = 999.98673$ m.

Balancing energy through the contraction transition from section 2 to 1 (neglecting friction loss within the transition), $z_2 + y_2 + V_2^2/2g + 0.2\{(V_2^2 - V_1^2)/2g\} = z_1 + y_1 + V_1^2/2g$.

Dimensions at trapezoidal sections 4 and 1 are the same. So, $999.98673 + 1.7 + 1.604^2/(2 \times 9.81) + 0.2\{(1.604^2 - 0.695^2)/(2 \times 9.81)\} = z_1 + 1.7 + 0.695^2/(2 \times 9.81)$.

So, $z_1 = 1,000.11454$ m.

The transition may start at a bed elevation of 1,000.11 at the upstream end of the expansion transition, sloping to 999.99 m at the upstream and 999.93 m at the downstream edge of the rectangular section, and rising to 1,000 m at the downstream end of the contraction.

Flood Control

Commonly used flood control methods include structural and nonstructural measures. Examples of structural measures of flood control include levees, groins, cutoffs, flood bypasses, flood proofing, and detention basins. Nonstructural measures include establishment of regulatory floodplains, flood zones, watershed management plans, and flood emergency planning.

Structural Measures

Levees

Levees are earth embankments constructed nearly parallel to the course of a stream to prevent inundation of large areas on the landward side of stream banks. Principal design considerations for levees include the following:

- Location: Usually distance between levees on the two sides of a river or the distance of the levee from the centerline of the river depends on the availability of land. If land is available or can be acquired without significant environmental, socioeconomic, or political problems, the levee may be located on natural high grounds or ridges such that sufficient waterway (or floodway) is available to pass the design-basis flood. Otherwise, other measures of flood control may have to be adopted. In some cases, alignment of levees has to be adjusted to protect or avoid existing historic, archaeological, or other significant features.

- Slope Stability: Levees are designed as earth dams so that the side slopes are stable under dry conditions and also when water is impounded up to the design flood elevation on the river side with dry conditions or low water levels on the landward side. Usual side slopes of levees vary from $2H:1V$ to $4H:1V$. Flatter slopes must be provided for embankment materials that have smaller angles of repose.

- Seepage: During high river stages, seepage through the levee section may result in piping unless appropriate measures for seepage control and/or piping control are implemented. It is desirable to incorporate such measures in the design of the levee. Commonly used methods include berms to minimize potential for piping, trenches to collect and transport seepage water to nearby tributaries, relief wells, and inverted filters.

- Interior Drainage: Construction of levees along the main channel may obstruct tributaries and natural overland flows entering the river through various reaches of its banks. This, along with seepage from the main river channel, may result in impoundment of water on the landward side of the levee. Measures for drainage of this water must be incorporated in the levee design. This may include pumping of water over the levee and diversion or discharge of impounded water into downstream tributaries through ditches or drain tiles nearly parallel to the levee.

- Top Width and Freeboard: To permit movement of equipment for maintenance and inspection, the top width of the levee should be greater than 3 m. However, smaller top widths may be used if other means of access are available (e.g., an existing road close to the levee). Freeboards of levees vary from 1.0 to 1.5 m, 1.0 m being the commonly adopted value (USACE 1994).

- Erosion and Scour Protection: Methods for erosion protection of levees are generally the same as for channel bank protection (see the section in this chapter entitled "Erosion Protection").

Hydraulic analyses for levees include floodplain delineation and floodway determination using computer models such as HEC2 and HEC-RAS (USACE 1991c, 1998) and interior drainage analysis using models such as HEC-IFH (USACE 1992). The floodplain is the area on the landward side of the riverbank that is inundated by floods in the river. The area inundated during the 100-yr flood is called the 100-yr floodplain. In flood insurance studies, floodway is defined as the channel of a river or other water course and the adjacent land areas that must be reserved in order to discharge the base flood (i.e., 100-yr flood) without cumulatively increasing the water-surface elevation more than a designated height. The designated maximum rise is usually taken to be 0.31 m above the 100-yr flood elevation under pre-floodway condition. In supercritical flow regime, the maximum allowable rise may be applied above the pre-floodway energy grade line rather than flood elevation. The area between the floodway and the boundary of the 100-yr flood is termed the floodway fringe. Thus, floodway fringe is the portion of the floodplain that could be obstructed without increasing the 100-yr flood elevation by more than 0.31 m at any point.

Floodway analysis may be performed using the channel encroachment methods available in steady-state water surface profile models (e.g., HEC-2; HEC-RAS). Normally, the width of the floodway is determined using equal loss of channel conveyance on opposite sides of the stream. If equal loss of conveyance is not practical or unusual flow patterns exist (e.g., interbasin flow, divided flow, etc.), unequal conveyances may be used subject to acceptance by the community, state and federal agencies, and insuring agency.

Groins

Groins are dikes extending from the bank of the river to a specified distance, which may usually be up to the normal waterline. They are constructed to protect the bank against erosion or to control channel meanders. Groins are more effective when constructed in series. They may be oriented perpendicular to the bank or at angles inclined slightly upstream or downstream. Groins oriented perpendicular to the bank or inclined upstream tend to deflect the main current away from the bank. Those inclined downstream cause scour closer to the bank and tend to maintain the deep current close to the bank. Groins inclined upstream are called deflecting or repelling, and those inclined downstream are called attracting groins. Depending on specific site conditions, the angles of inclination to the bank may vary from 10 to 30 degrees. Solid groins are constructed of earth with riprap protection and do not permit appreciable flow through them. Permeable groins, constructed of timber frames filled with brush and tree branches or rock, permit restricted flow through them. T-groins have a cross dike constructed at the riverside end of a normal groin. Usually, a greater length of the cross dike projects upstream and a smaller portion projects downstream of the main groin.

The configurations and designs of groins must be determined with the help of physical (hydraulic) models.

The length and spacing of groins depend on specific site conditions and objectives of the river training project. If the river is wide and the course of a major portion of the river flow is to be repelled toward the opposite bank, the repelling groin has to be fairly long. If the river bank has a curvature, groins in series may have varying lengths. Commonly used spacing between groins is 2 to 2.5 times the groin length for convex banks and is equal to the length of groins for concave banks (USAERDC 2003). Groins are usually spaced farther apart in a wide river than in a narrow one, if the two have nearly the same design discharge. Also, permeable groins may be spaced farther apart than solid groins.

Usual side slopes of groins are $2H:1V$ to $3H:1V$. Steeper slopes may be used for materials with higher angle of repose and flatter slopes for materials with smaller angle of repose. The nose of a straight, attracting or repelling groin may be approximately semicircular in plan. The section of the groin where the attack of the current is expected to be strongest must be well protected. In particular, the head or nose and toe must be armored with rock, concrete or soil cement blocks, or other erosion protection measures. In the case of repelling groins in series, the one at the upstream end of the protected river reach may require maximum protection. In the case of an attracting groin, the upstream side slope and toe may require maximum protection. Erosion protection on the side slope and toe may be designed using the methods described in "Erosion Protection." The design of a typical groin is illustrated in Figure 5-2.

Cutoffs

When the river meander develops into a horseshoe, there is potential for sudden formation of a straight channel or oxbow lake during high floods. This may result in channel

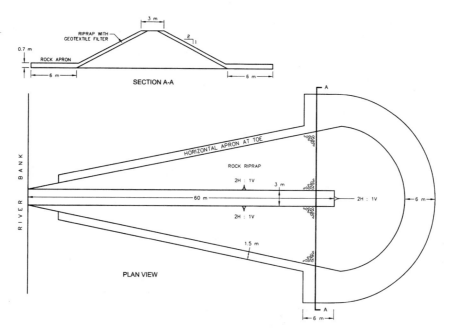

Figure 5-2. Typical groin

instability and significant scour on the upstream side of the bend. In anticipation of the above-mentioned situation, artificially excavated cutoffs may be used to straighten the channel. An artificial cutoff may be constructed to prevent a sharp meander in the river from endangering valuable property or to straighten the river course.

If the loop length exceeds 1.5 to 2 times the chord length of the meander or oxbow, an artificially induced cutoff may be a viable option. A relatively narrow pilot channel may be excavated along the course of the anticipated cutoff. Subsequent flood flows may gradually enlarge the pilot channel to the required size and may abandon the old curved alignment. The pilot channel should be excavated to a deeper section such that (Singh 1967)

$$R/L^2 > R_m/L_m^2 \tag{5-3}$$

where

R = estimated depth of excavation of the pilot channel

L = length of pilot channel

R_m = regime depth of original curved channel

L_m = loop length of original curved channel

Flood Bypass

Flood bypasses are channels, tunnels, or conduits designed to divert a portion of the flood flows of a river to another stream or to a downstream point on the same stream. If diversion is required only during floods larger than a specified magnitude, the invert of the bypass may be located higher than the bed of the river so that there is insignificant diversion during smaller flood events. Hydraulic analysis for different quantities of flow diversion may involve split flow computations using models such as HEC-2 and HEC-RAS (USACE 1991c, 1998). The diverted flows may be discharged to another stream, which may safely carry the additional flows. Alternatively, the diverted flows may be discharged downstream into the same channel if flooding conditions in the downstream channel permit.

Channelization

Channelization involves channel clearance, straightening, widening, deepening, and lining as required to improve hydraulic conveyance. Hydraulic impacts of channelization are evaluated using computer models such as HEC-2 and HEC-RAS (USACE 1991c, 1998). An economic analysis of the costs and benefits associated with each element of channelization should be conducted to identify the preferred course of actions.

Bridge Modification

Bridge modification involves removal, replacement, widening, or raising of existing bridges on the stream, which may be too narrow or too low. In some cases, the actual waterway of a bridge may be adequate, but the effective waterway may be too narrow because of skew with respect to the direction of flow, obstructions within the bridge waterway, or skewed approaches on the upstream and downstream sides.

Flood Proofing

Isolated units within the floodplain or flood hazard zones are flood proofed by a ring levee or floodwall or by raising the structure above the anticipated flood level. In some cases, lower levels of the affected units, which may be below the base flood elevation, may be made watertight and strengthened to withstand dynamic and hydrostatic forces of flood waters.

Detention Basins

Detention basins are relatively small impoundments designed to temporarily store storm runoff from tributaries of a river and release it gradually after peak flow in the main channel has passed. Computations for sizing detention basins are made using models for generating and routing surface runoff hydrographs (e.g., HEC-1, USACE 1991a). A dry detention basin is designed to remain dry most of the time except during storm events. This can be accomplished by providing an outlet at or below the elevation of the bottom of the basin. A wet detention basin or retention basin retains some quantity of water for most of the year and may be used for fishing and other shallow-water recreational activities. For such basins, the elevation of the outlet is kept above the bottom of the basin. Sedimentation basins are impoundments designed mainly to trap sediment load carried by surface runoff.

Nonstructural Measures

Nonstructural measures of flood control include flood insurance, enactment of storm water regulations and ordinances, and relocation of flood-prone units.

Flood insurance involves categorization of flood-prone areas in accordance with their flooding potential. For purposes of flood insurance, areas within the 100-yr floodplain boundary are termed special flood hazard areas; areas between the 100- and 500-yr flood boundaries are termed areas of moderate flood hazard; and the remaining areas outside the

500-yr floodplain boundary are termed areas of minimal flood hazard. In flood insurance studies, flooded areas are divided into several flood risk zones (FEMA 1993):

- Zone A: Areas that correspond to 100-yr floodplains which are determined by approximate hydraulic analyses. Base flood elevations and depths for this zone are not shown on flood insurance rate maps.

- Zone AE: Areas that correspond to 100-yr floodplains which are determined by detailed hydraulic analyses. Base flood elevations for this zone are shown on flood insurance rate maps.

- Zone AH: Areas that correspond to 100-yr shallow flooding with a constant water surface elevation (usually areas of ponding) where average depths are between 0.31 and 0.91 m. The flood elevations derived from detailed hydraulic analyses are shown at selected intervals within this zone.

- Zone AO: Areas that correspond to 100-yr shallow flooding (usually sheet flow on sloping terrain) where average depths are between 0.31 and 0.91 m. Average depths derived from detailed hydraulic analyses are shown within this zone. Alluvial fan flood hazards also are shown as Zone AO.

- Zone A99: Areas that correspond to areas of the 100-yr floodplains which will be protected by a federal flood protection system and for which no base flood elevations or depths are shown on flood insurance rate maps.

- Zone V: Areas that correspond to the 100-yr coastal floodplains determined by approximate hydraulic analyses that have additional hazards associated with storm waves for which no base flood elevations are shown on flood insurance rate maps.

- Zone VE: Areas that correspond to the 100-yr coastal floodplains determined by detailed hydraulic analyses that have additional hazards associated with storm waves for which base flood elevations are shown on flood insurance rate maps.

- Zone X: Areas that may be outside the 100-yr floodplain, within 100-yr floodplain where average flood depths are less than 0.31 m, areas of 100-yr flooding where the contributing drainage area is less than 2.59 km^2, or areas protected from the 100-yr flood by levees. No base flood elevations or depths are shown on flood insurance rate maps.

- Zone D: Unstudied areas where flood hazards are undetermined but possible.

Storm water regulations and ordinances are enacted by local agencies to regulate developments within their jurisdictions so that there are no adverse impacts on flooding conditions in the area. Examples of local flood control regulations include requirements to provide dry or wet detention basins such that peak flows released to the receiving stream during a prescribed storm (e.g., 10-, 25-, or 100-yr storm) do not exceed prescribed values.

Relocation includes moving isolated units from within the floodplain or flood hazard zone to safer areas with appropriate compensation for re-settlement.

Erosion Protection

Introduction

Erosion protection may be required on beds and banks of stream channels, areas disturbed by construction or mining, and slopes of earthen embankments. Channel bank erosion may be caused by tractive stress of water flowing along and across the bank slope, wave forces, and erosive forces at channel bends or meanders. Erosion on disturbed areas and earthen embankment slopes is normally caused by overland flow and may be in the form of sheet, rill, or gully erosion. Protection against erosion is required to prevent loss of land and shoreline property. Erosion protection measures include armoring the vulnerable area against erosive forces, structural measures to repel the main current and associated erosive forces away from the vulnerable portion of the channel bank, protection or avoidance of overland flow concentration, or a combination of two or more of these measures. Depending on economics and aesthetics, commonly used armoring measures may include riprap, gabions, fabriforms, vegetation cover, and biotechnical materials. Measures to divert the main current and associated erosive forces may include small groins, bamboo or wooden piles, and rock dumping along and in the vicinity of the bank toe. Protection against sheet and rill erosion usually includes vegetation cover, whereas that against gully erosion may include vegetation cover, rock protection, and grade control measures. The designs of erosion protection measures are generally based on empirical formulae, experimental data from physical models, practical experience, and judgment (Prakash 2000b).

Selection and design of erosion protection methods for specific stream reaches must include environmental and aesthetic considerations. This may involve evaluation of the feasibility of nonstructural erosion control measures. These may include biotechnical methods, watershed management, streambed stabilization, vegetation growth by seeding or plantation, and incorporation of sinuosity in stream channel alignment. The objective should be to minimize alterations to existing bars, pools, and other geomorphic features and maintain the natural riverine environment as far as possible.

Structural measures of erosion control include the following:

- Stream Bank Modification: This may include reshaping stream banks by flattening existing steep bank slopes or overexcavating erodible soils from banks and replacing with compacted, less erodible soils.

- Rock Riprap or Gabions: Rock riprap may be ungrouted or grouted or placed in steel or synthetic wire baskets commonly known as gabions. Ungrouted rock riprap is flexible and can be easily replenished or repaired. It imparts additional roughness to the protected reach and tends to reduce flow velocities. Over time, the void spaces may become filled with soil, which may support vegetation growth. In case of grouted riprap, the void space is filled with concrete. This type of riprap may be suitable for sites where available riprap sizes are smaller than required (refer to the following sections of this chapter entitled "Design of Riprap Protection" and "Riprap Sizing for Steep Slopes"). Grouted riprap is rigid and susceptible to damage due to settlement and freeze-thaw cycles. To minimize potential for damage due to settlement and slope failure, it is advisable to use grouted riprap on bank slopes flatter than the angle of repose of the bank material.

- Fabriform: This consists of bags of synthetic materials filled with concrete placed adjacent to one another. Fabriforms may be suitable for sites where riprap is scarce and sand and gravel are plentiful.

- Articulated Concrete Blocks: These are reinforced concrete blocks with reinforcement bars extending from their edges in the form of hooks and eyes. They are placed adjacent to one another and connected together by insertion of the hooks of one block into the eyes of the other.

- Concrete Bulkheads: These are vertical or sloping concrete retaining walls constructed to retain soils in steep banks. In some cases, vertical walls may be provided near the bank toe with sloping concrete, riprap, or vegetative protection above it.
- Grade Control: This includes construction of several grade control structures in series across the channel to reduce energy slope and flow velocities in the channel reach upstream of each grade control structure.
- Channelization: This includes measures to reduce erosive flow velocities by channel widening and increasing waterways of bridges.

The type of erosion protection for any particular channel reach should be determined after economic, hydraulic, geomorphologic, and environmental impact analyses (USAERDC 2003). Care must be taken to ensure that erosion protection in the study reach does not create unacceptable adverse flooding or erosion conditions in the upstream or downstream reaches. Measures such as grade control or those resulting in increased channel roughness may cause higher flood elevations upstream. Measures that may result in increasing channel capacity (e.g., channelization) could cause erosive velocities and higher flood elevations in the downstream reach.

Design of Riprap Protection

Design of flexible armors for channel beds and banks and embankment slopes requires determination of the minimum size of rock to withstand expected erosive forces. In practice, rock sizes are estimated using several different methods and the design value is selected by judgment, within the range of estimated values. Commonly used methods for riprap sizing include the following:

1. Maynord's Method (Maynord et al. 1989): In this method, the equation used to estimate riprap size is:

$$d_{30}/D = SF \cdot 0.30 \, [\{\gamma/(\gamma_s - \gamma)\}^{0.5} \cdot V/\sqrt{(gD)}]^{2.5} \tag{5-4}$$

where

d_{30} = riprap size than which 30% of the riprap material is finer by weight

D = average water depth in channel

SF = factor of safety suggested for this method to be 1.2

V = local depth-averaged velocity

g = acceleration due to gravity

γ = unit weight of water

γ_s = unit weight of stone taken to be 2,644 kg/m^3

For γ_s = 2,563 and 2,483 kg/m^3, the computed d_{30} should be multiplied by 1.06 and 1.14, respectively. The equation is valid for side slopes of 2H:1V or flatter. For side slopes of 1.5H:1V, a multiplying factor of 1.3 for d_{30} should be used.

2. U.S. Army Corps of Engineers Method 1 (USACE 1994): In this method, the equation to estimate d_{30} of the riprap size is

$$d_{30}/D_a = SF \cdot C_s \cdot C_v \, C_t \, [\{\gamma/(\gamma_s - \gamma)\}^{0.5} \cdot V/\sqrt{(g \, D_a \, K)}]^{2.5} \quad (5\text{-}5)$$

where

K = side slope correction factor = $[1 - (\sin^2 \theta / \sin^2 \varphi)]^{0.5}$

θ = angle of side slope to horizontal

φ = angle of repose for stone

SF = varying from 1.1 to 1.5

D_a = local depth of flow

C_s = stability coefficient = 0.30 for angular stones and 0.375 for rounded stones

C_v = vertical velocity distribution coefficient = 1.0 for straight channels and up to 1.283 inside channel bends

C_t = thickness coefficient = 1.0 for riprap thickness equal to d_{100}

For riprap protection on channel bed, $K = 1$, since $\theta \cong 0$.

3. U.S. Army Corps of Engineers Method 2 (USACE 1970): This method involves trial and error to estimate the value of d_{50} (m), using the equation

$$0.0122 \, (\gamma_s - \gamma) \, d_{50} \, K = SF \, \gamma \, V^2 / [32.6 \log_{10} (12.2 \, D/d_{50})]^2 \quad (5\text{-}6)$$

in which SF = safety factor, taken to be about 1.5, to account for nonuniformity of flow.

4. Simons and Senturk Method (Barfield et al. 1981; Nelson et al. 1986; Simons and Senturk 1976): This method computes SF corresponding to a trial value of d_{50}, using Eqs. (5-7) to (5-11). If the computed SF is not acceptable, then the trial value of d_{50} is modified until an acceptable SF is obtained.

$$\tau_{max} = \text{maximum shear stress} = 0.76 \, \gamma \, D \, S \quad (5\text{-}7)$$

$$\eta = 21 \, \tau_{max}/[(\gamma_s - \gamma) \, d_{50}] \quad (5\text{-}8)$$

$$\beta = \arctan [\cos \lambda / \{(2 \sin \theta / \eta \tan \varphi) + \sin \lambda\}] \quad (5\text{-}9)$$

$$\eta' = \eta \, [1 + \sin (\lambda + \beta)]/2 \quad (5\text{-}10)$$

$$SF = \cos \theta \tan \varphi / [\eta' \tan \varphi + \sin \theta \cos \beta] \quad (5\text{-}11)$$

where

S = channel bed slope

λ = angle of streamlines to horizontal $\cong$ angle of bed to horizontal

5. ASCE Equation (Vanoni 1977):

$$d_{50} = [6\,W/(\pi\,\gamma_s)]^{0.333} \tag{5-12}$$

$$W_{50} = 0.0232\,G_s\,V^6/[(G_s - 1)^3 \cos^3 \theta] \tag{5-13}$$

where

W_{50} = weight of stone (kg) of diameter, d_{50} (m)

G_s = specific gravity of stone

6. California Department of Transportation Equation (West Consultants 1996):

$$W_{33} = 0.0113\,G_s\,V_x^6/[(G_s - 1)^3 \sin^3 (\rho - \theta)] \tag{5-14}$$

$$d_{33} = [6\,W_{33}/(\pi\,\gamma_s)]^{0.333} \tag{5-15}$$

where

W_{33} = weight of stone (kg) of diameter, d_{33} (m)

V_x = 4/3 V_a for impinging flow and 2/3 V_a for tangential flow

V_a = average channel velocity (m/s)

ρ = 70° for randomly placed rubble stone

Several empirical equations for riprap sizing are based on average channel velocity only and do not specifically account for side slopes of channel banks. A few commonly used equations are presented here.

1. U.S. Bureau of Reclamation Equation (Peterka 1958):

$$d_{50} = 0.043\,V_a^{2.06} \tag{5-16}$$

2. U.S. Geological Survey Equation (West Consultants 1996):

$$d_{50} = 0.055\,V_a^{2.44} \tag{5-17}$$

3. Isbash Equation (Maynord et al. 1989; West Consultants 1996):

$$d_{50} = V_a^2/[2\,g\,C^2\,(G_s - 1)] \tag{5-18}$$

in which $C = 0.86$ for high- and 1.20 for low-turbulence zones.

4. HEC-11 Method (West Consultants 1996):

$$d_{50} = d'_{50}\,C_f\,C_s \tag{5-19}$$

$$d'_{50} = 0.005943\,V_a^3/[K^{1.5}\,\sqrt{D}] \tag{5-20}$$

Table 5-1. Riprap gradation

Size of stone	Percentage of total weight smaller than given size
2 d_{50}	100
1.7 d_{50}	85
1.0 d_{50}	50
0.42 d_{50}	15
0.10 d_{50}	0

Source: Barfield et al. (1981).

$$C_f = [SF/1.2]^{1.5} \qquad (5\text{-}21)$$

$$C_s = 2.12/[G_s - 1]^{1.5} \qquad (5\text{-}22)$$

From considerations of stability, stones of different sizes should be included in the riprap layer so that smaller pieces of stones may occupy voids between relatively larger pieces of rock. Different riprap gradations are recommended by different agencies for different classes of riprap sizes. An acceptable gradation may include the sizes of stones shown in Table 5-1 (Barfield et al. 1981).

The thickness of riprap layer, T, is usually taken to be equal to 2 d_{50} or equal to the largest size of stone in the riprap layer.

Example 5-3: Estimate riprap sizes for bank protection of two sandbed channels, A and B. Relevant hydraulic parameters for the two streams are shown in Table 5-2. Use a factor of safety of 1.5, where applicable, and assume $d_{50} = 1.5\ d_{30}$.

Solution: The riprap sizes estimated using Eqs. (5-4) to (5-22) with $D_a = D$, $C = 0.86$, and $V_x = 4/3\ V_a$ are shown in Table 5-3.

The adopted values of riprap sizes for the two streams are selected to be within the range of values estimated by Eqs. (5-4) to (5-22).

Table 5-2. Hydraulic parameters of channels

Parameter	Value	
	Stream A	Stream B
D (m)	20.73	9.3
V (m/s)	3.96	3.3
V_a (m/s)	3.05	2.5
θ (degrees)	21.8	21.8
φ (degrees)	40	40
γ_s (kg/m³)	2,404	2,404
G_s	2.4	2.4
S	0.000486	0.00154

Table 5-3. Estimated riprap sizes

Method	Estimated riprap size, d_{50} (m)	
	Stream A	Stream B
1. Maynord	0.30	0.23
2. U.S. Army Corps of Engineers Method 1	0.48	0.37
3. U.S. Army Corps of Engineers Method 2	0.15	0.13
4. Simons and Senturk	0.38[a]	0.38[a]
5. ASCE	0.43	0.30
6. USBR	0.43	0.28
7. USGS	0.84	0.51
8. Isbash	0.46	0.31
9. Cal. Dept. of Transportation	0.66	0.44
10. HEC-11	0.09	0.074
Adopted value	0.46	0.38

[a] $SF = 1.5$ for bank slopes for Stream A and 1.3 for Stream B.

Example 5-4: Design riprap protection for a groin whose nose is located in a channel segment where floodwater depth is 5 m and current velocity is 2 m/s (Figure 5-2). The side slopes of the groin are 2.5 H:1V. Use $\gamma_s = 2,400$ kg/m³, and angle of repose for riprap = 40°.

Solution: Use Eq. (5-5) for riprap sizing. Angle of side slope = $\theta = \tan^{-1}(1/2.5) = 21.8°$. So, $K = [1 - (\sin^2 21.8/\sin^2 40)]^{0.5} = 0.8162$. D_a = local depth of flow = 5 m.

Assume a safety factor of 1.5 for flow concentration. Use a stability coefficient of 0.375 for angular stones; a vertical velocity distribution coefficient of 1.25 for flow near the nose of the groin; and a thickness coefficient of 1.0, assuming riprap thickness to be equal to d_{max}. Also, to account for swirling and nonuniform velocity distribution at the nose of the groin, assume a velocity correction factor of 1.5. Then, V(max) = 1.50 × 2.0 = 3 m/s, and $d_{30}/D_a = 1.50 \times 0.375 \times 1.0 \times 1.25 \,[\{1,000/(2,400 - 1,000)\}^{0.5} \{3.0/\sqrt{(9.81 \times 5.0 \times 0.8162)}\}]^{2.5}$.

So, $d_{30} = 0.357$ m and $d_{50} = 0.357/0.70 = 0.51$ m. Use $d_{max} = 1.02$ m. Design a filter blanket for the riprap (see the section of this chapter entitled "Design of Filters"), and provide a horizontal apron along the toe of the dike as described in the section of this chapter entitled "Protection Against Scour at Bank Toe."

Riprap Sizing for Steep Slopes

In the case of flow along steep slopes (e.g., flow through mountainous channels, mining drainage channels, or ditches along embankments with slopes ranging from 2 to 20%), flow per unit width of the channel and depth of flow is generally low and flow velocity is relatively high. An acceptable equation for riprap sizing for such cases is as follows (USACE 1994):

$$d_{30} = 1.95 \, K \, S^{0.555} \, q^{2/3}/g^{1/3} \tag{5-23}$$

where

d_{30} = stone size (m) than which 30% of stones are finer by weight

K = flow concentration factor, usually taken to be 1.25

S = bed slope (m/m)

q = unit discharge (m²/s)

Design of Filters

Usually, the size of riprap is much larger than the particle sizes of the base material that constitutes the channel bank. To prevent loss of base material through voids in the overlying riprap, the two should be separated by geotextile filter, sand and gravel filter, or both. To ensure that the finest grains in the base material do not escape through the interstices of the filter, the pore spaces in the filter material should not be much bigger than the finest grains of the base material. Also, to avoid buildup of hydraulic pressure behind the filter and washing away of the filter material, the pores of the filter material should not be so small as to be blocked by the base material. Similarly, finer particles in the filter material should not be so small that they may escape through the voids in the riprap material. At the same time, particles in the filter material should not be so large as to block the void spaces of the riprap material.

Commonly adopted criteria to determine the gradation of filter material, based on size distributions of the riprap and base material, are

$$d_{50} \text{ (filter)}/d_{50} \text{ (base)} < 40 \quad \text{and} \quad d_{50} \text{ (riprap)}/d_{50} \text{ (filter)} < 40 \quad \text{(i)}$$

$$5 < d_{15} \text{ (filter)}/d_{15} \text{ (base)} < 40 \quad \text{and} \quad 5 < d_{15} \text{ (riprap)}/d_{15} \text{ (filter)} < 40 \quad \text{(ii)}$$

$$d_{15} \text{ (filter)}/d_{85} \text{ (base)} < 5 \quad \text{and} \quad d_{15} \text{ (riprap)}/d_{85} \text{ (filter)} < 5 \quad \text{(iii)}$$

These criteria include four filter sizes based on gradation of base material and four based on riprap gradation, and they provide three upper limit and two lower limit values for d_{15}, one upper limit and one lower limit value for d_{50}, and one lower limit value for d_{85} of the filter material. The values of d_{15}, d_{50}, and d_{85} of the filter material should be selected within these limits such that the gradation curve for the filter material is nearly similar to the gradation curves for the base material and riprap.

Protection Against Scour at Bank Toe

Due to channel bed scour, riprap at the junction of the bank slope with the channel bed (i.e., at the toe) may become dislodged. Design of riprap protection against scour at the bank toe requires estimation of potential scour depth below the channel bed. In practice, potential scour depth is estimated using several different methods and the design value is selected by judgment. Hydraulic model experiments may be required for certain special conditions. Some commonly used methods of estimating channel bed scour are listed below:

1. Regime Scour Depth (Davis and Sorensen 1970; Zipparro and Hansen 1993):

$$d_s = x \cdot 0.473 \, (Q/f)^{0.333} - D \quad (5\text{-}24a)$$

$$d_s = x \cdot 1.337 \, (q^2/f)^{0.333} - D \qquad (5\text{-}24b)$$

where

d_s = scour depth below channel bed (m)

x = multiplying factor varying from 1.25 to 2.0 depending on the severity of flow concentration in the vicinity of the bank toe

Q = design discharge (m³/s)

q = unit discharge per meter width of channel (m²/s)

f = Lacey's silt factor = $1.76 \sqrt{d_m}$

d_m = d_{50} of channel bed material expressed in millimeters

2. Neill's Equation (USBR 1984):

$$d_s = x \cdot d_i \, (q_f/q_i)^n \qquad (5\text{-}25)$$

where

x = multiplying factor varying from 0.5 to 0.7 depending on the possibility of flow concentration

d_i = average water depth in incised channel reach at bankfull discharge (m)

q_f = design discharge per unit channel width (m²/s)

q_i = bankfull discharge in incised reach per unit channel width (m²/s)

n = an exponent varying from 0.67 for sand bed to 0.85 for coarse gravel bed channels

3. USBR's Modification of Lacey's Equation (USBR 1984):

$$d_s = x \cdot 0.473 (Q/f)^{0.333} \qquad (5\text{-}26)$$

where x = multiplying factor varying from 0.25 to 1.25 depending on the severity of flow concentration near the bank toe.

4. USBR's Modification of Blench Equation (USBR 1984):

$$d_s = x \cdot (q_f^{0.67}/F_b^{0.33}) \qquad (5\text{-}27)$$

where

x = multiplying factor varying from 0.6 to 1.25

F_b = zero bed factor (m/s²) estimated to be 0.14 for d_{50} = 0.1 mm, 0.37 for d_{50} = 0.30 mm, 0.8 for d_{50} = 3.0 mm, 1.4 for d_{50} = 30 mm, and 3.0 for d_{50} = 1,000 mm

226 WATER RESOURCES ENGINEERING

5. Competent Velocity Method (USBR 1984):

$$d_s = D\,[(V_a/V_c) - 1] \qquad (5\text{-}28)$$

where V_c = competent velocity (m/s) varying from 0.6 to 1.8 m/s for $D = 1.5$ m, 0.65 to 2.0 m/s for $D = 3.0$ m, 0.7 to 2.3 m/s for $D = 6.0$ m, and 0.8 to 2.6 m/s for $D = 15$ m. Lower values of V_c pertain to easily erodible bed material and higher values to erosion-resistant bed material.

6. Field Measurements of Scour Method (USBR 1984):

$$d_s = 1.32\,(q_f)^{0.24} \qquad (5\text{-}29)$$

Example 5-5: Estimate scour depths for streams A and B of Example 5-3. Relevant hydraulic parameters are shown in Table 5-4.

Solution: Here, $f = 1.76\,\sqrt{(0.10)} = 0.56$; $n = 0.67$ for sandbed streams; $F_b = 0.14$ for $d_{50} = 0.10$ mm; and $V_c = 0.8$ m/s for easily erodible streambeds. Estimated scour depths using Eqs. (5-24) to (5-29) are shown in Table 5-5.

The adopted values of scour depths for the two streams are selected to be within the range of values estimated by Eqs. (5-24) to (5-29).

Launching Apron

To protect against sliding of bank riprap into scour holes near the toe, a launching apron of riprap is laid over the channel bed, extending from the toe of the sloping riprap toward the center of the channel. The length of the launching apron along the width of the channel is taken to be $p \times d_s$, where p = a multiplying factor varying from 1.5 to 2.0.

As a scour hole develops near the toe of the bank, stones from the launching apron slide along the side slope of the scour hole, armor the side slope of the scour hole against erosion, and minimize or prevent further scour. If the side slope of the scour hole is z:1 (horizontal:vertical) and riprap thickness required to armor the side slope of the scour hole is T_c, then the volume of stone, VS, required per unit length of the protected bank is

$$VS = d_s \cdot T_c \cdot \sqrt{(z^2 + 1)} \qquad (5\text{-}30)$$

Table 5-4. Parameters for estimation of scour depths

Parameter	Value	
	Stream A	Stream B
Q (m³/s)	35,390	770
d_{mm} (mm)	0.10	0.10
d_i (m)	12.0	4.1
q_f (m²/s)	82.1	13.6
q_i (m²/s)	43.9	9.4

Table 5-5. Estimated scour depths

	Estimated scour depth below bed (m)	
Method	Stream A	Stream B
1. Regime Scour Depth	2.8–17.0	0–1.2
2. Neill's Equation	9.1–12.8	2.6–3.7
3. USBR's Modification of Lacey's Equation	4.7–23.6	1.3–6.6
4. USBR's Modification of Blench Equation	22–45.8	6.6–13.8
5. Competent Velocity Method	58.3	19.8
6. Field Measurements of Scour Method	3.8	2.5
Adopted value	18	5

This volume of stone should be available within the launching apron. Thus, the thickness of stone in the launching apron, T_t, is given by

$$T_t = (T_c/p) \cdot \sqrt{(z^2 + 1)} \tag{5-31}$$

Usually, $z = 2$, $p = 1.5$, and $T_c = 1.24\ T$, where T = riprap thickness along bank slope. Therefore, $T_t = 1.85\ T$.

Example 5-6: Using the scour depths for streams A and B estimated in Example 5-5 and riprap sizes selected in Example 5-3, estimate the dimension of launching apron for each case.

Solution: The thickness of riprap along the banks of each stream is taken to be twice the estimated riprap size and $p = 1.5$. Thus, for stream A, $T = 2 \times d_{50} = 2 \times 0.46 = 0.92$ m, and $T = 2 \times 0.38 = 0.76$ m for stream B.

The dimensions of launching apron estimated using Eqs. (5-30) and (5-31) are shown in Table 5-6.

Conceptual Measures to Repel Main Current Away from Bank

In some cases, bank erosion may result due to the tendency of the main current to shift toward the riverbank, and a component of bank protection measures may be to repel the

Table 5-6. Design parameters for launching apron

	Value (m)	
Parameter	Stream A	Stream B
T	0.92	0.76
T_t	1.7	1.4
Length of launching apron	27	7.5

main current away from the convex (outer) side of the riverbank. Measures to accomplish this are listed below (Singh 1967; CBIP 1971):

1. Wooden piles may be driven along rectangular grids covering the channel portion occupied by the curved portion of the riverbank. These piles may be driven up to the bedrock or sufficiently deep into the riverbed material to be stable at least up to the bankfull flow conditions. Over time, the resistance to flow provided by these gridded piles causes silt deposition in the area.

2. Repelling groins may be provided that project from the bank at an acute angle with respect to the upstream side bank. These groins may be permeable, impermeable, submerged, or unsubmerged.

3. Rock may be dumped on the channel bed in the curved portion to raise the channel bed, reduce channel velocities, and minimize scour in that portion.

4. Attracting groins may be provided that project from the opposite bank at an obtuse angle with respect to the upstream side of that bank to attract the main current toward that bank. These groins also may be permeable, impermeable, submerged, or unsubmerged.

Biotechnical and Other Methods of Bank and Toe Protection

These methods include the use of vegetation or synthetic materials in place of rock riprap or gabions. The following alternative erosion protection materials may be used for bank erosion protection:

1. Fabriforms or Articulated Blocks: As defined previously, fabriforms are bags made of synthetic materials filled with concrete, cement, or soil cement and placed on bank slopes and toes. Green synthetic bags may be used to maintain a park-like visual setting. The bags may be placed adjacent to one another or with spaces left between them for plant growth or seeding. Articulated blocks are prefabricated concrete blocks with voids left in them for seeding and plant growth. These are placed on the bank slope and their voids are filled with soil and seeded with suitable plant species.

2. Rolled Erosion Control Products (RECPs): These are mats made of synthetic materials able to resist microbial and environmental degradation, and they contain open spaces for plant penetration. They are constructed from multiple layers of synthetic threads, which minimize migration of soil from areas covered by them. These mats may be reinforced by turf or wires and may have an organic component made of coir, excelsior, or wood chip fibers. Seeds are introduced into the soil prior to covering with RECP protection.

3. Vegetation Cover: This includes both the selection of vegetation type that will survive and grow under prevailing climatic conditions at the site and the seeding, sodding, or planting of such vegetation.

Alternative methods for bank toe protection (other than riprap and gabions) include:

1. Organic Logs: Logs manufactured from excelsior, coir, or straw materials. They are secured with rods or wood stakes at the anticipated low water level. The logs retard

flow velocities in their vicinity and allow development of vegetation behind them. After some time, the logs may biodegrade, leaving vegetation protection at the bank toe.

2. Lunkers: A hollow wooden box structure placed at the stream bank toe below the low water line, with the stream side of the box open for access by aquatic organisms. The top of the Lunker provides a bench for placement of riprap or fabriform to hold the Lunker in place. To provide further stability, steel reinforcing bars are driven through the Lunkers into the underlying soils, and stringers are buried into the sloping bank on the back side of the Lunker.

3. Root Wad Revetments: Grubbed sections of tree trunks and roots placed at the bank toe for erosion protection. The rough texture of the root wad retards flow velocities in the vicinity of the bank toe, promotes sedimentation, and allows development of vegetation at the toe.

Additional details of specific biotechnical methods of bank and toe protection may be obtained from individual manufacturers and vendors.

Drop Structures

Drops or grade control structures are provided to negotiate changes in the bed slope of a channel, to reduce existing channel bed slope, and as control points to limit retrogression due to potential scour in the downstream reach. There are several types of grade control structures suitable for different site conditions, as described in the following paragraphs.

Straight Vertical Drop

This drop is designed as a vertical wall with a broad crest and vertical downstream face. It may be suitable for small canals or drainages with relatively small drops (e.g., 1 to 1.5 m). The crest of the drop structure may be treated as a broad-crested weir if the length of the horizontal portion of the crest (parallel to the direction of flow) is greater than 2.5 H, where H is the head above crest. Thus,

$$q = 1.70 \ H^{1.5} \tag{5-32}$$

Empirical equations to estimate relevant dimensions for straight vertical drops are given below (Chow 1959):

$$D = \text{drop number} = q^2/g \ h^3 \tag{5-33a}$$

$$L_d/h = 4.30 \ D^{0.27} \tag{5-33b}$$

$$y_1/h = 0.54 \ D^{0.425} \tag{5-33c}$$

$$y_2/h = 1.66 \ D^{0.27} \tag{5-33d}$$

$$F_1 = q/\sqrt{(g \ y_1^3)} \tag{5-33e}$$

where

h = drop between upstream and downstream bed elevations

y_1 = water depth at the toe of the falling nappe or beginning of the hydraulic jump

y_2 = tailwater depth sequent to y_1

L_d = distance from the toe of the drop to the location of y_1

The length of the jump on the horizontal floor downstream of the drop may be taken to be about 6 y_2 or estimated from Table 3-15. Thus,

$$\text{approximate length of the downstream floor} \cong L_d + 6\, y_2 \tag{5-33f}$$

The downstream floor may be located at an elevation such that the design tailwater depth is greater than y_2. This may require the floor to be lower than the downstream channel bed. The thickness and length of the downstream floor must be checked to be safe against uplift and exit gradient. It is good practice to provide cutoff at the end of the horizontal floor that is about 0.75 to 1.0 m deep and 0.25 m thick. The depth of cutoff may be increased, if required, to protect against potential scour or critical exit gradient, as described in the following section of this chapter, "Drop with Sloping Apron."

A depressed stilling pool on the downstream side of the drop structure has been found effective for energy dissipation for small vertical drops. Preliminary design dimensions for such a stilling pool may be estimated by

$$L_s = 5\sqrt{(H \cdot H_L)} \tag{5-33g}$$

$$X = 0.25\,(H \cdot H_L)^{2/3} \tag{5-33h}$$

where

H = head above crest

H_L = drop between upstream and downstream water surface elevations

L_s = length of depressed floor

X = depth of depressed floor below downstream channel bed

Sometimes, from environmental considerations, it may be desirable to locate the top of the crest above the upstream channel bed to create a shallow pool on the upstream side. For flood control channels, a raised crest may cause high flood elevations on the upstream side and may not be desirable.

Drop with Sloping Apron

This drop may be suitable for drops of 1 to 3 m. Commonly used apron slopes vary from 2H:1V to 4H:1V. The structure may be constructed with concrete or rock. In the case of rock structures, the sizes of stones may be determined using the methods described in the previous section entitled "Erosion Protection." The crest may be a broad crest for rock structures and a broad or sharp crest for concrete structures. For a broad crest, the length of the horizontal portion of the crest should be greater than 2.5 times the head, H, above it. For a sharp crest, the crest length should be less than 2/3 H. The cross section at the crest may be rectangular or trapezoidal. If necessary, the width of the drop may be smaller than the width of the channel. However, any contraction in the flow section may result in higher water surface elevations on the upstream side. The downstream face is in the form of a sloping apron.

Computational steps to estimate design dimensions for a drop with a trapezoidal section and sloping apron are as follows:

- Determine design discharge, Q, bed width, B, and side slopes, z, and estimate critical depth, y_c, at the crest by trial and error:

$$Q = (\sqrt{g}) \, [B \, y_c + z \, y_c^2]^{1.5} / [B + 2 \, z \, y_c]^{0.5} \qquad (3\text{-}41)$$

- Assume that y_c is nearly equal to the average water depth over the sloping apron, and estimate flow per unit width of crest:

$$q \, (\text{average}) = Q/[B + z \, y_c] \qquad (5\text{-}34a)$$

- Estimate y_1 at the toe of the slope by trial and error using

$$H + Z = y_1 + q^2/(2 \, g \, y_1^2) + (q^2 \, n^2 \, L)/(y_c^{3.333}) \qquad (5\text{-}34b)$$

where

H = head above crest, including velocity of approach head

Z = height of crest above toe of slope

n = Manning's coefficient

L = length of slope

If required, further refinements may be made using $(y_c + y_1)/2$ in place of y_c in Eq. (5-34b) to compute a modified value of y_1.

- Estimate $V_1 = Q/[B \, y_1 + z \, y_1^2]$ and $F_1 = V_1/\sqrt{(g \, y_1)}$.
- Assume the jump is located on the horizontal floor downstream of the toe of the slope, and estimate sequent depth y_2 from

$$y_2/y_1 = (1/2) \, [\sqrt{(\{1 + 8 \, F_1^2\}} - 1] \qquad (3\text{-}52b)$$

- Set the horizontal floor (commencing from the toe of the sloping apron) at a depth greater than y_2 below design tailwater elevation such that the top of the floor is below the downstream channel bed.
- Obtain length of jump on a horizontal floor from Table 3-15 and adopt the length of horizontal floor equal to the length of jump. Usually, chute or floor blocks are not required for small drops of 1 to 2 m. For larger drops, St. Anthony Falls (SAF) type stilling basins may be used (see the section of this chapter entitled "Stilling Basins and Energy Dissipation Devices").
- Provide a vertical cutoff at the downstream end of the horizontal floor. Depth of cutoff below the tailwater elevation, R, may be estimated by

$$R = 1.25 \times 1.337 \, (q^2/f)^{1/3} \qquad (5\text{-}34c)$$

- For concrete structures founded on pervious soils, check for exit gradient and uplift pressures on the horizontal floor.

Exit gradient is the hydraulic gradient at the exit (downstream) end of the concrete floor or cutoff. Excessive exit gradient may result in formation of boils or piping in the saturated pervious soils at the downstream end. Under this condition, water seeping under the floor of the structure retains sufficient hydraulic head to lift up and wash off soil grains at the exit. This loss of soil grains may create a cavity under and at the downstream end of the floor and result in its subsidence. The critical exit gradient, G_E, is given by the following (Davis and Sorensen 1970; Zipparro and Hansen 1993):

$$G_E = (H/d)\,[1/(\pi\sqrt{\lambda})] \qquad (5\text{-}35a)$$

$$\lambda = [1 + \sqrt{\{1 + \alpha^2\}}]/2 \qquad (5\text{-}35b)$$

$$\alpha = L/d \qquad (5\text{-}35c)$$

where

H = difference of head between the upstream and downstream ends of the concrete floor or head causing seepage below the floor

d = depth of cutoff below downstream channel bed or thickness of floor below channel bed if no cutoff is provided

L = total horizontal length of impervious floor under which seepage occurs, including portions upstream of and under the crest

The hydraulic force of seepage at the exit is resisted by the submerged weight of soil:

$$dp \cdot dA = dA \cdot dl \cdot (1 - \varphi)\,(\gamma_s - \gamma), \text{ or}$$

$$\gamma\,dh/dl = (1 - \varphi)\,(\gamma_s - \gamma), \text{ or}$$

$$dh/dl = \text{exit gradient (resistance)} = (1 - \varphi)\,(G - 1) \qquad (5\text{-}36)$$

where

dA = area of soil column at the exit on which hydraulic force is exerted

dp = residual hydraulic pressure at the exit

dl = depth of submerged soil column resisting hydraulic force, i.e., depth of cutoff or floor below channel bed

γ = unit weight of water

γ_s = unit weight of soil grains

φ = porosity

G = specific gravity of soil grains

To minimize potential of piping at the downstream end, a factor of safety must be used for the exit gradient given by Eq. (5-36). With $\varphi = 0.35$ and $G = 2.6$, G_E (resistance, Eq. (5-36)) $\cong 1.0$. Suggested values for safe G_E (Eq. (5-35a)) are 1/4 to 1/5 for gravel; 1/5 to 1/6 for coarse sand, and 1/6 to 1/7 for fine sand.

With known depth of cutoff, d (required for scour protection), H, and required exit gradient, G_E, the required floor length, L, can be estimated using Eqs. (5-35a) to (5-35c). If the length of the sloping apron plus floor length estimated for energy dissipation is less than L, the balance may be provided on the upstream side of the crest. Note that there may be no uplift (see next paragraph) on the floor length located upstream of the crest requiring only a nominal floor thickness. Therefore, it may be desirable to provide as much of the floor length on the upstream side as practicable, provided sufficient impervious floor length is available on the downstream side for energy dissipation.

Water seeping under the structure floor exerts uplift at the bottom of the impervious floor equal to the residual hydraulic head at any location. Uplift pressure at any location may be estimated using Lane's weighted creep approach. A more refined estimate may be made using Khosla's design charts (Davis and Sorensen 1970; Zipparro and Hansen 1993). Lane's weighted creep length, L_c, is estimated as the sum of the depth of all vertical faces (steeper than 45 degrees) along the path of seepage and 1/3 of all horizontal lengths (flatter than 45 degrees) of the impervious floor. The average hydraulic gradient below the structure floor is estimated by

$$i = H/L_c \qquad (5\text{-}37)$$

The weighted creep length for the structure shown in Figure 5-3(a) is given by

$$L_c = d_1 + d_2 + d_3 + d_4 + d_5 + d_6 + 1/3\,[L_1 + L_2 + L_3 + L_4 + L_5 + L_6] \qquad (5\text{-}38)$$

The residual hydraulic head above the bottom of floor at a point A, located at a distance x from the downstream end of the floor, is

$$h = H\,(x/3 + \text{sum of vertical creep lengths downstream of point A})/L_c \qquad (5\text{-}39)$$

Thickness of concrete, d, required to withstand the uplift pressure at this location may be estimated by

$$\gamma\,h = \gamma_c\,d \quad \text{or} \quad h = G_c\,d \quad \text{or} \quad h - d = G_c\,d - d, \quad \text{or}$$
$$d = h/G_c = (h - d)/(G_c - 1) \qquad (5\text{-}40)$$

where G_c = specific weight of concrete, taken to be about 2.4. The last expression in Eq. (5-40) may be convenient because the ordinate between the hydraulic gradient line and the top of the floor $(h - d)$ can be easily estimated. Suggested values of i to check the adequacy of total weighted creep length are 1/3 to 1/4 for a mixture of sand, gravel, and boulder; 1/5 for coarse sands; and 1/6 to 1/8.5 for medium to fine sands.

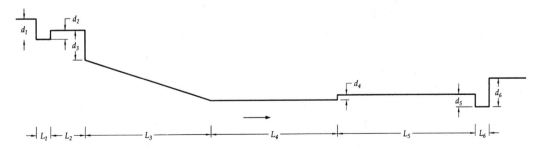

Figure 5-3(a). Weighted creep length

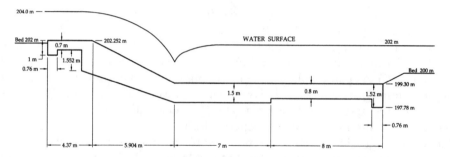

Figure 5-3(b). Line diagram of drop structure floor

Example 5-7: Design a drop structure with a sloping apron to negotiate a drop of 2 m between the upstream and downstream bed elevations of a stream that carries a design flood of 175 m³/s. The channel has a trapezoidal section with a bottom width of 55 m, side slopes of $2H{:}1V$, and water depth of 2 m. The upstream and downstream bed elevations are 202 and 200 m, respectively. The channel bed material consists of fine sand with $d_{50} = 0.30$ mm. At the drop structure, the channel is to be contracted to a bed width of 42 m and side slopes of $2H{:}1V$. Assume the apron slope to be $2H{:}1V$ (Figure 5-3(b)).

Solution: Assume there is a broad-crested weir at the contracted section; then, using Eq. (3-41), $175 = \sqrt{(9.81)} [42\,y_c + 2\,y_c^2]^{1.5}/[42 + 4\,y_c]^{0.5}$. By trial and error, $y_c = 1.19$ m. So, $A = 42 \times 1.19 + 2 \times 1.19 \times 1.19 = 52.81$ m². $V_c = 175/52.81 = 3.31$ m/s; $q = 175/(42 + 2 \times 1.19) = 3.94$ m³/s. Length of sloping apron = $L = 2\sqrt{(2^2 + 1)} = 4.472$ m. H (above crest) = $1.19 + 3.31^2/(2 \times 9.81) = 1.748$ m. Crest elevation = $202 + 2.0 - 1.748 = 202.252$ m or 0.252 m above the upstream channel bed. Length of broad crest along direction of flow = $2.5 \times 1.748 = 4.37$ m.

The bottom of the crest wall may be taken to El. 200 m or up to the bed of the downstream channel. Thus, height of crest wall = $202.252 - 200.0 = 2.252$ m.

Neglect head loss through channel contraction.

Equating energies upstream and downstream of the crest (with $n = 0.014$ and y_1 = water depth at toe of sloping apron) measured above downstream channel bed, $1.748 + 2.252 = y_1 + 3.94^2/(2 \times 9.81\,y_1^2) + (3.94 \times 0.014)^2 \times 4.472/(1.19)^{3.333}$. Or, $4.0 = y_1 + 0.791/(y_1^2) + 0.0076$. So, $y_1 = 0.474$ m; $V_1 = 175/[42 \times 0.474 + 2.0 \times 0.474^2] = 8.6$ m/s; and $F_1 = 8.6/\sqrt{(9.81 \times 0.474)} = 4.0$.

So, $y_2/y_1 = (1/2)[\sqrt{(1 + 8 \times 4.0^2)} - 1] = 5.18$ and $y_2 = 2.45$ m.

Set the top of the horizontal floor at El. 199.30 m [202 (downstream water surface or tailwater elevation) − 2.70 (a little greater than y_2)]. This will ensure that adequate tailwater depth is available for the formation of the jump near the toe of the sloping apron.

Length of jump (assuming it to form on the horizontal floor, Table 3-15) = $5.8 \times 2.45 = 14.2$ m. Provide a 15-m-long horizontal floor.

Lacey's silt factor = $1.76\sqrt{0.30} = 0.96$. So, $R = 1.25 \times 1.337\,(3.94^2/0.96)^{1/3} = 4.22$ m. Set bottom elevation of downstream cutoff at 202 (tailwater elevation) − 4.22 = 197.78 m or 1.52 m below the top of horizontal floor.

Depth of cutoff below top of horizontal floor = $d = 199.30 - 197.78 = 1.52$ m. Using $G_E = 1/6$ in Eq. (5-35a), $1/6 = (2/1.52)[1/\{\pi\sqrt{\lambda}\}]$, or $\lambda = 6.31 = [1 + \sqrt{1 + \alpha^2}]/2$ and $\alpha = 11.6 =$

L/d. This gives a desirable floor length of $11.6 \times 1.52 = 17.6$ m. The total floor length provided for the horizontal jump is greater, so no modification is necessary. As a first estimate, assume the thickness of the horizontal floor to be 0.8 m so that the depth of cutoff below the bottom of the horizontal floor is $1.52 - 0.80 = 0.72$ m.

Horizontal length of seepage path (Figure 5-3(b)) = $4.37 + 5.904 + 7 + 8 = 25.274$ m.

Vertical length of seepage path (Figure 5-3(b)) = $1 + (1 - 0.7) + 1.552 + (1.5 - 0.8) + (1.52 - 0.8) + 1.52 = 5.792$ m.

Lane's creep length = $5.792 + 25.274/3 = 14.2$ m. Total head differential = 2 m.

Average hydraulic gradient = $2/14.2 = 0.14$. This is acceptable for bed materials consisting of fine to medium sand.

Residual uplift at 8 m upstream from downstream end of concrete floor = $(8/3 + 1.52 + 0.72) \times 0.14 = 0.69$ m.

Required concrete thickness (using Eq. (5-40)) = $0.69/1.4 = 0.5$ m.

Residual uplift at 15 m upstream from downstream end of concrete floor = $(15/3 + 1.52 + 0.72 + 0.7) \times 0.14 = 1.11$ m.

Required concrete thickness (using Eq. (5-40)) = $1.11/1.4 = 0.8$ m.

The channel will require contraction and expansion designs (see the section entitled "Channel Transitions" in this chapter).

Example 5-8: Design a 1.25-m riprap drop structure with sloping apron in a trapezoidal channel with a design discharge of 22.65 m³/s. The bed width and side slopes of the channel are 6.1 m and $3H{:}1V$, respectively, and the water depth is 1.52 m. The d_{50} of channel bed material is 0.18 mm. Assume the apron slope to be $3H{:}1V$.

Solution: $A = 6.1 \times 1.52 + 3 \times (1.52)^2 = 16.2$ m²; $V = 22.65/16.2 = 1.398$ m/s.

Energy head above upstream channel bed = $1.52 + (1.398)^2/(2 \times 9.81) = 1.62$ m.

Assuming the falling water is to hit the toe of sloping apron with no energy loss with impingement on water in the downstream channel, H (above toe) = $1.62 + 1.25 = 2.87$ m.

Length of sloping apron = $1.25 \times \sqrt{(3^2 + 1)} = 3.95$ m.

As a first trial, assume average water depth between crest and toe of sloping apron to be 1.0 m, Manning's n of 0.04 for riprap, and apply conservation of energy equation between the upstream channel and toe of sloping apron: $2.87 = 3.95 \times (0.04^2 \times V_1^2)/(1.0^{4/3}) + V_1^2/(2 \times 9.81)$ or $V_1 = 7.077$ m/s.

$A = 22.65/7.077 = 3.2 = 6.1 y_1 + 3 y_1^2$. Subscript 1 refers to values at the toe before the hydraulic jump.

Thus, $y_1 = 0.43$. Average depth between crest and toe of sloping apron = $(1.52 + 0.43)/2 = 0.975$ m, which is nearly equal to the assumed value of 1.0 m. So, $F_1 = 7.077/\sqrt{(9.81 \times 0.43)} = 3.45$; $y_2/y_1 = 1/2 [\sqrt{(1 + 8 \times 3.45^2)} - 1] = 4.40$; and $y_2 = 1.89$ m. From Table 3-15, length of jump on a horizontal apron = $5.4 \times 1.89 = 10.2$ m.

The riprap apron will provide more friction and energy loss than a smooth horizontal floor. Provide a 10-m-long horizontal apron. The top of the horizontal apron may be set at $1.89 - 1.52 = 0.37$ m below the downstream channel.

Energy head above upstream channel bed $= H = 1.62$ m. Length of upstream horizontal apron $= 2H = 3.24$ m.

Lacey's silt factor $= 1.76 \sqrt{0.18} = 0.75$. Potential scour depth below downstream channel bed (see the previous section entitled "Protection Against Scour at Bank Toe") $= 1.5 \times 0.473 \times (22.65/0.75)^{1/3} - 1.52 = 0.69$ m. Provide 0.7-m-deep cutoff at the upstream and downstream ends of the riprap aprons below the upstream and downstream channel beds.

The size and thickness of riprap may be estimated using methods given previously (see the section entitled "Design of Riprap Protection") with values of V_1 and y_1 estimated previously.

Usually, riprap would be pervious enough to minimize uplift pressures and piping. If significant clogging is expected, then the total length of the riprap apron including cutoffs should be checked to provide safe creep ratio and exit gradient as in Example 5-7.

Dams and Reservoirs

Planning and Investigations

This section describes preliminary planning and investigations required for siting and designs of dams and reservoirs. Depending on the objective, dams and reservoirs may be classified as single- or multipurpose and, depending on size, they may be categorized as minor, medium, or major projects. A single-purpose project is constructed to serve a single purpose (e.g., water supply, hydropower, recreation, flood control, etc.). A multipurpose project is designed to accomplish two or more such purposes. Common steps in the planning and investigation of such projects include the following:

1. Identification of project objectives including approximate magnitudes (e.g., water supply for a specified community or agricultural area, hydropower generation to meet a specified demand, and flood control for a specified community).

2. Selection of dam and reservoir site. This is conducted by a multidisciplinary team, usually composed of a water resources engineer, geologist, geotechnical engineer, and community leader. Additional members may be added depending on specific circumstances. Generally, more than one site is identified on topographic maps, and the preliminary choice is narrowed down after field visits. The salient items to be observed include:

    (a) Suitability of foundations for an earth, earth and rockfill, or gravity dam

    (b) Existence of a relatively narrow valley to avoid unduly large embankments and inundated areas

    (c) Suitability of reservoir bottom to hold the anticipated volume of water without undue seepage losses

    (d) Availability of construction materials within reasonable hauling distances (e.g., embankment fill materials, rockfill, concrete aggregate)

(e) Proximity to the service area (e.g., agricultural area and community to be served)

(f) Availability of suitable site for spillways

3. Preliminary sizing and determination of dam type. Preliminary estimation of the dam height and length and reservoir capacity may be made using available contoured topographic maps of the area. This estimation may be used to reduce the number of alternative sites identified in the previous steps. Suitable type of dam (e.g., earth, earth and rockfill, and gravity) for each promising site is determined by a multidisciplinary team using preliminary geological, geotechnical, and economic analyses.

4. Preliminary surveys. Preliminary field surveys are conducted for selected promising sites. These include preliminary geological surveys to assess the rock and soil conditions for the dam, reservoir, spillway, and borrow areas, and topographic surveys to estimate reservoir capacity with different dam heights. The topographic surveys include cross sections across the valley covering potential reservoir area. These cross sections are used to prepare elevation-area and elevation-capacity tables or curves for the reservoir.

5. Hydrologic investigations. These include demarcation of the selected sites and estimation of drainage areas upstream of each location from available topographic maps of the respective watersheds. Available data on streamflows for the stream intended to serve as the source of water and rainfall data for precipitation gauges in its watershed are assembled and data gaps are identified. At this time, a hydrologic monitoring plan is prepared that includes installation of stream and precipitation gauges at suitable locations. Available streamflow and precipitation data for monitoring stations in adjacent watersheds are collected, along with information such as the hydraulic characteristics of the respective watersheds, location and datum of existing stream gauges, and location and altitude of existing precipitation gauges.

6. Hydrologic analyses. These include flood routing computations for several combinations of dam heights and spillway widths.

The reservoir capacity is divided into several segments:

1. Dead or Inactive Storage: The capacity at the bottom of the reservoir that is reserved for sediment accumulation during the anticipated life of the project.

2. Conservation Storage: The storage between the top of the dead storage pool and normal reservoir water surface elevation. This storage is available for various project uses (e.g., water supply and hydropower generation).

3. Flood Control Storage: The storage between the top of the conservation pool and maximum permissible water level in the reservoir. This storage is reserved for flood control. Usually, it is kept unused or empty for most of the time to be utilized for temporary storage of floodwater during storm events.

4. Freeboard or Surplus Storage: The storage available within the freeboard of the dam, between the top of the flood control pool and top of the dam. This storage is not used except during abnormal conditions. It is reserved to prevent dam overtopping due to wind wave setup and runup, potential landslide-generated waves, and extreme flood events larger than the design basis flood.

Reservoir Sedimentation

Methods to estimate sediment yield of watersheds and rates of reservoir sedimentation include:

1. Periodic bathymetric surveys in the case of an existing reservoir, usually at an interval of 10 yr or less.
2. Periodic measurement of total sediment load inflowing into an existing reservoir during storm events and snowmelt periods.
3. Estimation of sediment load using the Universal Soil Loss Equation (USLE) and sediment delivery ratio (SDR) for proposed reservoirs or where data are not available.
4. Estimation based on regression equations for different regions.
5. Estimation using data for other reservoirs (e.g., Chow 1964; USDA 1969).

If data for inflowing discharges and sediment loads for other reservoirs in similar climatic, hydrologic, and geomorphologic regions are available, then a regression equation of the following form may be developed

$$Q_s = a\, Q^b \tag{5-41}$$

where

Q_s = inflowing sediment load (t/day)

Q = discharge (m^3/s)

a and b = regression coefficients.

The following is a commonly used regression equation for reservoirs in semi-arid climate (USBR 1987):

$$Q_s = 1{,}098\, A^{-0.24} \tag{3-84}$$

where

Q_s = sediment deposition in reservoir (m^3/km^2/yr)

A = watershed area (km^2)

Other empirical equations to estimate the weight of sediment likely to be deposited in a reservoir are the Dendy-Bolton equations, shown below as Eqs. (5-42) and (5-43) (USACE 1989):

- For watersheds with mean annual surface runoff equal to or less than 5 cm,

$$T = 292.32\, R^{0.46}\, [1.537 - 0.26 \log A] \tag{5-42}$$

where

T = sediment load likely to be deposited in the reservoir (t/km^2/yr)

R = mean annual surface runoff (cm)

Table 5-7. Reservoir sedimentation rates

Location	Drainage area (km²)	Sedimentation rate (t/km²/yr)
Northeastern United States	808–37	76–217
Southeastern United States	23,509–11.76	41–783
Midwestern United States	11,930–414	90–693
South Central United States	252.5–59.8	96–1,333
Northern Great Plains, United States	1,199–3.6	43–1,862
Southwestern United States	12.7–282.3	51–857
Northwestern United States	10.6–481.7	11–375

Source: Chow (1964).

- For watersheds with mean annual surface runoff greater than 5 cm,

$$T = 686.58 \, [\exp(0.02165 \, R)] \, [1.537 - 0.26 \log A] \qquad (5\text{-}43)$$

Eqs. 5-42 and 5-43 are the same as Eqs. 3-82 and 3-83 except that R is expressed in centimeters.

Typical rates of sedimentation for reservoirs in humid regions vary from 190 to 714 m³/km² for a drainage area of 26 km² and 86 to 333 m³/km² for a drainage area of 12,950 km². The corresponding rates for reservoirs in semi-arid regions vary from 381 to 1,667 m³/km² and 143 to 476 m³/km², respectively (Golze 1977).

Sedimentation rates for some reservoirs with high and low rates of deposition per square kilometer of drainage area are shown in Table 5-7 (Chow 1964).

Example 5-9: Bathymetric survey of a reservoir indicated a total sediment deposition of 41,000 m³ in 62 yr. Based on particle-size distribution of the sediments in the vicinity of the reservoir shoreline, it is estimated that the deposited sediments may have 41% of clay, 22% of silt, and 37% of sand by weight, with initial unit weights of 561, 1,138, and 1,554 kg/m³, respectively. Other relevant data include:

- Reservoir capacity = 474,821 m³
- Drainage area = 18.8 km²
- Inflowing annual surface runoff = 4,254,889 m³/yr
- Length of watershed = 7,322 m
- Average daily streamflow = 51 to 453 l/s
- USLE estimate of soil erosion from the watershed = 5,978 t/yr

Regression of the inflowing water and suspended sediment load gives

$$Q_s = 14.74 \, Q^{1.7853} \qquad (5\text{-}44)$$

where

Q_s = suspended sediment load (t/day)
Q = average daily streamflow (m³/s)

Using the bathymetric data, estimate the rate of sediment deposition in t/yr and rate of sediment that may have entered the reservoir in t/yr. Also, use other empirical methods to estimate the rate of sediment deposition in the reservoir.

Solution:

1. Sediment yield using bathymetric survey data: Initial composite unit weight of sediments = $561 \times 0.41 + 1{,}138 \times 0.22 + 1{,}554 \times 0.37 = 1{,}055.4$ kg/m³.

   Unit weight of sediments after 62 yr of deposition may be estimated using Miller's equations (USBR 1987):

   $$K_a = K_s p_s + K_m p_m + K_c p_c \tag{5-45}$$

   where

   K_s, K_m, and K_c are empirical coefficients with values of 0.0, 29, and 135

   p_s, p_m, and p_c are fractions of sand, silt, and clay, respectively

   K_a = weighted average coefficient for the deposited material

   Using the given data, $K_a = 0.0 + 29 \times 0.22 + 135 \times 0.41 = 61.73$. Then,

   $$W_T = W_i + 0.4343\, K_a\, [\{T/(T-1)\} \ln(T) - 1] \tag{5-46}$$

   where

   W_T = unit weight (kg/m³) of sediment after T years of deposition
   W_i = initial unit weight (kg/m³)

   Thus, for $T = 62$ yr, $W_T = 1{,}055.4 + 61.73 \times 0.4343\, [\{62/61\} \ln(62) - 1] = 1{,}141$ (kg/m³).

   Thus, average sedimentation rate for the reservoir = $41{,}000 \times 1{,}141/(62 \times 1{,}000) = 755$ t/yr. This does not represent the entire sediment entering the reservoir because the trap efficiency of the reservoir may not be 100%. Trap efficiency is defined as E = quantity of sediment deposited in reservoir/quantity of sediment entering reservoir.

   Trap efficiency of the reservoir may be estimated using Brown's, Brune's, and Churchill's methods (USACE 1989). Brown's equation gives

   $$E = 1 - [1/\{1 + 0.0021\,(K C/W)\}] \tag{5-47}$$

   where

   K = a coefficient ranging from 0.046 to 1.0 with a median value of 0.1
   C = reservoir capacity = 474,821 m³
   W = drainage area = 18.8 km²

   This gives E = trap efficiency = 0.84.

The USBR has developed trap efficiency curves based on Brune's and Churchill's equations. Approximate trap efficiencies for different C/I ratios are shown in Table 5-8, where I = average annual inflow (m^3).

C/I ratio for the reservoir = 474,821/4,254,889 = 0.11. Therefore, Brune's trap efficiency = 0.87 and Churchill's trap efficiency = 0.73. Adopt an average trap efficiency of 0.81.

Sediment load entering the reservoir = 755/0.81 = 932 t/yr.

2. Sediment yield using inflowing sediment load: Assuming bed load transport to be 15% of suspended load (Simons and Senturk 1992), Eq. (5-44) may be modified to

$$Q_T = 16.95 \ Q^{1.7853} \qquad (5\text{-}48)$$

where Q_T = total daily sediment load entering the reservoir (t/day).

$Q_T = 16.95 \times (0.051)^{1.7853} = 0.0835$ t/day or 30.5 t/yr and $Q_T = 16.95 \times (0.453)^{1.7853} = 4.123$ t/day or 1,505 t/yr. Average sediment inflow to reservoir = (30.5 + 1505)/2 = 768 t/yr. Assuming $E = 0.81$, sediment load expected to deposit in the reservoir = $0.81 \times 768 = 622$ t/yr.

3. Sediment yield based on USLE: In this case,

Sediment deposited in the reservoir = sediment erosion from the watershed
(USLE estimate) $\times$ SDR $\times$ E \qquad (5-49)

in which SDR = sediment delivery ratio, which accounts for sediment deposited in portions of the watercourse or watershed during the process of transport to the reservoir. To estimate the average annual sediment load reaching the reservoir, the USLE estimate has to be multiplied by the SDR.

Some empirical methods to estimate SDR are given below (USACE 1989; Vanoni 1977):

$$\text{SDR} = 0.30 \ (A)^{-0.20} \qquad (3\text{-}81)$$

$$\text{SDR} = 0.76 \ (L)^{-0.23} \qquad (5\text{-}50)$$

where

A = watershed area = 18.8 km^2

L = length of watershed = 7,322 m

Eqs. (3-81) and (5-50) give SDR = 0.17 and 0.10, respectively. In addition to these empirical equations, generalized charts between SDR and watershed have also been developed by various agencies (e.g., USACE 1989). For this case, an average value of about 0.14 is adopted for SDR.

Thus, annual sediment inflow to the reservoir = $5,978 \times 0.14 = 837$ t/yr. Using $E = 0.81$, sediment load expected to deposit in the reservoir = $837 \times 0.81 = 678$ t/yr.

4. Sediment deposition rate using Dendy-Bolton equation (USACE 1989): Mean annual surface runoff = $R = 4,254,889/(18.8 \times 1,000 \times 1,000) = 0.226$ m or 22.6 cm. So, using Eq. (5-43), $T = 686.58 \ [\exp(0.02165 \times 22.6)] \ [1.537 - 0.26 \times \log (18.8)] = 1,350$ t/km^2/yr.

For $A = 18.8$ km^2, rate of reservoir sedimentation = $1,350 \times 18.8 = 25,386$ t/yr.

Assuming a unit weight of deposited sediment of about 1,100 kg/m^3, this gives a total deposition (in 62 yr) of $(25,386/1.1) \times 62 = 1,430,000$ m^3. This is about three times the reservoir capacity and appears unrealistic. Apparently, this method is not applicable to this site.

5. Sediment deposition rate using USBR equation (USBR 1987): Using Eq. (3-84), Q_s = 1,098 $(18.8^{-0.24})$ = 543 m³/km²/yr. For A = 18.8/km², reservoir sedimentation = 543 × 18.8 = 10,209 m³/yr and results in a sediment deposition of 632,958 m in 62 yr, which is more than the reservoir capacity. Apparently, this method is also not applicable to this site.

The results of these computations suggest that the sedimentation rate for the reservoir is about 622 to 755 t/yr.

Wind Wave Heights and Erosion Protection Design

Freeboard in reservoirs includes the sum of wind wave setup and wave runup. Wave setup is the tilting of reservoir water surface caused by wind-induced movement of the surface water toward the shore. Due to this, the reservoir water surface elevation on the leeward or downwind side is higher than the still water elevation and lower on the windward or upwind side. Generally, wind wave setup is larger in shallow reservoirs with rough bottoms. Wave runup computations also include wave setup components, except in situations involving complex shore configurations (USACE 1984). Wave runup is the vertical height up to which a wave will run up a slope. It is a function of the wave height, wave length, slope of the embankment, and permeability and roughness of the surface along which the wave runs up. The wave height used to estimate freeboard and wave forces on structures is the significant wave height, H_s. It is the average of the heights of the one-third highest waves. Significant wave period is the average period of 10 to 15 successive prominent waves. It is approximately the average of all waves whose troughs are below and crests are above the mean water level. Usually,

$$H_{10} = \text{average height of the highest 10\% of all waves} \cong 1.27\, H_s \quad (5\text{-}51)$$

$$H_1 = \text{average height of the highest 1\% of all waves} \cong 1.67\, H_s \quad (5\text{-}52)$$

$$H_s = \sqrt{(2)} \cdot H_{\text{rms}} \quad (5\text{-}53)$$

$$H_{\text{rms}} = \text{root mean square wave height} = \sqrt{[(1/N)\, \Sigma\, H_j^2]} \quad (5\text{-}54)$$

Table 5-8. Reservoir trap efficiencies

C/I	Brune's trap efficiency	Churchill's trap efficiency
0.01	0.45	0.47
0.10	0.86	0.72
1.0	0.98	0.88
10.0	0.98	0.96

Source: USBR (1987).

where N = number of waves in the record and the summation is from $j = 1$ to $j = N$. A dominant parameter that affects wind wave characteristics is the length of water surface along the direction of wind, known as fetch length, F. Fetch is subjectively defined as a region in which the wind speed and direction are reasonably constant. A recommended procedure for estimating fetch length, F, is to construct nine radials from the point of interest at 3-degree intervals and extend these radials until they first intersect the shoreline. The central radial extends up to the farthest point on the opposite shoreline along the dominant wind direction. Then, four radials are drawn on either side of the central radial starting from the same point. The length of each radial is measured and arithmetically averaged. If necessary, angular spacings other than 3 degrees may be used (USACE 1984).

The following are steps for approximating significant wave height (H_s or H_m):

1. Estimate fetch length, F (m).

2. Estimate design over-land wind speed, U_{land} (m/s), at a height of about 10 m above ground and design wind duration, t_d, for the wind at the site.

3. Convert over-land wind speed to over-water wind speed, U_{water} (m/s), using empirical adjustment factors indicated in Table 5-9. If $F < 16$ km, use $U_{water}/U_{land} = 1.2$.

4. Compute wind stress factor (i.e., adjusted wind speed):

$$U_A = 0.71 \, U_{water}^{1.23} \tag{5-55}$$

5. Assume deep water waves and estimate fetch-limited wave parameters:

$$H_m = \text{wave height (m)} = 5.112 \times 10^{-4} \times U_A F^{1/2} \tag{5-56}$$

$$T_m = \text{wave period (s)} = 6.238 \times 10^{-2} (U_A F)^{1/3} \tag{5-57}$$

$$t = \text{wave duration (s)} = 32.15 (F^2/U_A)^{1/3} \tag{5-58}$$

If computed t is less than t_d, the wave height and period are fetch-limited and are determined by Eqs. (5-56) and (5-57), respectively. If t is greater than t_d, the fetch-limited values will not be reached for the given t_d. In this case, set $t = t_d$, compute limiting fetch, F_l, using Eq. (5-58), and use this value in Eqs. (5-56) and (5-57) to estimate wave height and wave period.

Table 5-9. Ratio of over-water to over-land wind speed

Wind speed (m/s)	U_{water}/U_{land}
5	1.45
7.5	1.27
10	1.13
15	1.0
20	0.98
25	0.97

Source: USACE (1984).

6. Check if the values are less than or equal to the fully developed values:

$$H_m \leq 2.482 \times 10^{-2} \times U_A^2 \tag{5-59}$$

$$T_m \leq 0.830 \, U_A \tag{5-60}$$

$$t \leq 7{,}296 \, U_A \tag{5-61}$$

Otherwise, use the fully developed values given by Eqs. (5-59), (5-60), and (5-61).

7. Estimate wave length:

$$L_0 = \text{wave length (m)} = 1.56 \, T_m^2 \tag{5-62}$$

8. Check if waves are deep, transitional, or shallow water waves:

If $d/L_0 > 1/2$, it is a deep water wave; if $1/25 < d/L_0 < 1/2$, it is a transitional wave; and if $d/L_0 < 1/25$, it is a shallow water wave, where d = average depth from bottom of reservoir to still water level along the fetch (m).

9. If waves are in shallow waters, use the following shallow water wave equations to estimate wave parameters:

$$H g/U_A^2 = 0.283 \tanh [0.530 \{g d/U_A^2\}^{3/4}]$$
$$\tanh [0.00565 \, (F g/U_A^2)^{1/2}/\tanh \{0.530 \, (g d/U_A^2)^{3/4}\}] \tag{5-63}$$

$$T g/U_A = 7.54 \tanh [0.833 \{g d/U_A^2\}^{3/8}]$$
$$\tanh [0.0379 \, (F g/U_A^2)^{1/3}/\tanh \{0.833 \, (g d/U_A^2)^{3/8}\}] \tag{5-64}$$

$$g t/U_A = 537 \, (g T/U_A)^{7/3} \tag{5-65}$$

Wave runup computations for different embankment slopes, depths of water above base, and types of bank surface roughness should be made using experimental charts developed by USACE (1984). For preliminary estimates, wave runup on graded riprap laid on a $2H{:}1V$ embankment slope may be taken to be about 1.7 times the significant wave height.

It is commonly assumed that a structure (e.g., rubble mound and revetment) may be subjected to attack by breaking waves if $d_s \leq 1.3 \, H$, where d_s = water depth above the structure toe and H = design wave height. The breaking point is defined as the point where foam first appears on the wave crest, the front face of the wave first becomes vertical, or the wave crest first begins to curl over the face of the wave. Experimental charts to estimate the multiplying factors for H to obtain the breaker wave height are available only for slopes of $10H{:}1V$ or flatter. If wave breaking is suspected, it may be advisable to use a factor of safety by judgment based on available experimental charts (USACE 1984).

The size of rock, riprap, or other armor material required to protect the reservoir embankment or lake or sea shore against wind wave forces may be estimated by

$$W = \gamma_s \, H^3/[K_D \, (G-1)^3 \cot \theta]] \tag{5-66}$$

where

W = weight of armor unit than which 50% of the armor material is finer by weight (kg)

γ_s = unit weight of armor material (kg/m^3)

G = specific gravity of armor material

θ = angle of bank slope with horizontal

K_D = stability coefficient, which varies with the shape of the armor units (e.g., rocks and blocks), roughness of the armor unit surface, sharpness of edges, and degree of interlocking of adjacent armor units

For nonbreaking waves, a K_D of 1.9 may be used for smooth rounded quarry stones and 2.3 for rough angular quarry stones, with appropriate factors of safety. The corresponding values for breaking waves may be 1.1 and 1.3, respectively. Note that angular stones may tend to become somewhat rounded after aging. For graded riprap, W(max) = maximum weight of graded rock = $4W$, and W(min) = minimum weight of graded rock = $0.125 W$. Assuming individual armor units to be approximately spherical in shape, the diameter of the armor unit may be estimated by

$$d \text{ (m)} = 1.24 \, (W/\gamma_s)^{1/3} \qquad (5\text{-}67)$$

If armor units are placed in layers, the coarsest units must be placed in the cover (top) layer. The rock sizes in different layers may be designed so that d_{15} of cover layer $\leq 5 \times$ (d_{85} of underlying layer).

It is advisable to use a filter blanket or bedding layer to protect rubble mound structures, riprap, or other armor units against differential wave pressures on the underlying soils and seepage or groundwater flow, which may remove the underlying soils through voids in the armor units. The bedding layer may consist of quarry spalls, crushed stone, gravel, or gabions. A geotextile filter may be used in place of the filter blanket or with a thinner bedding layer. A geotextile, coarse gravel, or crushed stone filter may be placed directly over a sand layer, if one exists. However, silty and clayey soils and fine sands in the embankment or seashore should be covered with coarser sand before placement of the bedding layer. Typically, quarry spalls may range from 0.45 to 23 kg in weight if placed over a geotextile, coarse gravel, or crushed stone filter. The thickness of the bedding layer or filter blanket should not be less than 0.3 m. Gradation of the filter layer may be designed using the criteria applicable to filter blankets for erosion protection on riverbanks (see the section entitled "Erosion Protection").

Example 5-10: Estimate wave height, wave runup, freeboard, and riprap size for a reservoir with average design water depth of 10 m, maximum expected wind speed of 22.97 m/s with a duration of 1.0 h, and fetch length of 3,200 m. The upstream slope of the dam embankment is $2H{:}1V$.

Solution: Assume over-water wind speed = $22.97 \times 0.975 = 22.4$ m/s. Using Eq. (5-55), estimate wind stress factor = $U_A = 0.71 \, (22.4)^{1.23} = 32.51$ m/s.

Assume deep water waves and compute wave parameters using Eqs. (5-56) to (5-58):

H_m = wave height (m) = $5.112 \times 10^{-4} \times 32.51 \times 3{,}200^{1/2} = 0.94$ m.

T_m = wave period (s) = $6.238 \times 10^{-2} \, (32.51 \times 3{,}200)^{1/3} = 2.93$ s.

t = wave duration (s) = $32.15 \, (3{,}200^2/32.51)^{1/3} = 2{,}187.4$ s = 0.607 h. The waves would become fetch-limited after 0.607 h. Since $0.607 < 1.0$ h (t_d), the fetch-limited equations are valid.

From Eq. (5-62), L_0 = wave length (m) = $1.56 \times (2.93)^2$ = 13.39 m. So, d/L_0 = 10/13.39 = 0.747 > 0.5. This suggests that these are deep water waves.

Check with the limiting values estimated by Eqs. (5-59) to (5-61) that the equations used are valid. Thus, $H_m = 2.482 \times 10^{-2} \times (32.51)^2 = 26.23$ m; $T_m = 0.830 \times 32.51 = 26.98$ s; $t = 7,296 \times 32.51 = 237,193$ s.

The previously computed values are less than these limiting values. So, the estimated values are valid.

Use wave runup = $1.7 \times 0.94 = 1.6$ m = freeboard above the maximum flood elevation in the reservoir.

Assume $K_D = 1.9$ for smooth rounded quarry stone with unit weight of 2,400 kg/m³, $G = 2.4$, and cot $\theta = 2.0$. Then, using Eq. (5-66), $W_{50} = 2,400 \times 0.94^3/[1.9 \times (2.4 - 1)^3 \times 2.0] = 191$ kg; $W_{max} = 4 \times 191 = 764$ kg, and $W_{min} = 0.125 \times 191 = 23.9$ kg.

Using Eq. (5-67); d_{50} (m) = $1.24 (191/2,400)^{1/3} = 0.54$ m; $d_{max} = 0.85$ m; and $d_{min} = 0.27$ m.

Spillways

Spillways are structures designed to pass excess flood flows that cannot be stored in a reservoir. Usually, designs of spillways on major dams are finalized based on hydraulic model experiments. The water resources engineer is often required to perform preliminary hydraulic design computations for alternative types and sizes of spillways and to identify the preferred type and size. Commonly used types of spillways are described below (USBR 1987; ASCE 1989):

Earth-Cut Spillway with Grass or Riprap Protection

These spillways are excavated in a portion of the dam, in the abutment, or in a suitable location on the shoreline where spillway outflows may be discharged to an existing gulch. They are designed as broad-crested weirs with a relatively flat downstream slope ranging from $3H:1V$ to $10H:1V$ with riprap or vegetative protection. Erosion protection for the crest and downstream slope may be designed as described in the previous section entitled "Erosion Protection." The crest length and head above crest for a design discharge, Q, may be estimated by

$$Q = 1.48\, L\, H^{1.5} \tag{5-68}$$

The discharge coefficient, 1.48, may be increased up to 1.70 for very smooth crests and reduced up to 1.46 for rough crests. For a given design inflow hydrograph and reservoir storage-elevation data, flood routing computations may be made using computer programs such as HEC-1 and HEC-HMS (USACE 1991a, 2002) with spillway rating for various values of L given by Eq. (5-68) to obtain an acceptable combination of L and maximum H. Water depth and velocity at the toe of the downstream slope may be estimated assuming uniform flow is established over the length of the slope. Dimensions of the stilling basin at the toe of the slope may be estimated as for a drop with sloping apron (see the previous section entitled "Drop Structures").

Fuse-Plug Spillway

A fuse-plug spillway consists of a section of an earth dam, generally lower than the crest of the main dam. When overtopped, this section is designed to fail due to rapid erosion so that the remaining portion of the dam is not damaged. Because of difficulties in designing a section, which would fail at the desired reservoir water surface elevation, fuse-plug spillways are not very common. When required, they are located on an abutment, on the reservoir rim, or at a low spot on the embankment crest. The selected section is designed as a pilot channel, which is overtopped when the reservoir reaches a predetermined elevation. Rapid erosion is ensured by placing erodible materials in the pilot channel. The dimensions of the ultimate fuse plug are controlled by providing a sill or nonerodible foundation at the bottom and nonerodible sides for the pilot channel. Usually, fuse plugs are designed not to operate for floods with recurrence intervals of less than 100 yr. They are designed as dams stable for all reservoir conditions except the design flood elevation that should cause it to be overtopped and breached.

Free Overfall (Straight Drop) Spillway

Straight drop spillways are suitable for drops of less than 6 m. In these spillways, water drops freely from the spillway crest on a horizontal apron. The discharge over the spillway may be estimated using sharp- or broad-crested weir equations depending on the length of the crest parallel to flow. To direct the flow away from the vertical face of the spillway during low discharges, an overhanging lip may be provided at the edge of the crest. If sufficient tailwater depth is available, a hydraulic jump may form on the horizontal apron. For small drops (less than 6 m or so), the velocity, V_1, and water depth, y_1, at the toe for a unit discharge, q, per unit width of the crest may be approximated by

$$V_1 = C_d \sqrt{(2\,g\,H)} \tag{5-69a}$$

$$y_1 = q/V_1 \tag{5-69b}$$

where

$C_d = 0.8$ to 0.9

$H =$ head above the crest

Hydraulic jump analysis may be performed using the methods described in "Drop Structures."

Ogee (Overflow) Spillway

These are spillways where the crest and downstream face conforms to the lower nappe of the sheet of water falling from a sharp-crested weir. The discharge equation for an ogee spillway is given in the section of Chapter 3 entitled "Ogee Crest." The effects of piers and abutments on the spillway crest may be accounted for by modifying the crest length as

$$L_e = L - 2\,(n\,K_p + K_a)\,H \tag{5-70}$$

where

$L_e =$ effective crest length

$L =$ total unobstructed crest length

$n =$ number of piers

K_p = pier contraction coefficient

K_a = abutment contraction coefficient

H = head over crest

For square-nosed piers, $K_p = 0.02$; for round-nosed piers, $K_p = 0.01$; and for pointed-nose piers, $K_p = 0.0$. For square abutments, $K_a = 0.20$; and for rounded abutments, $K_a = 0.0$ to 0.10. The discharge coefficient is affected by the height of spillway above upstream bed, slope of upstream face, heads other than design head, and degree of submergence. Variation with height of upstream vertical face is shown in Table 3-12(a), with heads other than design head in Table 3-12(b), and with degree of submergence in Table 5-10 (USBR 1987).

The equation to determine the profile of the downstream face of an ogee spillway is given in "Ogee Crest" in Chapter 3:

$$X^n = K H_d^{n-1} Y \qquad (3\text{-}49)$$

The slope of the downstream face may be determined by judgment or hydraulic model tests. Commonly used slopes vary from $0.6H{:}1V$ to $0.8H{:}1V$. The profile may be determined using the above-mentioned equation up to a point when its slope becomes equal to the selected slope of the downstream face (i.e., $z{:}1$). The coordinates of this point may be estimated by setting

$$dY/dX = n X^{n-1}/[K H_d^{n-1}] = 1/z \qquad (5\text{-}71)$$

The spillway profile upstream of the origin of coordinates (apex of the crest) consists of three connected circular arcs. The radii and coordinates of centers of these arcs developed by the US Army Engineer Waterways Experiment Station are shown in Table 5-11 (USAEWES 1977). The radii are denoted by R and the origin is at the highest point on the crest.

In Table 5-11, X is positive to the right and Y is positive downward. Using the information in Table 5-11, the coordinates of various points on the spillway profile upstream of the crest up to the upstream vertical face are shown in Table 5-12.

Table 5-10. Variation of discharge coefficient of ogee crest with submergence (H_s = crest elevation − tailwater elevation and C_s = modified discharge coefficient)

H_s/H_d	C_s/C
1	0
0.9	0.65
0.8	0.86
0.7	0.93
0.6	0.96
0.5	0.98
0.4	0.99
0.3	0.995
0.2	1

Source: USBR (1987).

Table 5-11. Radii and coordinates of centers of arcs forming upstream spillway profile

	Coordinates of center		
	X/H_d	Y/H_d	R/H_d
	0.0	0.50	0.50
	−0.105	0.219	0.20
	−0.2418	0.136	0.04

Source: USAEWES (1977).

Example 5-11: Develop the profile of an ogee spillway with $H_d = 5$ m, 80-m-high upstream vertical face, and downstream face slope of $0.7H{:}1V$. Use $K = 2.0$ and $n = 1.85$ (see Figure 3-2 in Chapter 3).

Solution: For the downstream face, coordinates of the point where the slope $0.7H{:}1V$ begins are given by Eq. (5-71): $1.85 \, X^{0.85}/(2.0 \times 5^{0.85}) = 1/0.7$. This gives $X = (6.06575)^{1.1765} = 8.338$ m.

From Eq. (3-49), $Y = 8.338^{1.85}/(2 \times 5^{0.85}) = 6.438$ m.

The coordinates of selected points on the downstream face above the point ($X = 8.338$, $Y = 6.438$) are estimated by Eq. (3-49) and are shown in Table 5-13.

Using Table 5-12, coordinates of points on the upstream profile are shown in Table 5-14.

The profiles estimated by the USBR (1987) and the USAEWES (1977) approaches are slightly different. The design must be finalized based on hydraulic models.

Table 5-12. Coordinates of points on upstream spillway profile

X/H_d	Y/H_d
0	0
−0.05	0.0025
−0.1	0.0101
−0.15	0.023
−0.175	0.0316
−0.2	0.043
−0.22	0.0553
−0.24	0.0714
−0.26	0.0926
−0.276	0.1153
−0.278	0.119
−0.28	0.1241
−0.2818	0.136

Source: USAEWES (1977).

Table 5-13. Computed coordinates of points on downstream spillway face

X	Y
1	0.127
2	0.459
3	0.972
4	1.654
5	2.5
6	3.503
7	4.659
8.338[a]	6.439
20	23.099 = 6.439 + [(20 − 8.338)/0.70]

[a]The spillway profile has a slope of 0.7H:1V below this point.

Table 5-14. Computed coordinates of points on upstream spillway face

Y	Y/H_d	X/H_d	X
0	0	−0	−0
0.0505	0.0101	−0.10	−0.50
0.158	0.0316	−0.175	−0.875
0.2765	0.0553	−0.22	−1.10
0.463	0.0926	−0.26	−1.30
0.595	0.1190	−0.278	−1.390
0.680	0.1360	−0.2818	−1.409
1.0	—	—	−1.409
2.0	—	—	−1.409
80	—	—	−1.409

Side Channel Spillway

In this spillway, the crest is nearly parallel to the downstream channel (usually a narrow trough) into which water is discharged. After flowing parallel to the spillway crest, the downstream channel turns at an angle to join the main discharge channel. The spillway crest may be designed the same way as an ogee crest or as a sharp- or broad-crested weir. If the trough is designed to have supercritical flow, velocities in the trough will be high and water depths will be low, resulting in high drop from the spillway crest to the water surface in the trough. This will tend to sweep the flow to the far side of the trough and result in violent turbulence and vibrations due to intermixing of supercritical longitudinal flow and high-velocity transverse flow from the spillway. On the other hand, if the trough is designed with too flat a slope, then it may not be able to carry the overflow from the spillway crest without submerging the crest, particularly during high floods. If the crest submergence exceeds 3/4 H or so, the spillway may be choked. H is the total head above crest. Under these conditions, the discharge capacity of the trough may control the outflow, rather than the crest, and the spillway may become hydraulically inefficient. Nevertheless, where long overflow crests are needed to limit water surface rise in the reservoir and the abutments are steep, or where the control must be connected to a narrow discharge channel or tunnel, a side channel may be a desirable alternative to be evaluated. For satisfactory hydraulic performance of a side channel spillway, it is desirable to maintain subcritical flow in the trough. This may be achieved by establishing a control section at the downstream end of the trough.

Hydraulic analysis of side channel spillways is based on conservation of momentum along the axis of the trough (USBR 1987):

$$\Delta y = -[Q_1(v_1 + v_2)/\{g(Q_1 + Q_2)\}][(v_2 - v_1) + \{v_2(Q_2 - Q_1)/Q_1\}] \quad (5\text{-}72)$$

where

Δy = change in water surface elevation from section 1 (downstream) to section 2 (upstream)

Q_1, v_1 = flow and velocity at section 1

Q_2, v_2 = flow and velocity at section 2

Because of finite-difference approximation, the smaller the length between sections 1 and 2, the better the accuracy of Eq. (5-72). Computational steps to estimate design dimensions are provided below:

1. Assume crest submergence of about $2/3H$ and estimate discharge coefficient for spillway crest. Develop spillway rating curve and perform reservoir routing computations for design inflow hydrograph to estimate preliminary spillway crest length and elevation.

2. Assign preliminary dimensions to the trough, i.e., bottom width, side slopes, and bottom slope. The trough is nearly parallel to the spillway crest and extends along its entire length.

3. Assume preliminary dimensions of the trough at the control section, i.e., bottom elevation, bottom width, and side slopes. A control section may be provided by contracting the trapezoidal section to a rectangular one or by a raised sill with a trapezoidal or rectangular section. A transition length is required between the end of the trough at the trapezoidal section and the control section.

4. Estimate water depth, velocity, and velocity head at the control section using critical flow equations (see "Critical Flow" in Chapter 3).

5. Estimate water depth and velocity at the downstream end of the trapezoidal section using energy equations between this section and the control section.

6. Using Eq. (5-72), estimate water depth and water surface elevations at various sections of the trough by trial and error.

7. Estimate crest submergence, if any, and determine if the assumed discharge coefficient is valid. If not, make appropriate adjustments to trough dimensions, bottom elevations, and slope, and repeat the computations.

Example 5-12: Inflow design flood routing computations for a reservoir indicated a desirable spillway crest elevation and length of 2,404.3 m and 35 m, respectively, and a peak outflow of 70 m³/s. Allowing for some submergence, the spillway discharge coefficient for flood routing was taken to be 1.98. Estimate design dimensions for a side channel spillway. Use a trapezoidal section for the trough with bottom width, side slopes, and bottom slope of 4.5 m, $0.5H:1V$, and 0.010, respectively.

Solution: Assume that a rectangular control section with a width of 4.5 m is located at a horizontal transitional length of 8 m downstream from the downstream end of the spillway or trapezoidal section of the trough.

By discharge equation for the crest, $70 = 1.98 \times 35 \times H^{1.5}$; so, $H = 1.0$ m.

Unit discharge at control section = $q = 70/4.5 = 15.56$ m^3/s. So, critical depth = $y_c = [15.56^2/9.81]^{1/3} = 2.91$ m; $v_c = q/y_c = 15.56/2.91 = 5.347$ m/s; and velocity head = $h_{vc} = 5.347^2/(2 \times 9.81) = 1.457$ m.

By trial and error, estimate water depth and velocity at the downstream end of the trapezoidal section using the energy equation $y_c + v_c^2/2g + 0.2(h_{vc} - h_{v1}) = y_1 + v_1^2/2g$.

Section 1 is located 8 m upstream of the control section, i.e., at the downstream end of the trapezoidal trough.

By trial and error, $y_1 = 4.295$ m. So, $A_1 = 4.5 \times 4.295 + 0.5 \times (4.295^2) = 28.551$ m^2, $v_1 = 70/28.551 = 2.4517$ m/s, and $h_{v1} = 2.4517^2/(2 \times 9.81) = 0.306$ m. So, right-hand side = $4.295 + 0.306 = 4.601$ m.

Left-hand side = $2.91 + 1.457 + 0.2(1.457 - 0.306) = 4.597$ m.

Discharge per unit width of spillway crest = $70/35 = 2$ m^2/s.

Estimate water surface elevations by trial and error using Eq. (5-72). The computations made on an Excel spreadsheet are shown in Table 5-15.

Values for station 0 pertain to computed values for the downstream end of the trapezoidal trough. The value in column (8) is spillway outflow over the crest length from the upstream edge up to the end of the station indicated in column (1) computed at 2 m^2/s per meter crest length. Values in column (4) are trial values used to compute values in other columns. These values are changed again and again until they closely match the computed values in column (18). Subscript 1 refers to the downstream and 2 to the upstream one of two adjacent stations.

The estimated maximum water surface elevation 2 m downstream from the upstream edge of the trough is 2,404.875 m. Thus, the submergence above the crest is $2,404.875 - 2,404.3 = 0.575$ m. This is less than $2/3 H$. So reduction in discharge coefficient due to submergence is expected to be insignificant, and the value of 1.98 used in flood routing computations is valid.

Chute or Trough Spillways

These spillways are similar to a drop structure with inclined downstream slope. The chute may be an open channel or may have baffles installed for energy dissipation and may be designed to have subcritical or supercritical slope for the design flow. It is suitable for dams where sufficient width (e.g., a saddle) is available to locate a relatively wide, broad crest. The flow section may be contracted downstream to a chute or trough of width and slope adequate to safely pass the design flood. The spillway crest is sized using the discharge equation for a broad-crested weir (Eq. (3-45)). The main distinguishing feature from a side channel spillway is that the crest is located normal or nearly normal to the axis of the chute. Usually, a chute spillway consists of an approach channel, a control section (saddle or broad crest), a relatively steep discharge channel (trough), a stilling basin, and an outlet channel (USBR 1987).

Table 5-15. Hydraulic computations for side channel spillway

(1)	(2)	(3)	(4)	(5)	(6)	(7)	(8)	(9)	(10)
Station	D_x	Bed El.	dy	W.S. El.	y	A	Q	v	$Q_1 + Q_2$
0	0	2,400	0	2,404.295	4.295	28.551	70	2.45175	—
7	7	2,400.07	0.2344	2,404.5294	4.4594	30.01042	56	1.866018	126
14	7	2,400.14	0.1587	2,404.6881	4.5481	30.80906	42	1.363236	98
21	7	2,400.21	0.1057	2,404.7948	4.5838	31.13271	28	0.899376	70
28	7	2,400.28	0.0618	2,404.8566	4.5756	31.05826	14	0.450766	42
33	5	2,400.33	0.019	2,404.8756	4.5446	30.77739	4	0.129966	18

(11)	(12)	(13)	(14)	(15)	(16)	(17)	(18)
$Q_1/(g \times \text{col. (10)})$	$v_1 + v_2$	$v_1 - v_2$	$Q_1 - Q_2$	Col. (14)/Q_1	$v_2 \times \text{col. (14)}/Q_1$	Col. (13) + col. (16)	dy
—	—	—	—	—	—	—	—
0.056632	4.317768	0.585732	14	0.2	0.373204	0.958935	0.234481
0.05825	3.229254	0.502783	14	0.25	0.340809	0.843592	0.158682
0.061162	2.262611	0.46386	14	0.333333	0.299792	0.763652	0.105679
0.067958	1.350141	0.44861	14	0.5	0.225383	0.673993	0.061841
0.079284	0.580731	0.3208	10	0.714286	0.092833	0.413633	0.019045

Conduit or Tunnel Spillways

These are vertical or inclined shafts or horizontal tunnels designed to flow partly full throughout their length with the exception of drop inlets. For drop inlets, the tunnel may be designed to flow full for a short length near the drop and partly full thereafter.

Morning Glory or Glory Hole Spillway

These spillways consist of a funnel-shaped bell mouth inlet in the form of a weir, a vertical shaft, and a closed discharge conduit carrying flows to the downstream channel. Usually, the crest profile and transition to the vertical shaft are designed to conform to the shape of the lower nappe of a jet flowing over a sharp-crested circular weir. The profile of the crest may be estimated using experimentally developed tables for the coordinates of points on the profile (USBR 1987). For lower heads, the discharge is controlled by weir flow over the crest. For intermediate heads, orifice flow in the transition through the vertical shaft controls. For high heads, pipe flow through the vertical shaft and conduit governs.

The discharge over the crest of a morning glory spillway with a nappe-shaped profile is given by

$$Q = C(2\pi R)H^{1.5} \qquad (5\text{-}73)$$

where

Q = outflow

R = crest radius

H = head over crest

C = discharge coefficient given in Table 5-16 (USBR 1987)

For very low heads, the discharge coefficient may be reduced to about 87% of the value for the design head.

Transitional flow through the vertical shaft is given by

$$Q_a = \pi R_a^2 \sqrt{[2g(H_a - 0.1 H_a)]} \qquad (5\text{-}74)$$

Table 5-16. Discharge coefficients for morning glory spillway crest

	Discharge coefficient, C		
H/R	$P/R = 2.0$	$P/R = 0.30$	$P/R = 0.15$
0.2	2.15	2.20	2.21
0.4	1.97	2.04	2.04
0.8	1.36	1.44	1.47
1.2	0.94	0.99	1.02
1.6	0.71	0.73	0.76
2.0	0.55	0.60	0.61

P = height of crest above upstream bed.
Source: USBR (1987).

where

R_a = radius of jet in transition

0.1 H_a = jet contraction and other losses

H_a = distance between reservoir water surface and elevation corresponding to jet radius R_a

The jet radius decreases as it falls down. Thus, the jet profile in the transition may be estimated by

$$R_a = 0.275 \, Q_a^{1/2}/(H_a^{1/4}) \tag{5-75}$$

The pipe flow begins at an elevation in the shaft where R_a becomes equal to the radius of the conduit. This location is called the throat. Usually, the conduit is designed to flow 75% full at the design discharge to minimize potential for fluctuations from pressure to open-channel flows in the conduit. From tabulated values of dimensions of circular conduits, for 75% full flows (i.e., for 75% of conduit area), water depth = y = 0.702 D and hydraulic radius = r_h = 0.2964 D, where D = conduit diameter (USBR 1987). Flow through a conduit of length L is governed by the energy equation

$$\text{water surface elevation at throat} + \text{velocity head at throat} = \\ V^2 \, n^2 \, L/(r_h^{4/3}) + V^2/2g + y + \text{invert elevation at exit} \tag{5-76}$$

The rating curve for a morning glory spillway includes three components:

1. A curve based on weir flow over the crest applicable for low heads
2. A curve based on orifice flow in the transition zone through the shaft applicable for intermediate heads
3. A curve based on pipe flow applicable for high heads

The intersections of these curves are joined by approximate smooth transitions curves. Hydraulic computations for a morning glory spillway involve trial and error.

Example 5-13: Develop preliminary dimensions for a morning glory spillway that will operate under a head of 3.05 m but will limit the outflow to 57 m³/s. The crest elevation is 1,000 m. The length of the downstream conduit is 82 m and its invert elevation at the exit is 988.60 m. Manning's roughness coefficient for the conduit is 0.014. Assume $P/R > 2.0$.

Solution: The crest radius is determined by trial and error. Assume a trial R = 2.13 m. So, H/R = 3.05/2.13 = 1.43. From Table 5-16, C = 0.80.

Q = 0.80 (2 π × 2.13) (3.05$^{1.5}$) = 57 m³/s. So, use R = 2.13 m.

To find the throat diameter, use Eq. (5-75): R_a = 0.275 × (57$^{0.5}$)/($H_a^{0.25}$) = 2.076/($H_a^{0.25}$).

Computations for the jet profile through the shaft are shown in Table 5-17.

The conduit diameter is also estimated by trial and error. Assume R_a = 1.37 m. From Table 5-17 of jet profile computations, this corresponds to El. 997.777 m in the shaft, which becomes the throat elevation. Adopt a conduit radius of 1.37 m. Then, H_a = difference between reservoir water surface elevation and throat elevation = 1,000 + 3.05 − 997.777 = 5.273 m. For the conduit flowing 75% full, $A = 0.75 \pi (1.37^2) = 4.422$ m^2; $V = 57/4.422 = 12.89$ m/s; $y = 0.702 \times 2 \times 1.37 = 1.9235$ m; and $r_h = 0.2964 \times 2 \times 1.37 = 0.8121$ m.

Velocity head at throat $\cong 5.273 - 0.1 \times 5.273 = 4.746$ m.

For conduit flow, $997.777 + 4.746 = 12.89^2 \times (0.014^2) \times 82/(0.8121^{4/3}) + 12.89^2/(2 \times 9.81) + 1.9235 +$ conduit invert elevation at exit.

Invert elevation of conduit at exit = $1,002.523 - (3.5245 + 8.47 + 1.9235) = 988.605$ m. This is nearly equal to the given exit elevation.

Computations for the rating curve for weir and orifice flows are shown in Table 5-18.

Values of the discharge coefficient C are taken from Table 5-16. The coefficients should be reduced by about 10% for other than design heads (USBR 1987). Q in column (5) is computed using Eq. (5-73) with $R = 2.13$ m. In column (6), H_a = column (1) − 997.777. Q_a in column (7) is computed using Eq. (5-74) with $R_a = 1.37$ m.

A plot of the rating curve for weir and orifice flows is shown in Figure 5-4. It may be seen that the weir and orifice flow curves intersect when reservoir water surface elevation is 1,003.05 m.

Siphon Spillway

This spillway operates on the principle of a siphon. It consists of an inlet, a crest, a crown, a step, and an air inlet to break the siphon when desired. Siphon spillways discharge relatively large volumes of water with a relatively narrow crest width. These spillways are suitable for situations where available space is limited and the design discharge is small. The discharge of a siphon spillway is given by the following orifice equation (USBR 1978):

$$Q = C A_T \sqrt{(2 g H)} \quad \text{or} \quad q = C D \sqrt{(2 g H)} \tag{5-77}$$

Table 5-17. Computations of jet profile trough shaft of morning glory spillway

Water surface elevation in shaft	H_a	$R_a = 2.076/H_a^{0.25}$
1,000	3.05	1.571
999.5	3.55	1.512
999	4.05	1.463
998.5	4.55	1.421
998	5.05	1.385
997.5	5.55	1.352
997	6.05	1.324
996.5	6.55	1.298
996	7.05	1.274
997.6	5.45	1.359
997.7	5.35	1.365
997.8	5.25	1.371
997.9	5.15	1.378

Table 5-18. Rating curve for weir and orifice flows

(1)	(2)	(3)	(4)	(5)	(6)	(7)
		Weir flow			Orifice flow	
Reservoir water surface elevation	H	H/R	C	Q	H_a	Q_a
1,000	0	0	0	0	2.223	—
1,000.5	0.5	0.2347	2.1	9.9365	2.723	—
1,001	1	0.4695	1.88	25.1604	3.223	—
1,001.5	1.5	0.7042	1.55	38.1091	3.723	—
1,002	2	0.9390	1.16	43.910	4.223	50.91
1,002.5	2.5	1.1737	0.95	50.2567	4.723	53.84
1,003	3	1.4084	0.81	56.3283	5.223	56.62
1,003.05	3.05	1.4319	0.8	57.0295	5.273	56.89
1,003.5	—	—	—	—	5.723	59.27
1,004	—	—	—	—	6.223	61.80
1,004.5	—	—	—	—	6.723	64.24
1,005	—	—	—	—	7.223	66.58
1,005.5	—	—	—	—	7.723	68.85
1,006	—	—	—	—	8.223	71.04

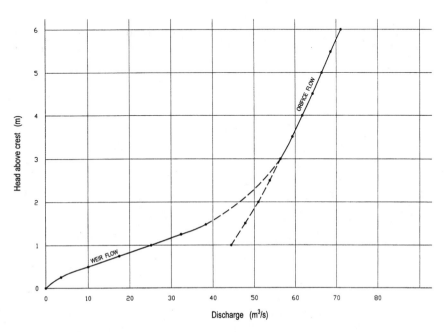

Figure 5-4. Morning glory spillway discharge curve

where

Q = design discharge

q = discharge per unit width of crest

D = throat depth

A_T = throat area

H = available head between reservoir water surface elevation and tailwater elevation

C = orifice discharge coefficient varying from 0.57 to 0.90 depending on depth of water and radius of curvature at the throat (Davis and Sorensen 1970; Zipparro and Hansen 1993)

The maximum discharge is limited by cavitation potential at the throat due to subatmospheric pressures. Assuming free vortex flow, the limiting discharge is estimated by

$$q_{max} \leq R_c \sqrt{\{0.7 \ (2 \ g \ h)\}} \ln (R_s/R_c) \qquad (5\text{-}78)$$

where

R_c = radius of curvature at crest

R_s = radius of curvature at summit

h = atmospheric pressure in terms of height of water under design conditions at the site

0.7 = a coefficient that provides for limitation on permissible subatmospheric pressure at the throat

In a siphon spillway, when reservoir water surface elevation rises above the spillway crest, water spills over the crest (Figure 5-5). The siphonic action starts and the spillway gets

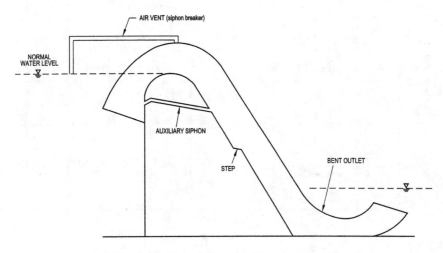

Figure 5-5. Siphon spillway schematic

primed (i.e., starts flowing full) only after air trapped in the siphon is ejected. The following are some devices that initiate or facilitate priming:

1. A step in the inner wall of the downstream leg of the siphon to deflect water so as to hit the outer wall, fill the outlet pipe, and seal the lower leg preventing entry of air from the downstream side
2. An auxiliary (smaller) siphon provided nearly parallel to and underneath the main siphon to release a small quantity of water to seal the lower leg
3. A nearly horizontal or concave outlet to create a water depth at the exit sufficient to seal the lower leg

To break the siphonic action at the required elevation, an air vent, consisting of a bent (downward) pipe, is connected to the siphon. The air vent is located above the siphon hood (cover) with its inlet slightly below the reservoir water surface elevation and outlet at the summit. As soon as water rises above the design water surface elevation in the reservoir, the air vent is sealed and priming can occur. When water level drops below the spillway crest, the air vent is exposed and air enters the siphon. This breaks the siphonic action. The lip (entrance of the siphon tube) is submerged below the design reservoir water surface elevation. The siphon outlet is a diverging duct, usually but not necessarily submerged below the tailwater elevation.

Preliminary dimensions of a siphon spillway include the following:

- A crest slightly above the design reservoir water surface elevation
- A rectangular inlet with area, A (inlet) $\cong$ 2 to 3 times A_T
- Throat width, $B_T \cong$ 2 to 3 times D
- Radius of curvature at crest = $R_c \cong 1.5\ D$ and radius of curvature at summit = $R_s \cong 2.5\ D$
- Area of vent pipe $\cong A_T/24$

The final design dimensions should be determined by hydraulic model tests.

Example 5-14: Determine preliminary dimensions of a siphon spillway with a design discharge of 85 m³/s and head drop of 4.9 m between design reservoir water surface elevation and tailwater elevation. Use a battery of five siphon spillways. Atmospheric pressure under design conditions is 9.14 m of water.

Solution: Q (per siphon spillway) = 85/5 = 17 m³/s; H = 4.9 m; and h = 9.14 m. For preliminary design, assume $B_T = 2D$ and $C = 0.6$.

From Eq. (5-77), $Q = 17 = 0.6 \times 2D \times D \sqrt{(2 \times 9.81 \times 4.9)}$. So, D = 1.2 m; B_T = 2 × 1.2 = 2.4 m; and A_T = 2.88 m².

$q = Q/B_T = 17/2.4 = 7.08$ m²/s.

$R_c = 1.5 \times 1.2 = 1.8$ m, and $R_s = 2.5 \times 1.2 = 3.0$ m.

Using Eq. (5-78), $q_{max} = 1.8 \sqrt{\{0.7\ (2 \times 9.81 \times 9.14)\}} \ln (3.0/1.8) = 10.3$ m²/s > 7.08 m²/s. So, D = 1.2 m is acceptable.

A (inlet) = 2.5 A_T = 2.5 × 2.88 = 7.2 m².

Provide the same area at the exit of the downstream diverging duct as at the inlet.

A (vent pipe) $= A_T/24 = 2.88/24 = 0.12$ m^2 or air vent diameter $= 0.39$ m.

Provide a step in the inner wall of the downstream leg of the siphon as a priming device.

Stepped Spillway

Stepped spillways are designed to have an ogee crest with the downstream face consisting of a number of vertical steps to maximize energy dissipation. For convenience of construction, stepped spillways are specially suitable for roller-compacted-concrete dams. Flow over stepped spillways can be divided into nappe flow and skimming flow regimes (Chanson 1994). Nappe flow is characterized by a succession of free-fall jets impinging on the next step followed by a partially or fully developed hydraulic jump. Nappe flow regime occurs on relatively flat spillways with large steps and small discharges. In skimming flow regime, water flows as a coherent stream skimming over the steps. On short chutes with flatter slopes, nappe flow is reported to dissipate more energy than skimming flow. On the other hand, for large dams, skimming flow can dissipate more energy than nappe flow. For preliminary designs, energy losses for ungated stepped spillways can be estimated by the following (Chanson 1994):

1. Nappe Flow:

$$\Delta H/H_{max} = 1 - \{[0.54\,(d_c/h)^{0.275} + 1.715\,(d_c/h)^{-0.55}]/[1.5 + H_{dam}/d_c]\} \quad (5\text{-}79)$$

2. Skimming Flow:

$$\Delta H/H_{max} = 1 - \{[(f/8\sin\theta)^{1/3}\cos\theta + 0.5\,(f/8\sin\theta)^{-2/3}]/[2/3 + H_{dam}/d_c]\} \quad (5\text{-}80)$$

where

H_{dam} = height of spillway crest above toe
d_c = critical depth = $(q^2/g)^{1/3}$
q = spillway discharge per unit length of crest
$H_{max} = H_{dam} + 1.5\,d_c$
h = height of step
ΔH = loss of head (i.e., difference between H_{max} and residual head at the bottom of spillway)
f = friction factor for flow over steps, varying from 0.5 to 4, with a mean value of 1.3
θ = angle of spillway slope to horizontal

It is reported that skimming flow occurs when $d_c/h > 0.8$ (approximately), and nappe flow occurs for $d_c/h < 0.8$ (approximately) (Chamani and Rajaratnam 1994). For a typical stepped spillway, approximate energy loss in nappe flow regime may be estimated by

$$\Delta E/E_0 = \Delta H/H_{max} = 1 - \{(1-\alpha)^N [1 + 1.5\,(d_c/h)] + \Sigma\,(1-\alpha)^i\}/\{N + 1.5\,(d_c/h)\} \quad (5\text{-}81)$$

Table 5-19. Toe velocities for stepped and unstepped spillways

Spillway discharge (m²/s)	Toe velocity (m/s)	
	Stepped spillway	Unstepped spillway
1.9	4.6	≥16.8
4.6	7.6	≥20.7
7.4	9.1	≥22.6
9.3	11.3	≥22.9
13.9	12.2	≥24.4

Source: Campbell and Johnson (1984); Sorensen (1985).

where

N = number of steps

E_0 = total energy

ΔE = loss of energy for stepped spillway (i.e., difference between E_0 and residual energy at toe of stepped spillway)

Σ is from $i = 1$ to $i = N - 1$

$\alpha = a - b \log (d_c/h)$; $a = 0.30 - 0.35 \, (h/l)$, and $b = 0.54 + 0.27 \, (h/l)$, where l = length of step

Approximate toe velocities with a typical stepped and unstepped spillway are indicated in Table 5-19 (Sorensen 1985; Campbell and Johnson 1984).

Preliminary dimensions for stepped spillways for the Upper Stillwater Dam in Utah (Young 1982) and Monksville Dam in New Jersey (Sorensen 1985) are indicated in Table 5-20. The dimensions may provide guidance to estimate preliminary dimensions for similar stepped spillways. The design dimensions must be finalized by model experiments for specific site conditions.

Example 5-15: Estimate head loss in an ungated stepped spillway under nappe flow conditions with $h = 0.61$ m, $l = 0.48$ m, $d_c = 0.18$ m, and $H_{dam} = 27.4$ m.

Solution: Here, $h/l = 0.61/0.48 = 1.27$ and $d_c/h = 0.18/0.61 = 0.295$. Using Eq. (5-79), $\Delta H/H_{max} = 1 - \{[0.54 \, (0.295)^{0.275} + 1.715 \, (0.295)^{-0.55}]/[1.5 + 27.4/0.18]\} = 1 - [(0.386 + 3.356)/(1.5 + 152.22)] = 0.976$ or 97.6%.

Labyrinth Spillways and Spillways with Semicircular or Double-Sided Entry

These are zigzag-shaped (or folded) spillways in plan such that a larger crest length is available within a limited space for spillway location. The total crest length may be three to five times the available width, and the discharge capacity may be about twice that of the standard overflow crest located within the same available width. The labyrinth is formed by a series of walls vertical on the upstream face and sloping on the downstream face. The slope

Table 5-20. Preliminary dimensions of stepped spillways

Variable	Stillwater Dam, Utah (Young 1982)	Monksville Dam in New Jersey (Sorensen 1985)
Height of spillway (m)	61	27.4 to 36.6
Width of spillway (m)	183	61
Design head above crest (m)	1.07	2.6
Design discharge (m²/s)	2.39	9.3
Downstream slope of spillway face	$0.6H{:}1V$[a]	$0.78H{:}1V$
Height of steps (m)	0.61	0.61[b]
Length of steps (m)	0.37	0.48[b]
Crest shape	Nappe-shaped (curved) up to 6.1 m below crest	Ogee-shaped near crest
Velocity at toe (m/s)	8–11	9.2
$\Delta E/E_0$	0.75	—
Length of stilling basin (m)	7.6 (approx. one-half of conventional stilling basin)	—

[a] Spillway slope $0.32H{:}1V$ up to 15.8 m below curved portion.
[b] Step height = 0.46 m, step length = 0.36 m and step height = 0.30 m, step length = 0.24 m within the ogee-shaped portion near the top.
Source: Young (1982); Sorensen (1985).

of the downstream face may be $1H{:}10V$ to $1H{:}16V$. These spillways are suitable for situations where space available for locating the full length of the spillway is inadequate due to topographic constraints or structural elements of existing facilities. Generally, they are used as ungated service spillways or auxiliary spillways for reservoirs or as control or diversion weirs on canals. The hydraulic performance of labyrinth spillways is fairly complex and depends on crest length per cycle, number of cycles forming the zigzag shape, crest height, angle of labyrinth with flow, head above crest, crest shape, wall thickness, and apex configuration. The design must be finalized by hydraulic (physical) modeling.

Preliminary dimensions of a labyrinth spillway may be estimated using empirical equations based on model experiments (see Figure 5-6). One set of such equations gives the fol-

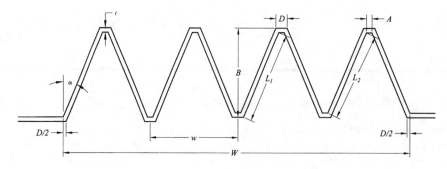

Figure 5-6. Plan view of four cycles of labyrinth spillway

lowing (Tullis et al. 1995): Wall thickness = $t = P/6$; inside width at apex = $A = t$ to $2t$; outside width at apex = $D = A + 2\ t \tan(45 - \alpha/2)$; effective crest length = $L = 1.5\ Q/[C_d H^{1.5} \sqrt{(2\ g)}]$; apron length parallel to flow = length of labyrinth = $B = [\{L/(2N)\} + t\tan(45 - \alpha/2)]\cos \alpha + t$; actual length of side leg = $L_1 = (B - t)/\cos \alpha$; effective length of side leg = $L_2 = L_1 - t\tan(45 - \alpha/2)$; total length of walls = $L_3 = N(2\ L_1 + D + A)$; distance between cycles = $w = 2L_1 \sin \alpha + A + D$; width of labyrinth (normal to flow) = $W = N\ w$; length of equivalent linear weir for same flow = $1.5\ Q/[C_L\ H^{1.5} \sqrt{(2\ g)}]$; and ratio of distance between cycles to crest length = $w/P \cong 3$ to 4. Herein, P = height of spillway above upstream floor; H = total head above crest; total length of crest = $N[2\ L_1 + A + D]$; N = number of cycles; W = straight width between abutments; α = labyrinth angle; and C_L = discharge coefficient for linear (standard) overflow weir varying from about 0.61 for $H/P = 0.1$ to 0.76 for $H/P = 0.9$. Empirical equations to estimate the discharge coefficient C_d for rounded crests are as follows:

- $C_d = 0.49 - 0.24\ (H/P) - 1.20\ (H/P)^2 + 2.17\ (H/P)^3 - 1.03\ (H/P)^4$ for $\alpha = 6°$
- $C_d = 0.49 + 1.08\ (H/P) - 5.27\ (H/P)^2 + 6.79\ (H/P)^3 - 2.83\ (H/P)^4$ for $\alpha = 8°$
- $C_d = 0.49 + 1.06\ (H/P) - 4.43\ (H/P)^2 + 5.18\ (H/P)^3 - 1.97\ (H/P)^4$ for $\alpha = 12°$
- $C_d = 0.49 + 1.00\ (H/P) - 3.57\ (H/P)^2 + 3.82\ (H/P)^3 - 1.38\ (H/P)^4$ for $\alpha = 15°$
- $C_d = 0.49 + 1.32\ (H/P) - 4.13\ (H/P)^2 + 4.24\ (H/P)^3 - 1.50\ (H/P)^4$ for $\alpha = 18°$

The design dimensions are based on a labyrinth with upstream and downstream horizontal aprons on the same elevation assuming that the downstream channel has a supercritical slope.

Example 5-16: Estimate preliminary dimensions for a labyrinth spillway for a design discharge of 1,538 m³/s; head above crest = 1.975 m; number of cycles = $N = 13$; crest height = $P = 3.05$ m; and angle of side legs = $\alpha = 8°$. This example is taken form Tullis et al. (1995).

Solution: $H = 1.975$ m; $H/P = 0.648$; $t = P/6 = 0.508$ m; $A \cong 0.95$ m; $D = 0.95 + 2 \times 0.508 \tan 41° = 1.83$ m; $C_d = 0.49 + 1.08\ (0.648) - 5.27\ (0.648)^2 + 6.79\ (0.648)^3 - 2.83\ (0.648)^4 = 0.3255$; $L = 1.5 \times 1{,}538/[0.3255 \times 1.975^{1.5} \sqrt{(2 \times 9.81)}] = 576.5$ m; $B = [\{576.5/(2 \times 13)\} + 0.508 \tan 41°]\cos 8° + 0.508 = 22.90$ m; $L_1 = (22.9 - 0.508)/\cos 8° = 22.61$ m; $L_2 = 22.61 - 0.508 \tan 41° = 22.17$ m; $L_3 = 13[2 \times 22.61 + 1.83 + 0.95] = 624$ m; $w = 2 \times 22.61 \sin 8° + 0.95 + 1.83 = 9.07$ m; $W = 13 \times 9.07 = 118$ m; $w/P = 9.07/3.05 \cong 2.97$; length of linear overflow spillway for same flow (with $C_L = 0.76$) = $1.5 \times 1{,}538/[0.76 \times 1.975^{1.5} \sqrt{(2 \times 9.81)}] = 247$ m.

Stilling Basins and Energy Dissipation Devices

Stilling basins are devices to dissipate the kinetic energy of water falling at the toe of a spillway in order to minimize its scouring potential before it reaches the downstream channel. Experimental charts are available to estimate velocities at the toe of spillways with downstream face slopes of $0.6H{:}1V$ to $0.8H{:}1V$ (Peterka 1978; Chow 1959). Alternatively, toe velocities may be estimated by

$$V = C\sqrt{[2\ g\ (Z - H/2)]} \qquad (5\text{-}82)$$

where

V = toe velocity (m/s)

Z = drop from reservoir water surface elevation to spillway toe (m)

H = head above spillway crest (m)

C = coefficient to account for friction and other losses along the spillway face or ratio of actual to theoretical velocity

Values of C for different values of Z and H may be obtained from the aforementioned experimental charts (Peterka 1978). Toe velocities for selected drops and heads above spillway crest are shown in Table 3-13.

Where practicable, stilling basin designs must be finalized by hydraulic model tests. Based on hydraulic model tests, various types of stilling basins have been developed for different flow conditions. The Soil Conservation Service has developed the St. Anthony Falls (SAF) stilling basin for use on small drainage structures with Froude numbers (F) varying from 1.7 to 12, which may not justify individual hydraulic model studies (USDA 1959). The SAF basin includes chute blocks, baffle blocks, an end sill, and a level floor. The dimensions of various elements of the SAF stilling basin are

$$F = V_1/\sqrt{(g\, y_1)} \tag{5-83a}$$

$$\text{Length of basin} = L_B = 4.5\, y_2/F^{0.76} \tag{5-83b}$$

$$\text{Height of chute blocks} = y_1 \tag{5-83c}$$

$$\text{Width of chute blocks} = 3\, y_1/4 \tag{5-83d}$$

$$\text{Spacing of chute blocks} = 3\, y_1/4 \tag{5-83e}$$

$$\text{Distance of floor blocks from upstream end of stilling basin} = L_B/3 \tag{5-83f}$$

$$\text{Minimum distance of floor blocks from side walls} = 3\, y_1/8 \tag{5-83g}$$

$$\text{Height of floor blocks} = y_1 \tag{5-83h}$$

$$\text{Width of floor blocks} = 3\, y_1/4 \tag{5-83i}$$

$$\text{Spacing of floor blocks} = 3\, y_1/4 \tag{5-83j}$$

$$\text{Height of end sill} = 0.07\, y_2 \tag{5-83k}$$

where

V_1 = pre-jump velocity

y_1 = pre-jump water depth

y_2 = post-jump, conjugate, or sequent water depth

The floor blocks are placed in a staggered pattern with respect to the chute blocks and are square in plan with vertical upstream and downstream faces or with inclined downstream and vertical upstream faces.

The tailwater elevation or depth in the channel downstream of a stilling basin may be estimated using water surface profile computations for the design discharge (USACE 1991c, 1998).

Example 5-17: Estimate preliminary dimensions of a SAF stilling basin for a spillway with $Z = 9.2$ m; $H = 1.6$ m; and discharge coefficient of the ogee crest = 2.22. The riverbed consists of fine sand with $d_{50} = 0.30$ mm, and the design tailwater elevation is at El. 100 m.

Solution: $q = 2.22\ (1.6)^{1.5} = 4.5$ m²/s. From Table 3-13, toe velocity = pre-jump velocity = $V_1 \cong 12$ m/s; $y_1 = 4.5/12 = 0.375$ m; $F = [12/\sqrt{(9.81 \times 0.375)}] = 6.26$; and $y_2 = (0.375/2)$ $[\sqrt{\{1 + 8 \times 6.26^2\}} - 1] = 3.138$ m.

To account for some lowering of tailwater elevation due to scour in the downstream channel, locate the stilling basin floor about 3.75 m below tailwater elevation or at El. 96.25 m.

Basin floor length = $L_B = (4.5 \times 3.138)/(6.26^{0.76}) = 3.5$ m. Height of chute blocks = height of floor blocks = 0.375 m; width and spacing of chute and floor blocks = $3 \times 0.375/4 = 0.28$ m; distance of floor blocks from the upstream edge of stilling basin = $3.5/3 = 1.17$ m; and height of end sill = $0.07 \times 3.138 = 0.22$ m.

Lacey's silt factor = $1.76\sqrt{0.30} = 0.96$ and potential scour depth below tailwater elevation = $2.0 \times R = 2.0 \times 1.337\ (4.5^2/0.96)^{1/3} = 7.39$ m (see Eq. (5-24b)).

Provide a sheet pile cutoff up to about 7.5 m below tailwater elevation at the end of the stilling basin.

The thickness of the stilling basin floor must be checked to be safe against uplift pressures.

The USBR has developed several stilling basins applicable to different hydraulic conditions (Peterka 1978):

1. *Basin I* is the simplest elementary stilling basin with a flat floor whose length is equal to the length of the jump. The length of jump, L_j, for different values of F may be estimated from Table 3-15. This basin has no appurtenances and so the required basin length is relatively large.

2. *Basin II* has been tested for $F = 4.1$ to 14.38. It includes chute blocks of height, width, and spacing equal to y_1 and a dentated sill of height $0.2\ y_2$ at the end of the basin, but no floor blocks. The width and spacing of dentations is $0.15\ y_2$. The basin floor elevation is set to utilize full conjugate tailwater depth, plus an added factor of safety. As an approximation, the basin floor may be set at a depth of $1.05\ y_2$ below the design tailwater elevation. The length of basin may be approximated from Table 5-21. The basin is designed using experimental charts that have been developed based on hydraulic model tests and is suitable for high dams and spillways up to 61 m in height and for flows up to 46.4 m³/s per meter width of the basin.

Table 5-21. Length of stilling basins II and III

F	L/y_2 (basin II)	L/y_2 (basin III)
4	3.6	—
4.25	3.78	2.2
6	4	2.48
8	4.2	2.64
10	4.35	2.75
12	4.38	2.78
14	4.36	2.78
14.25	4.32	2.78
16	—	2.75

Source: Peterka (1978).

Example 5-18: Estimate preliminary dimensions for USBR stilling basin II for a spillway with $Z = 40$ m (from upstream water surface elevation to existing riverbed); $H = 5.4$ m; tailwater elevation of 100 m; and spillway discharge coefficient of 2.23.

Solution: $q = 2.23\ (5.4)^{1.5} = 28$ m²/s; from Table 3-13, toe velocity = $V_1 \cong 22$ m/s; $y_1 = 28/22 = 1.27$ m; $F = [22/\sqrt{(9.81 \times 1.27)}] = 6.233$; and $x\ y_2 = (1.27/2)\ [\sqrt{\{1 + 8 \times 6.233^2\}} - 1] = 10.58$ m.

To account for some lowering of tailwater elevation due to scour in the downstream channel, locate the stilling basin floor about 12 m below tailwater elevation or at El. 88 m.

From Table 5-21, $L/y_2 = 4.1$; $L = 43.4$ m = length of basin; height of chute blocks = width of chute blocks = $y_1 = 1.27$ m; height of dentated sill = 0.2 $y_2 = 2.12$ m; width of dentated sill = spacing = 0.15 $y_2 = 1.59$ m.

If the basin floor elevation (El. 88 m) is found to be significantly below existing riverbed, then the total drop, Z, from the upstream water surface elevation to the basin floor may be more than 40 m, and the velocity, V_1, estimated from Table 3-13 may be more than 22 m/s. In this case, the computations may be repeated with the modified value of Z. The depth of sheet pile cutoff and thickness of basin floor may be estimated as described previously.

3. *Basin III* is a short stilling basin for canal structures, small outlet works, and small spillways, and is reported to be effective for F values above 4.5. It includes chute blocks, floor blocks, and an end sill. The height, width, and spacing of chute blocks are equal to y_1, except that the height of chute blocks should not be less than 0.2 m. The height of the baffle blocks, h_B, and end sill, h_s, may be taken from Table 5-22. The upstream face of the baffle blocks should be placed at a distance of 0.8 y_2 from the downstream face of the chute blocks; the downstream face of these blocks slopes at 1H:1V; the upstream face is vertical; and the cross section is trapezoidal with the top width along the direction of flow equal to 0.2 times the height. The upstream face of the end sill slopes at 2H:1V, and the downstream face is vertical. Details of the basin are finalized using experimental charts that have been developed based on hydraulic model tests.

Table 5-22. Heights of baffle blocks and end sill for basin III

F	h_B/y_1	h_s/y_1
4	1.25	1.2
6	1.6	1.35
8	1.95	1.45
10	2.3	1.55
12	2.6	1.65
14	2.95	1.8
16	3.3	1.9
17	3.45	1.95
18	—	2

Source: Peterka (1978).

4. *Basin IV* is suitable for canal structures, diversion dams, or small drops with values of F between 2.5 and 4.5 and is applicable to rectangular cross sections only. It includes chute blocks and an end sill but no baffle blocks. This basin design reduces the height of waves created in imperfect or unstable jumps.

5. *Basin V* has a sloping apron. If proper tailwater depth is available, horizontal and sloping aprons are reported to perform equally well for high values of F. Steps for the preliminary design of Basin V are as follows:

   (a) Select a suitable configuration for the sloping apron.

   (b) Position the apron so that the front of the jump is at the upstream end of the slope for the maximum or design discharge and corresponding tailwater elevation using information from published experimental charts (Peterka 1978). It may be necessary to raise or lower the apron or change its slopes after several trials.

   (c) Obtain length of jump, L_j, for the design discharge from tabulated values (Peterka 1978) and adopt an apron length equal to the length of jump. A short horizontal apron may be added at the end of the sloping apron.

   (d) Provide a triangular end sill with height between 0.05 y_2 to 0.10 y_2 and slope of upstream face between $2H:1V$ and $3H:1V$.

   For high spillways with design discharges exceeding 46 m³/s, a hydraulic model study should be conducted.

   The conjugate depth and length of jump may be estimated from Table 5-23 or published charts (Peterka 1978).

6. *Basin VI* is an impact-type energy dissipator contained in a relatively small box-like structure, which requires no tailwater for successful performance. It is commonly used for pipe outlets. Design dimensions for the basin are selected from tabulated values based on model experiments (Peterka 1978).

7. *Basin VII* includes slotted and solid buckets for high, medium, and low dam spillways. Design dimensions for the basin are selected from tabulated values based on model experiments (Peterka 1978).

Table 5-23. Conjugate depths and length of jump on sloping aprons

F	Apron slope 5H:1V		Apron slope 4H:1V		Apron slope 3H:1V	
	y_2/y_1	L/y_2	y_2/y_1	L/y_2	y_2/y_1	L/y_2
2	4.8	2.55	5.5	2.2	6.8	—
4	10	3.15	12	2.8	14	—
6	15.6	3.38	18.5	2.95	22	—
8	21	3.4	25	3	30	—
10	26.5	3.4	—	—	—	—

— indicates test results are not available.
Source: Peterka (1978); Chow (1959).

8. *Basin VIII* is used for energy dissipation downstream of hollow jet valves.

9. *Basin IX* is suitable for canals or small spillway drops up to a unit discharge of 5.6 m³/s per meter width of the chute. It includes a chute on a slope of 2H:1V or flatter to negotiate the drop in channel bed. Staggered blocks are provided on the slope for energy dissipation. Steps to determine preliminary dimensions of the basin are as follows:

    (a) Provide a short sill at the upstream edge of the slope to create a stilling pool. The height of the sill may be about 0.61 m.

    (b) Estimate $y_c = (q^2/g)^{1/3}$, where $q = Q/B$; Q = design discharge and B = width of the chute.

    (c) Provide baffle blocks with height = $H = 0.5\ y_c$, width of blocks = spacing = 1.5 H. Partial blocks with widths equal to 1/3 to 2/3 H should be placed against one wall in rows 1, 3, 5, 7, etc., and against the other wall in rows 2, 4, 6, 8, etc. The distance between adjacent rows of chute blocks should be 2 H. Blocks may be placed with their upstream faces normal to the chute surface or with vertical faces.

    (d) Riprap protection should be placed at the downstream end of the chute.

10. *Basin X* is intended for tunnel spillways but is used for open chutes as well. It consists of a flip bucket at the downstream end of the tunnel, chute, or spillway face. Stilling basins of this type are known as ski jump, deflector, diffuser, flip, or trajectory buckets and are applicable to F values varying from 6.8 to 10.3. The inclination of the bucket flip varies from about 15° to 35° with the horizontal. The radius of the bucket should be at least four times the depth of flow in the bucket. The length of the trajectory of the jet (i.e., horizontal distance from the lip or downstream edge of the bucket to the point where the jet falls vertically down), X, and the rise of the jet above the bucket invert may be approximated by

$$X = V^2 \sin 2\theta/g \qquad (5\text{-}84a)$$

$$r = \text{rise} = (V \sin \theta)^2/2g \qquad (5\text{-}84b)$$

where θ = angle of lip to the horizontal and V = velocity of jet exiting the bucket. The end of the bucket lip may have a short horizontal sill or may have an inclined edge followed by a vertical downstream face. Downstream of the vertical face of the bucket, a plunge pool is provided to dissipate the energy of the jet exiting the bucket. When the jet falls vertically on the riverbed, it creates a scour hole. The ultimate scour depth may be approximated by the following (USBR 1987):

$$D_s = 1.897\, H^{0.225}\, q^{0.54} \tag{5-85}$$

where

D_s = ultimate scour depth below tailwater elevation (m)

H = elevation drop from the reservoir elevation to the tailwater elevation (m)

q = unit discharge m^3/s per meter width of chute or bucket

Unless the riverbed is comprised of nonerodible hard rock, a reinforced concrete plunge pool should be provided to protect against scour. The floor of the plunge pool may be located at a depth, D_s, below the tailwater elevation. The length of the plunge pool may be slightly greater than the length of the jet trajectory, X.

Preliminary dimensions for comparative evaluation of alternative stilling basin designs or preliminary cost estimates may be developed using the information presented in this section. Detailed designs for specific conditions must be finalized using published design charts for individual types of basins and hydraulic model tests.

Example 5-19: Estimate preliminary dimensions of a ski jump bucket type stilling basin and plunge pool for a spillway with $H = 6.1$ m; $Z = 100$ m; and $q = 33$ m^2/s. The design tailwater elevation is 95 m below the design water surface elevation in the reservoir, which is at El. 1,000 m. Use $\theta = 25°$.

Solution: From Eqs. (5-82) and (5-84a), it may be seen that the velocity at the bottom of the drop varies approximately as the square root of the net drop and the length of plunge pool varies as the square of the velocity. To optimize the dimensions of the plunge pool, several trial computations may be required to determine the elevation of the ski jump bucket. For this case, it has been determined that the end sill of the ski jump bucket is located at 90 m below the design reservoir water surface elevation or at El. 910 m. For $Z = 90$ m and $H = 6.1$ m, $V \cong 34$ m/s (Table 3-13); $y = 33/34 = 0.97$ m; and $R \cong 4 \times 0.97 \cong 4$ m. There are no specific guidelines to estimate the length and depth of the bucket. But they can be estimated once the radius of the bucket and lip angle are determined; for example, length of bucket arc = $2\,R\,\theta°\,\pi/180$ and bucket depth = $R - R\cos\theta$.

For preliminary estimates, the length, L, beyond the low point of the bucket may be taken to be about 0.25 to 0.60 times the radius. For this case, use $L = 2$ m and depth of bucket between its end sill and low point = 0.80 m. Provide a 0.61-m-wide horizontal sill at the downstream edge of the bucket (see Figure 5-7). These preliminary dimensions must be refined based on hydraulic model tests or test data for similar types of flip buckets.

270 WATER RESOURCES ENGINEERING

Using Eqs. (5-84a) and (5-84b), $X = (34^2 \times \sin 50°)/9.81 = 90$ m, and rise of jet = $(34 \times \sin 25°)^2/(2 \times 9.81) = 10.5$ m above the end sill or up to El. 920.5 m.

The kinetic energy with which the jet strikes the riverbed corresponds to a drop from the top of the jet trajectory to the tailwater elevation (El. 905 m); i.e., $H = 920.5 - 905 = 15.5$ m.

Using Eq. (5-85), $D_s = 1.897 \times 15.5^{0.225} \times 33^{0.54} = 23.2$ m.

Provide a 95-m-long plunge pool at an elevation of about 24 m below design tailwater elevation or at El. 881 m. The floor slab of the plunge pool must be checked to be safe against uplift pressures. Depending on site conditions, the length of the plunge may be reduced by selecting a higher elevation for the ski jump bucket. Alternatively, other types of stilling basins may be evaluated.

Hydroelectric Power

Power Plant

A hydroelectric power plant has several components:

1. Tunnel or conduit or power channel: Conveys water from the source (e.g., a reservoir or river intake) to the forebay or penstocks in the vicinity of the power plant.

2. Forebay: A short open channel with a relatively small storage capacity to absorb short-term (e.g., hourly or diurnal) fluctuations in flows. It connects the power channel to an intake, which controls entry of water to the penstocks or turbines. Depending on site conditions, a forebay may or may not be necessary.

3. Intake Structure: Controls the flow of water from the forebay to the penstocks or turbines. A trash rack is provided just upstream of the intake structure to control the entry of debris and fish into the turbine.

4. Penstocks: For medium- or high-head hydroelectric facilities, water from the forebay or power tunnel (if forebay is not provided) is conveyed to the turbine entrance through pressure pipes, usually made of steel. These pressure pipes are called pen-

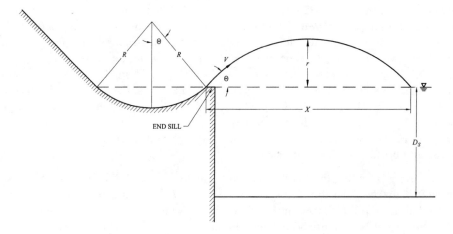

Figure 5-7. Schematic of ski jump bucket

stocks. They are laid on steep slopes without sharp bends to minimize losses. The entrance of the penstocks should be sufficiently submerged to prevent air entry. For long tunnels, a surge tank is provided between the junction of the tunnel and the penstocks, close to the power plant, to protect the tunnel and penstocks against water hammer pressures in case of sudden stoppage of flow through the turbines. In some low-head installations, where water can be supplied to the turbines through open flumes, penstocks may not be necessary.

5. Scroll Case: From the end of the penstocks, water is transitioned to the turbine through a scroll case, which is spiral in plan. The main purpose of the scroll case is to maintain nearly uniform velocity at the entrance of the guide vanes (or wicket gate) at the entrance to the turbines.

6. Wicket Gate: Water enters the turbine through the wicket gate.

7. Turbine-Generator: Various types of turbines are available from different manufacturers suitable for different site conditions. These are machines used to convert the potential and kinetic energy of water to useful work. The generator transforms this energy into electric energy.

8. Draft Tube: From the turbines, water is discharged downstream through a draft tube. The draft tube has a gradually expanding cross-sectional area so that the exit velocity into the downstream channel is significantly reduced. Usually, the design of the draft tube is provided by the manufacturers.

9. Tailrace: The channel into which water enters from the draft tube. It conveys water discharged from the power plant to the natural channel on the downstream side.

Depending on the installed capacity, hydroelectric power plants are classified as follows:

1. Conventional: installed capacity is greater than 15 MW.
2. Small-scale: installed capacity is between 1 and 15 MW.
3. Mini: installed capacity is between 100 KW and 1 MW.
4. Micro: installed capacity is less than 100 KW.

Hydraulic structures related to a hydroelectric facility include intake structure, power canal, tunnel, penstocks, surge tank, draft tube, and tailrace channel. Guidelines for the design of these structures are available in various publications (e.g., ASCE 1989; Monsonyi 1963; USACE 1985; Davis and Sorensen 1970; Zipparro and Hansen 1993; Creager and Justin 1950).

The installed capacity or hydropower generation potential for a given site is estimated by

$$P = 9.80 \, Q H \eta \qquad (5\text{-}86)$$

where

P = power in kilowatts

Q = flow (m^3/s)

H = available net head (m)

η = turbine-generator efficiency, usually in the range of 0.80 to 0.90

272 WATER RESOURCES ENGINEERING

Design values of Q and H are estimated from the flow duration and head-discharge tables or curves for the water supply source. To maximize use of available water, a value of Q which is exceeded 10 to 30% of times may be selected to estimate the installed capacity of the facility. Several values of Q and the corresponding H may be examined to estimate the values that result in optimum installed capacity from water use and economic considerations. For preliminary planning, optimum installed capacity may be that beyond which relatively large increases in Q are required to obtain relatively small increases in P.

To estimate the annual hydropower generation potential of a facility, a reservoir and power plant operation study has to be conducted using models such as HEC-3 (USACE 1981) and HEC-5 (USACE 1982) with sequences of available flows and corresponding heads for relatively long periods of time (e.g., 10 to 50 yr or more). For preliminary estimates, flow duration and head-discharge tables may be used. For these estimates, the overall water to wire efficiency should be used, which varies from about 0.70 to 0.86 and includes efficiencies of the turbine, generator, transformers, and other equipment. In addition, adjustments to the efficiency may be made for tailwater fluctuations and unscheduled down time.

Example 5-20: The average annual flow duration and head discharge relationship for a run-of-river power plant are shown in Table 5-24. Estimate the installed capacity and annual energy generation potential for the facility.

Solution: It may not be advisable to install equipment to operate at flows more than 405 m³/s because flows in excess of this are available only for about 11% of the times. The efficiency of the units will vary with discharge and head according to the characteristics provided by the manufacturer. For this example, assume a constant overall efficiency of 0.86. The units are assumed to have an overload factor of 1.15. So the generation may be up to 1.15 times the rated installed capacity.

With maximum operating $Q = 405$ m³/s, head (from given head-discharge Table 5-24) = 31.53 m.

From Eq. (5-86), installed capacity = $9.80 \times 405 \times 31.53 \times 0.86 = 108{,}000$ KW.

Use four units of 30 MW each capable to generate up to 120 MW with no overload and 138 MW with an overload factor of 1.15. Specifications provided by the manufacturer are used to determine the minimum discharge below which it may not be efficient to operate the turbine generation unit. For this example, it is assumed that it may not be efficient to operate any one of these turbine-generator units when flows are less than about 116 m³/s. The corresponding head is 16.3 m. The generated power is about $9.80 \times 116 \times 16.3 \times 0.86/1{,}000 = 15.9$ MW, or 53% of the installed capacity of a single unit. It is assumed that the load can be evenly distributed on the machines so that a reasonable minimum efficiency is maintained on each operating unit. One, two, three, or four units will be operated depending on available discharge and head at any given time.

Using a plot of the flow duration table (Q on y-axis and percentage exceedance on x-axis), divide the flows into several convenient flow intervals and, by interpolation, estimate the mean of each flow interval, corresponding head, and the percentage of times covered by the flow interval. For example, the flow interval 29 to 57 m³/s occurs for $(90 - 70) = 20\%$ of the times. The average annual energy generation potential may be estimated by

$$\text{average annual energy (KWh)} = E = 8{,}760 \times 9.8 \times \Sigma \, (Q_i \, p_i \, H_i \, \eta_i)/100 \qquad (5\text{-}87)$$

where

> i represents the number of flow intervals
> Q_i = mean flow in interval i
> H_i = head corresponding to Q_i
> η_i = overall efficiency corresponding to Q_i and H_i
> p_i = the percentage of times covered by the flow interval
> Σ is over all flow intervals

Selected flow intervals, percentage of times covered by each flow interval, mean of the flow interval, and head corresponding to the mean flow are shown in Table 5-25.

The annual energy generation potential of the facility is about 232,246 MWh without overload factor and 248,337 MWh with overload factor.

Intakes

Various types of intake designs are used in hydroelectric facilities (ASCE 1989). Commonly used types of intakes include the following:

1. A dry vertical well that contains a hoisting arrangement to control valve openings in a horizontal inlet located at the bottom of the shaft conveying water from the reservoir or river to the point of delivery.

Table 5-24. Flow duration and head-discharge table

Q (m³/s)	% exceedance	Head (m)
0	100	0
5	99.99	3.5
28	90	9.74
57	70	11.06
86	54	15.74
115	41	16.17
144	33.5	19.92
173	29	20.13
202	25	23.32
231	22	23.47
260	19.5	26.25
289	17.5	26.42
318	15.5	28.87
347	13.5	29.08
376	12	31.28
405	11	31.53
434	10.5	33.5
463	10	33.81
492	9.5	35.59
521	9	35.92
522	0	36.29

Table 5-25. Computations of annual energy generation

Flow interval		% time covered by flow interval	Mean of flow in the interval (m^3/s)	Head corresponding to mean flow (m)	Energy (MWh)	
Q_1 (m^3/s)	Q_2 (m^3/s)				Using actual installed capacity	With overload factor of 1.15
0	28	10	14	5.94	0	0
29	57	20	43	10.42	0	0
58	86	16	72	13.48	0	0
87	115	13	101	15.96	0	0
116	144	7.5	130	18.11	13,036	13,036
145	173	4.5	159	20.03	10,581	10,581
174	202	4	188	21.78	12,092	12,092
203	231	3	217	23.4	11,247	11,247
232	260	2.5	246	24.91	11,310	11,310
261	289	2	275	26.34	10,696	10,696
290	318	2	304	27.69	12,429	12,429
319	347	2	333	28.98	14,250	14,250
348	376	1.5	362	30.22	12,115	12,115
377	405	1	391	31.41	9,067	9,067
406	434	0.5	420	32.55	5,047	5,047
435	463	0.5	449	33.66	5,256[a]	5,579
464	492	0.5	478	34.73	5,256[a]	6,044[a]
493	521	0.5	507	35.76	5,256[a]	6,044[a]
522		9	522	36.29	94,608	108,799[a]
Total		100			232,246	248,337

[a]The machines will be able to produce 120 MW with no overload factor and 138 MW with an overload factor of 1.15 during these periods.

2. A solid horizontal pipe intake with a gate or valve at its entrance.

3. An inclined conduit supported along the face of the dam with ports located at different elevations and connected to a nearly horizontal pipe extending through the body of the dam.

4. A vertical shaft with a velocity cap on top to prevent vertical influx of water into the intake and connected to a nearly horizontal pipe extending through the body of the dam.

5. A vertical shaft where water enters the vertical shaft from its top and also from the sides through inlet ports located at different elevations and the control valve is located at the outlet pipe connected to the vertical shaft. The outlet pipe extends through the body of the dam. This type of intake is called a wet well intake because the vertical shaft always contains water.

The hydraulic designs of intakes are based on flow equations described in Chapter 3. It is important to avoid vortex formation at the entrances of power intakes. The minimum submergence required to avoid vortex formation at horizontal bell mouth entrances of intakes is given by

$$S = c V d^{0.5} \tag{5-88}$$

where

S = minimum required submergence above the upper lip of the bell mouth (m)

V = velocity through the intake entrance (m/s)

d = diameter of the intake pipe (m)

c = 0.54 for symmetrical and 0.72 for lateral approach flow conditions

For horizontal intakes with a vertical bell mouth inlet,

$$S > 62 V^2 \tag{5-89}$$

where

S = minimum required submergence above the centerline of the bell mouth (m)

V = intake velocity at the bell mouth

Alternative criteria for submergence at such horizontal intakes are

$$S/d \leq 0.7 \quad \text{and} \quad F = V/\sqrt{(g\,d)} \leq 0.5 \tag{5-90}$$

For vertical intakes,

$$2.14\, F^{0.04} \leq S/d \leq F^{0.12} \tag{5-91}$$

The head loss through the entrance, exit, bends, and conduit of an intake may be estimated using the methods described in Chapter 3. Head loss through the trash rack installed at the entrance of the intake may be estimated by the following (Davis and Sorensen 1970; Zipparro and Hansen 1993; Chow 1959; ASCE 1989):

$$h_L = k_T\, (t/b)^{4/3}\, (V^2/2\,g)\, \sin \alpha \tag{5-92}$$

where

h_L = head loss through the trash rack

k_T = a coefficient with a value of 2.42 for trash rack bars with square nose and 1.79 for round bars

t = thickness of bars (cm)

b = clear spacing between bars (cm)

α = angle of bar inclination to horizontal

V = velocity of approach ahead of trash rack

WATER RESOURCES ENGINEERING

An alternative method to estimate head loss through the trash rack is to use a head loss coefficient given by the following (USBR 1987):

$$k_T = 1.45 - 0.45\, a_n/a_g - (a_n/a_g)^2 \tag{5-93}$$

Then,

$$h_L = k_T\, (V_n^2/2g) \tag{5-94}$$

where

V_n = velocity through the open area of the trash rack

a_n = net open area through the trash rack

a_g = gross area of the trash rack and supports

For practical cases, head loss through the trash rack should be estimated assuming the trash rack to be 50% clogged.

The contraction loss in pipe flow may be estimated by the following (ASCE 1989):

$$h_L = k_c\, (V_2^2/2g) \tag{5-95}$$

where V_2 = velocity in the smaller pipe, and typical values of k_c for different contraction ratios are given in Table 5-26.

Typical values of loss coefficients, k_b, for a 90 degree bend with a radius, R, in a pipe of diameter, d, are indicated in Table 5-27 (USBR 1987).

Example 5-21: Water levels in a reservoir fluctuate between 300 and 310 m. The dead storage elevation is 295 m. Water is to be supplied to a point 200 m away from the dam at El. 289.70 m. The required water supply discharge from the reservoir is 2.85 m³/s. Estimate the dimensions of a pipe intake for this case. Use an average friction coefficient of 0.01.

Solution: The intake could be an inclined pipe laid along the embankment slope or a vertical tower located near the embankment. For this case, a vertical tower is adopted. To prevent fish from getting sucked into the intake shaft, a vertical cylindrical fish screen will be provided at the entrance, and the velocity of approach toward the screen will be limited to 0.15 m/s. A velocity cap will be provided at the top of the screen so that water enters only through the height of the screen. The screen bars may occupy approximately 20% of the gross area of the screen. However, for design purposes, assume that blockage of the screen area due to screen bars, debris, and fouling is 50%. Thus,

$$A_n = (1 - 0.20)\, A_g = 0.80\, A_g \quad \text{and} \quad A_n\,(\text{unclogged}) = 0.50\, A_n = 0.40\, A_g$$

To pass the design discharge,

$$0.15 \times (1 - 0.20)\, \pi\, D_s\, H_s \times 0.50 \geq 2.85 \quad \text{or} \quad D_s\, H_s \geq 15.10 \text{ m}^2$$

where

D_s = diameter of the cylindrical screen cage (m)

H_s = height of screen (m)

Table 5-26. Typical values of contraction coefficients for pipes

A_2/A_1	k_c
0.1	0.363
0.2	0.339
0.4	0.268
0.6	0.164
0.8	0.053
1	1

A_2 = cross-sectional area of smaller pipe and A_1 = cross-sectional area of larger pipe.
Source: ASCE (1989).

Table 5-27. Typical values of bend loss coefficients for pipes (90° bend)

R/d	K_b
0.8	0.3
1	0.23
1.5	0.18
2	0.13
3	0.1
≥4.0	0.08

Source: USBR (1987).

Assume $D_s = 3.7$ m and $H_s = 4.1$ m. Assume square-cornered entrance for the circular crest of the intake pipe at El. 296 m and $D = 3.7$ m, which gradually reduces to the pipe diameter, d. The intake must carry 2.85 m³/s with the lower reservoir elevation of 300 m. Then,

$$300 - 289.7 = [(k_T + k_e + k_b + fL/d + 1)\, V^2/2g + k_c\{(V^2/2g) - (V_b^2/2g)\}]$$

where

- k_T = trash rack loss coefficient = $1.45 - 0.45 \times 0.40 - 0.16 = 1.11$ (Eq. (5-93))
- k_e = entrance loss coefficient = 0.70
- k_b = bend loss coefficient for the elbow where the vertical pipe becomes nearly horizontal with a bend radius of twice the pipe diameter or more = 0.13
- k_c = loss coefficient for contraction from the diameter of the bell mouth to the pipe diameter = 0.25 (USBR 1987)
- V_b = velocity through the circular intake entrance ≅ $2.85/(\pi \times 3.7^2/4) = 0.27$ m/s

Thus, $10.3 = [1.11 + 0.7 + 0.13 + (0.01 \times 200/d) + 1]\, V^2/2g + 0.25\, [V^2/2g - 0.27^2/2g) = [2.94 + 2/d]\, V^2/2g + 0.25\, V^2/2g - 0.0009$.

Assume a trial $d = 1.0$ m so that $V = 2.85/(\pi \times 1.0^2/4) = 3.63$ m/s. Then, $h_L = (3.19 + 2.0) [3.63^2/(2 \times 9.81)] - 0.0009 = 3.485$ m.

Similarly, try $d = 0.8$ m; then $V = 5.67$ m/s, and $h_L = 9.32$ m.

For $d = 0.7822$ m, $V = 5.931$ m/s, and $h_L = 10.3$ m, which is nearly equal to the available head. Use a pipe diameter of $d = 0.80$ m to allow some variation in the assumed loss coefficients.

For reservoir elevations higher than 300 m, flow through the intake pipe will be controlled by a valve located at the intake tower.

Surge Tank

Surge tank is a vertical tank located at the downstream end of the power tunnel and upstream end of the penstocks. The function of the surge tank is to attenuate pressure increases in the tunnel and penstocks due to sudden load rejection at the turbine and consequent reduction or complete stoppage of flow through the turbines. A simple surge tank may be a vertical shaft connected to the tunnel. The minimum area of a surge tank may be estimated by the following (Davis and Sorensen 1970; Zipparro and Hansen 1993):

$$A_s = A L / (2 g c H) \tag{5-96}$$

where

A_s = minimum required area of surge tank

A = area of tunnel

H = difference in elevations between normal reservoir water surface and centerline of the tunnel

L = length of tunnel from the reservoir to the surge tank

c = head loss coefficient for the tunnel such that friction and other losses equal $c V^2$, where $c \cong [f L/(2 g d) + \{\text{coefficients of minor losses}/(2 g)\}]$, d = diameter of tunnel, and f = Darcy-Weisbach friction coefficient (Figure 5-8)

In practice, the diameter of the simple surge tank estimated by Eq. (5-96) should be increased by 50% or more. Unless overflow from the surge tank is acceptable, its height should be higher than the maximum anticipated surge height above the steady-state water level due to sudden stoppage of flow through the turbine. The surge height is estimated using the continuity and Eulerian equations:

$$dz/dt = (A/A_s) V \tag{5-97}$$

$$(L/g) \, dV/dt = -(z - c V^2) \tag{5-98}$$

where z = change in water level in the surge tank from the water level in the reservoir, measured positive in the upward direction. These equations may be solved using finite increments, with starting values (at $t = 0$) of $z = c V^2$ and V = steady-state velocity in the tunnel:

$$\Delta z = (A/A_s) V \Delta t \tag{5-99}$$

$$\Delta V = -(g/L)(z - c V^2) \Delta t \tag{5-100}$$

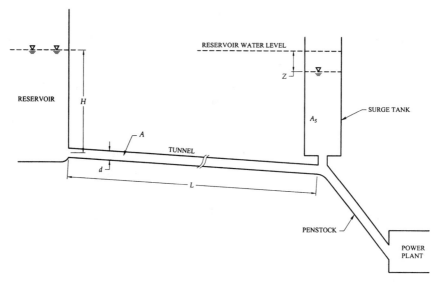

Figure 5-8. Simple surge tank

For preliminary estimates for a simple surge tank, cV^2 may be neglected (ASCE 1989); then,

$$z = (Q_0/A_s) \ [\sqrt{\{A_s L/(A\ g)\}}] \ \sin\ [\sqrt{\{A\ g/(A_s L)\}}\ t] \qquad (5\text{-}101)$$

$$z_{max} \cong (Q_0/A_s) \ \sqrt{\{A_s L/(A\ g)\}} \qquad (5\text{-}102)$$

$$T = 2\ \pi \sqrt{\{A_s L/(A\ g)\}} \qquad (5\text{-}103)$$

where

z_{max} = maximum water surface rise above steady-state level in surge tank

T = time period of water level fluctuations

Example 5-22: Estimate preliminary dimensions of a simple cylindrical surge tank with circular cross section for a power plant that is supplied water from a reservoir through a 1,000-m-long, 1.75 diameter concrete-lined tunnel. Steady-state discharge through the tunnel is 6 m³/s. The average head above the centerline of the conduit from the reservoir to the surge tank is 30 m. Assume the design condition to be the sudden closure of a valve at the penstock.

Solution: For concrete-lined tunnel, assume $f = 0.015$, $V = 6/(\pi \times 1.75^2/4) = 2.4945$ m/s. $c \cong fL/(2\ g\ d) = 0.015 \times 1,000/(2 \times 9.81 \times 1.75) = 0.437$ and z (initial or steady-state value) = $cV^2 = 0.437 \times 2.4945^2 = 2.72$ m.

Cross-sectional area of tunnel = $A = \pi \times 1.75^2/4 = 2.405$ m².

Using Eq. (5-96), minimum area of surge tank = $A_s = 2.405 \times 1,000/(2 \times 9.81 \times 0.437 \times 30) = 9.35$ m², and minimum required diameter of surge tank = $\sqrt{(4 \times 9.35/\pi)} = 3.45$ m. From practical considerations, use $d_s = 5.5$ m and $A_s = \pi \times 5.5^2/4 = 23.758$ m².

$A/A_s = (1.75/5.5)^2 = 0.1012$.

Ignoring friction loss and using Eq. (5-102), $z_{max} \cong (6/23.578) \sqrt{\{(5.5/1.75)^2 \times 1{,}000/(9.81)\}} = 8.1$ m.

Considering friction loss, the maximum surge may be estimated using Eqs. (5-99) and (5-100) on an Excel spreadsheet (Rouse 1950). The computations are illustrated in Table 5-28. The steps of computations are:

1. Determine time, t, at which computations are made (column (1)).
2. Determine time step of computations, Δt (difference in two consecutive values of t in column (1)).
3. Set initial values of previously computed V, z, and $c\,V^2$ in columns (3), (5), and (6), and compute initial values in other columns. Note that initial z is negative because steady-state water level in surge tank is lower than reservoir water level due to losses in the tunnel.
4. Determine Δz (column (4)) using Eq. (5-99).
5. Determine z in column (5) (adding Δz to z computed in the previous time step).
6. Compute $c\,V^2$ using value of V from previous time step, making sure that the sign of this value in column (6) is opposite that of the velocity for the previous time step in column (3); i.e.,

$$c\,V^2 = -c\,\text{abs}\cdot(V)\cdot V$$

where abs · represents absolute value and V refers to the previous time step.

7. Compute $z - c\,V^2$ and enter result in column (7).
8. Compute ΔV using Eq. (5-100) and enter result in column (8).
9. Compute V (new time step) = V (previous time step) + ΔV and enter result in column (3) for current time step.

The estimated maximum surge height above the reservoir water level is 6.21 m at $t = 55$ s, and the minimum height below the reservoir level is 4.36 m at $t = 159$ s. The effect of friction in the tunnel is to reduce the maximum surge height from $z_{max} \cong 8.1$ m to 6.21 m. Use a surge tank diameter of 5.5 m and a height of 7.25 m above the reservoir water surface level, allowing a freeboard of about 1.0 m.

Draft Tube

The design of the draft tube is usually provided by the manufacturer. However, its setting with respect to the turbine and tailrace may have to be checked to ensure that there is no potential for cavitation at the junction of the turbine and the draft tube. This is done using the following energy equation (Davis and Sorensen 1970; Zipparro and Hansen 1993):

$$Z_1 + p_1/\gamma + V_1^2/2g = V_3^2/2g + fL\{(V_1+V_2)/2\}^2/[2g(d_1+d_2)/2] + \{(V_2-V_3)^2/2g\} + p_{atm} \quad (5\text{-}104)$$

Table 5-28. Computations of surge height

(1)	(2)	(3)	(4)	(5)	(6)	(7)	(8)
t	Δt	V	Δz	z	CV^2	$z - cV^2$	ΔV
0	0	2.4945	0	−2.72	−2.72	0	0
1	1	2.4920	0.2524	−2.4676	−2.7192	0.2517	−0.0025
2	1	2.4871	0.2522	−2.2154	−2.7139	0.4985	−0.0049
5	3	2.4506	0.7551	−1.4603	−2.7032	1.2430	−0.0366
10	5	2.3326	1.24	−0.2203	−2.6243	2.4040	−0.1179
15	5	2.1689	1.1803	0.9600	−2.3778	3.3378	−0.1637
20	5	1.9672	1.0975	2.0575	−2.0557	4.1133	−0.2018
25	5	1.7345	0.9954	3.0529	−1.6911	4.7440	−0.2327
30	5	1.4772	0.8776	3.9305	−1.3147	5.2452	−0.2573
35	5	1.201	0.7475	4.6780	−0.9536	5.6316	−0.2762
40	5	0.9108	0.6077	5.2857	−0.6303	5.916	−0.2902
45	5	0.6111	0.4609	5.7465	−0.3625	6.1091	−0.2997
50	5	0.3061	0.3092	6.0558	−0.1632	6.2190	−0.3050
55	5	−0.0005	0.1549	6.2107	−0.041	6.2516	−0.3066
56	1	−0.0615	−5.5E-05	6.2106	1.29E-07	6.2106	−0.0609
57	1	−0.1223	−0.0062	6.2044	0.0017	6.2027	−0.0608
60	3	−0.3036	−0.0371	6.1673	0.0065	6.1607	−0.1813
65	5	−0.5966	−0.1536	6.0136	0.0403	5.9733	−0.293
75	10	−1.1121	−0.6038	5.4098	0.1556	5.2543	−0.5154
90	15	−1.5802	−1.6881	3.7217	0.5404	3.1813	−0.4681
105	15	−1.6143	−2.3987	1.323	1.0912	0.2318	−0.0341
110	5	−1.5833	−0.8168	0.5061	1.1388	−0.6327	0.03103
115	5	−1.5151	−0.8011	−0.295	1.0954	−1.3904	0.06820
120	5	−1.4138	−0.7666	−1.0616	1.0031	−2.0647	0.10127
125	5	−1.2838	−0.7154	−1.777	0.8735	−2.6505	0.1300
130	5	−1.1294	−0.6496	−2.4266	0.7202	−3.1468	0.1544
135	5	−0.9550	−0.5715	−2.9981	0.5574	−3.5555	0.1744
140	5	−0.7647	−0.4832	−3.4813	0.3986	−3.8799	0.1903
145	5	−0.5625	−0.3869	−3.8683	0.2556	−4.1238	0.20228
150	5	−0.352	−0.2846	−4.1529	0.1382	−4.2911	0.2105
155	5	−0.1369	−0.1781	−4.331	0.0541	−4.3851	0.2151
156	1	−0.0942	−0.0138	−4.3448	0.0082	−4.3530	0.0427
157	1	−0.0514	−0.0095	−4.3544	0.0039	−4.3582	0.0428
158	1	−0.0086	−0.0052	−4.3596	0.0012	−4.3607	0.0428
159	1	0.0341	−0.0009	−4.3604	3.27E-05	−4.3605	0.0428
160	1	0.0769	0.0035	−4.357	−0.0005	−4.3565	0.0427
161	1	0.1195	0.0078	−4.3492	−0.0026	−4.3466	0.0426
162	1	0.162	0.0121	−4.3371	−0.0062	−4.3309	0.0425
163	1	0.2043	0.0164	−4.3207	−0.0115	−4.3093	0.0423
164	1	0.2463	0.0207	−4.3001	−0.0182	−4.2818	0.0420
165	1	0.288	0.0249	−4.2751	−0.0265	−4.2486	0.0417
166	1	0.3292	0.0291	−4.246	−0.0362	−4.2098	0.0413
167	1	0.3701	0.0333	−4.2127	−0.0474	−4.1653	0.0409
168	1	0.4105	0.0375	−4.1752	−0.0599	−4.1154	0.0404
169	1	0.4503	0.0415	−4.1337	−0.0736	−4.0601	0.0398
170	1	0.4895	0.0456	−4.0881	−0.0886	−3.9995	0.0392
175	5	0.6728	0.2477	−3.8404	−0.1047	−3.7357	0.1832
180	5	0.8347	0.3404	−3.5	−0.1978	−3.3022	0.162

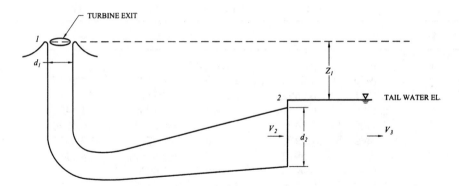

Figure 5-9. Line diagram of draft tube

where

Z_1 = elevation difference between turbine exit and minimum tailwater elevation

p_1 = absolute pressure at turbine exit

V_1 = velocity at turbine exit

V_2 = velocity at draft tube exit

V_3 = velocity in tailrace

d_1 = diameter of draft tube at turbine exit

d_2 = diameter of draft tube at exit to tailrace

L = length of draft tube

p_{atm} = atmospheric pressure (m) (Figure 5-9)

Example 5-23: Examine whether there is potential for cavitation in a draft tube where the following conditions exist: elevation difference between the turbine exit and expected minimum tailwater elevation is 8 m, velocity of flow at the turbine exit is 7 m/s, diameter of the draft tube at the turbine exit is 1.5 m, and that at the exit to tailrace is 3 m, velocity of flow in tailrace is 1 m/s, and length of the concrete draft tube is 20 m. Atmospheric pressure at the tailrace is 10.34 m (of water) and vapor pressure of water at the turbine exit is 0.125 m (of water).

Solution: Using the notation of Eq. (5-104), Z_1 = 8 m, V_1 = 7 m/s, d_1 = 1.5 m, d_2 = 3 m, V_3 = 1 m/s, L = 20 m, p_1 = 0.125 m, and p_{atm} = 10.34 m. For the concrete draft tube, assume f = 0.015. Also, for continuity of discharge through the draft tube, $V_2 = V_1 \times (d_1/d_2)^2 = 7 \times (1.5/3)^2 = 1.75$ m/s.

Using Eq. (5-104), $p_1/\gamma = -8 - 7^2/(2 \times 9.81) + 1/(2 \times 9.81) + [0.015 \times 20 \{(7 + 1.75)/2\}^2]/[2 \times 9.81 \{(1.5 + 3)/2\}] + (1.75 - 1)^2/(2 \times 9.81) + 10.34 = 0.053$ m.

The estimated pressure at the turbine exit is lower than the vapor pressure, so there is potential for cavitation. To allow for unforeseen fluctuations in tailwater elevations, the turbine setting needs to be lowered by about 1.0 m.

Dissolved Oxygen Content of Power Plant Releases

In some cases, water released from hydroelectric facilities contains low dissolved oxygen (DO), which is detrimental to fish and other aquatic organisms and reduces the waste assimilative capacity of the downstream channel. Therefore, DO enhancement becomes an important environmental consideration. In particular, this occurs at hydroelectric facilities located on stratified reservoirs. Deep reservoirs tend to stratify, particularly during summer months when the top layer of water (epilimnion) becomes warmer and less dense due to exposure to sunlight than the bottom layer (hypolimnion). The processes that use oxygen dissolved in the reservoir water include respiration of living organisms, decomposition of organic matter, and chemical reactions between dissolved matter and reservoir sediments. These processes occur in both the epilimnion and hypolimnion. However, the oxygen depleted from the epilimnion is replenished due to exposure to atmosphere and sunlight. But in the isolated hypolimnion in a stratified reservoir, there is continuous DO depletion with little replenishment. As a result, water withdrawn from the lower levels of a stratified reservoir has lower DO concentrations. When this water passes through the low-pressure regions of the turbine and draft tube, it releases air and oxygen, resulting in further reduction in DO. Thus, the water released from the turbine and draft tube contains low DO.

Methods to enhance DO in waters from hydropower plants include the following:

1. Selective Withdrawal Intakes: These intakes are designed with multiple inlets to simultaneously withdraw proportional quantities of water from the epilimnion (with high DO) and hypolimnion (with low DO), which are mixed before entering the turbine.

2. Turbine Venting and Compressed Air Injection in Draft Tube (USAEWES 1983): In this case, air is aspirated or drawn into the regions of subatmospheric pressure in the draft tube below the turbine exit. These regions of subatmospheric pressure may occur as described in "Draft Tube," or may be created by design. Vacuum is created through the installation of baffles or deflector plates near vent holes, which cause flow separation and localized low-pressure areas near the vent openings. Vents are often provided on hydroturbines as part of a vacuum breaker system to protect the turbine during rapid shutdown and to reduce vibration and cavitation during normal operation. Compressed air injection into the draft tube is used when hydraulic conditions do not permit the use of the aspirating type of turbine venting.

3. Reservoir Aeration or Destratification: This involves oxygen injection using diffusers or by introduction of a diffused air bubble plume in the water column. The rise of air bubbles from the hypolimnion to the epilimnion tends to destratify the reservoir and add oxygen to the lower water column.

The design and economics of the DO enhancement devices are based on model experiments, experience at similar installations, and consultation with manufacturers of specific devices. Care must be taken to ensure that DO enhancement does not result in supersaturation of water with gases, which may cause gas-bubble disease in fish. Gas-bubble disease results when water becomes supersaturated with nitrogen, oxygen, and argon, and the dissolved gases come out of solution in the blood of fish. If the intake is located at a great depth in the reservoir, water may be released to the tailrace, where pressure may be lower and temperature may be higher. This may create potential for gases coming out of solution, resulting in gas-bubble disease (ASCE 1989).

Infiltration into Rock Tunnels

In addition to evaluating the size and lining requirements of tunnels, the water resources engineer has to estimate potential infiltration through rock tunnels so that the contractor may make appropriate arrangements for groundwater control during the construction and operation phases. Measured and estimated rates of infiltration through the dolomite formations encountered in Chicago's tunnel and reservoir plan (TARP) system varied from 61 to 1,176 m^3/day per kilometer length of tunnel (Dalton and La Russo 1979). In many field situations, adequate data are not available for the calibration and implementation of sophisticated dual-porosity or fracture flow models to simulate saturated–unsaturated groundwater flow through various rock formations. Usually, infiltration will occur from the top surface or overburden to the location of the tunnel through several intermediate rock formations. Because of the empirical nature of the estimation procedures, it is advisable to estimate infiltration rates using several different methods and select the design value by judgment. Some approximate methods are described here.

Continuum Approach

This approach involves representation of the joints and fractures in rock formations overlying the tunnel segment by equivalent homogeneous porous media using hydraulic conductivities obtained by slug tests or packer tests in different formations. This approach is a reasonable approximation of the hydraulic characteristics if the spacing of the joints, bedding planes, and fractures is fairly dense and the hydraulic behavior of the rock masses is similar to granular porous media (Freeze and Cherry 1979).

In this approach, it is assumed that the hydraulic conductivity of each formation obtained from slug and packer tests is the one causing infiltration into the tunnel. This means the estimated average hydraulic conductivity represents the vertical hydraulic conductivity of the respective formations. Further, it is conservatively assumed that the joints, fractures, or porous media in the overburden and underlying formations are hydraulically connected.

Each formation may have a different hydraulic conductivity and thickness. An equivalent vertical hydraulic conductivity for all formations above a tunnel segment is estimated by

$$K \text{ (equivalent)} = \Sigma H_n / \Sigma (H_n / K_n) \qquad (4\text{-}2)$$

where

K = equivalent vertical hydraulic conductivity for infiltration into a tunnel segment (m/day)

H_n = thickness of formation n (m)

K_n = hydraulic conductivity of formation n (m/day)

Σ is over all formations overlying the tunnel segment

The infiltration will be under a unit hydraulic gradient in the vertical direction along the perimeter of the tunnel. The tunnel will act as a sink at atmospheric pressure. Thus, there may be some concentration of streamlines in its vicinity. To account for this, a flow concentration factor of 1.5 may be used. Thus,

$$Q = 1.5 \times K \times i \times L \times P \qquad (5\text{-}105)$$

where

Q = infiltration rate (m³/day)

L = length of tunnel segment (m)

i = hydraulic gradient for vertical flow

P = perimeter of tunnel cross section (m)

Actually, the major portion of infiltration would occur through the top half of the tunnel. Eq. (5-105) is approximate because it assumes infiltration to be uniformly distributed on all sides of the tunnel.

Infiltration through Soil-Filled Joints

In some cases, there may be identifiable soil-filled joints in rock formations. Usually, most of these joints may be filled with fine sand, silt, and clay particles transported from the surface or overburden with infiltrating rainwater. Based on published data for hydraulic conductivities, the fill materials may be assigned lower and higher limit hydraulic conductivities (e.g., 10^{-3} and 10^{-2} cm/s).

Based on site-specific data on distribution of joint sizes, weighted average width and spacings of rock joints are estimated. The spacings are used to estimate the number of potential joints in the tunnel segment. The number of joint sets in a tunnel segment is given by

$$N = [(L_T/NW) + (L_T/NE)] \qquad (5\text{-}106)$$

where

L_T = length of tunnel segment (m)

NW = weighted average spacing of one set of joints (m)

NE = weighted average spacing of the orthogonal set of joints (m)

If required, other sets of joints may be added.

Since the rock matrix is relatively tight, infiltration through the rock matrix may be assumed to be negligible and the total infiltration may be assumed to occur only through soil-filled joints under saturated conditions. The flow is assumed to be through vertical slits filled with clay, silt, and/or sand/gravel; there may be little pressure flow through the bottom of the tunnel sections. The slits are assumed to open along the ceiling and side walls of the tunnel. Therefore,

$$Q = N \times (W + 2T) \times b \times K \times i \qquad (5\text{-}107)$$

where

Q = infiltration in the tunnel segment (m³/day)

N = estimated number of total joints or fractures in the tunnel segment

W = width of tunnel (m)

T = height of tunnel (m)

b = weighted average width of a typical joint (m)

K = hydraulic conductivity of fill material (m/day)

i = hydraulic gradient for vertical infiltration

Infiltration through Fractures

In some cases, infiltration may be through narrow fractures in the formations. Flow through a narrow fracture is given by the following (Streeter 1971):

$$q = g\, b^3\, i/(12\nu) \qquad (5\text{-}108)$$

where

q = flow in m³/s per meter length of the fracture

b = plausible width of a water-filled joint (m)

i = hydraulic gradient

g = gravitational acceleration (m/s³)

ν = kinematic viscosity of water (m²/s)

The range of widths of open apertures in rocks used in some previous analyses is 0.001 to 0.002 cm (Long et al. 1982; Long and Witherspoon 1985; Freeze and Cherry 1979); however, widths of rock joints may be significantly greater. If field data are not available, then a relatively high fracture width (e.g., 0.002 cm) may be assumed.

Because of the zigzag patterns of interconnected fractures, water flowing through interconnected fractures follows a tortuous path. Tortuosity, τ, is defined as the square of the ratio of actual length of flow along zigzag flow paths in a porous medium to the straight length. This gives the following approximate hydraulic gradient (Brooks and Corey 1964; Delleur 1999):

$$i \cong 1/\sqrt{(\tau)} \qquad (5\text{-}109)$$

Tortuosity may be estimated by laboratory experiments on rock cores. Where no data are available, an approximate value of $\tau = 5$ may be used for preliminary estimates and the results may be adjusted with appropriate factors of safety.

The length of fractures per square meter area of rock may be estimated from fracture mapping. For preliminary estimates in cases where such maps are not available, fracture maps for other similar formations may be used with appropriate safety factors. Then,

$$Q = q\, L_f\, W L_T \qquad (5\text{-}110)$$

where L_f = length of fractures per square meter of rock. If, in addition to the ceiling of the tunnel, fracture openings are anticipated along the side walls as well, then W in Eq. (5-110) should be changed to $W + 2T$.

Table 5-29. Thicknesses and hydraulic conductivities of formations above tunnel bottom

Formation	Thickness (m) H_n	Hydraulic conductivity (m/day) K_n	H_n/K_n
A	12.2	0.006736	1,811.2
B	26.7	0.071807	371.8
C	7.6	0.056751	133.9
D	13.7	0.038951	351.7
E	11.4	0.020482	556.6
Sum	71.6	—	3,225.2

Example 5-24: Make a preliminary estimate of infiltration through a tunnel segment for inclusion in a bid package for tunnel construction. There are five rock formations above the tunnel. The thicknesses and hydraulic conductivities (obtained by packer tests) of these formations are given in Table 5-29. Inspection of cores and fracture mapping indicates that there may be a mix of narrow joints and fractures with average aperture length of 3 m/m² of rock surface. The average width of apertures (with narrow fractures and some relatively wide air- or water-filled joints) is 0.01 cm. The tunnel segment has a width of 10 m, height of 7 m, and length of 300 m. Assume $\tau = 5.0$ and a kinematic viscosity of water of 1.31×10^{-6} m²/s at about 10° C.

Solution: Using the continuum approach, with $W = 10$ m and $T = 7$ m, $P = 2 \times (10 + 7) = 34$ m.

Using Eq. (4-2), equivalent vertical hydraulic conductivity of the formations above the tunnel segment, $K = 71.6/3225.2 = 0.0222$ m/day.

Using Eq. (5-105), infiltration rate for this tunnel segment, $Q = 1.5 \times 0.0222 \times 1.0 \times 300 \times 34 = 340$ m³/day.

Using the fracture flow approach with $b = 0.0001$ m, $\tau = 5.0$, and $\nu = 1.31 \times 10^{-6}$ m²/s, $W = 10$ m, $L_T = 300$ m, and $L_f = 3$ m.

$i = 1/\sqrt{5.0} = 0.447$, and $q = 9.81 \times (0.0001)^3 \times 0.447 \times 10^6/(12 \times 1.31) = 0.00000028$ m²/s $= 0.0241$ m²/day and $Q = 0.0241 \times 300 \times 10 \times 3 = 216.91$ m³/day.

The estimated range of infiltration rates into the tunnel segment is about 217 to 340 m³/day. These estimates must be updated when site-specific data become available.

CHAPTER 6

Economic Analysis

Estimates of Costs and Benefits of Water Resources Engineering Projects

The following costs are associated with a typical water resources engineering project:

1. Land acquisition

2. Mobilization/demobilization of equipment and labor

3. Buildings for storage, workshops, and housing

4. Civil structures and other facilities associated with the project (e.g., diversion or coffer dam, tunnels, intake structures, dam and its appurtenant structures, levees, erosion protection, canals, wells, and excavation of contaminated soils or other remediation activities)

5. Access roads and communication systems

6. Rehabilitation, relocation, and resettlement of evacuees

7. Permitting and environmental compliance, including monitoring

8. Management during construction

9. Operation and maintenance during life of the project

The benefits may include primary benefits (e.g., municipal water supply, irrigation, electric energy, flood control, environmental enhancement, and remediated site); secondary benefits (e.g., developments occurring because of the availability of the water supply, flood control facilities, electric energy, and environmental enhancement); and tertiary benefits (e.g., development of other businesses, utilities, and industries incidental to project-related developments).

Both costs and benefits may be expressed in tangible (dollar-denominated) and/or intangible (non–dollar-denominated) terms. Intangible terms include expression of project-related impacts as minor, medium, or major and insignificant or significant. In addition, they may be expressed as number of people or households, trees, archaeological features, and acreage (e.g., areas of fisheries and wildlife habitat) affected by the project.

Benefit-Cost Analysis

Economic analyses of water resources engineering projects involve comparison of costs and benefits of projects or alternatives of a project. Costs and benefits are expressed as annualized or capitalized values over the lifetime of the project with appropriate rates of discount (interest) and price escalation over time. For simplistic comparative evaluation of alternatives, price escalation is usually ignored. The annual costs may include annualized values of the capital cost, annual operation and maintenance cost, and annual value of liability for damages and indemnification/insurance costs associated with the construction and operation of the project. The annual benefits may include monetary values of primary and secondary benefits attributable to the project. The costs of liability and indemnification or insurance may include annual payments to cover damages to utilities, facilities, or communities attributable to the operation of project facilities (e.g., malfunction of a warning system) or in the event of potential failure of project structures (e.g., damages resulting from failure of a dam). In addition, these costs may include capital cost to reconstruct or rehabilitate the structure or provide equivalent benefits for the projected residual life of the project. These costs are estimated using probabilities of failure, lifetime of the project, discount rate, and estimated monetary value of failure consequences (ASCE 1988; Prakash 1992a, 1992b).

The capital cost (present-day cost) may be converted to an equivalent annual cost by

$$CA = CP\, i\, (1 + i)^n / [(1 + i)^n - 1] \qquad (6\text{-}1)$$

where

CA = equivalent annual cost

CP = capital cost (present cost)

i = discount rate per annum per dollar

n = estimated life of the project

All costs are in present-day dollars

The capital recovery factor, CRF, is the annual value that, after n years, will yield the equivalent of one dollar invested today. CRF is given by

$$CRF = i(1 + i)^n / [(1 + i)^n - 1] \qquad (6\text{-}2)$$

The benefit-cost ratio, B/C, is computed as

$$B/C \text{ ratio} = BA / (CA + OM) \qquad (6\text{-}3)$$

where

BA = annual benefits

OM = annual operation and maintenance cost

Future price escalation in present-day annual costs and benefits may be incorporated by the following (USACE 1979; Prakash 1992a):

$$A = CA\,(1 + j)\,i\,[(1 + i)^n - (1 + j)^n] / [\{(1 + i)^n - 1\}(i - j)] \qquad (6\text{-}4)$$

where

A = annual value adjusted for price escalation

CA = estimated annual value of benefits or costs

j = rate of price escalation per dollar per year

Usually, rate of price escalation, j, will be less than the discount rate, i.

Average annual damage that is protected by a flood control project can be estimated by the expected value approach:

$$ED = \Sigma D(P) \times \Delta P \qquad (6\text{-}5)$$

where

ED = expected damage

$D(P)$ = damage likely to occur due to a flood of probability, P

ΔP = incremental probability or frequency

Σ = from 1 to the total number of incremental probabilities

The steps of computations are listed below:

- List return periods, T, for different floods from 1 yr to the upper limit flood (column (1), Table 6-1).

Table 6-1. Computation of expected damage for a flood control project

(1) Return period (yrs)	(2) Probability (P)	(3) Frequency or incremental probability (ΔP)	(4) Damage, $D(P)$ (thousand dollars)	(5) Expected damage $D(P) \times \Delta P$ (thousand dollars)
1.005	0.995	0.005	0	0
2.000	0.5	0.495	300	148.5
2.500	0.4	0.100	400	40
3.333	0.3	0.100	500	50
5.000	0.2	0.100	550	55
10.000	0.1	0.100	600	60
20.000	0.05	0.050	650	32.5
25.000	0.04	0.010	700	7.0
33.333	0.03	0.010	750	7.5
50.000	0.02	0.010	800	8.0
100.000	0.01	0.010	850	8.5
200.000	0.005	0.005	1,000	5.0
>200	<0.005	0.005	1,100	5.5
Total		1.000		427.5

- Compute cumulative probabilities for each return period; $P = 1/T$ (column (2), Table 6-1).

- Compute incremental probabilities between adjacent cumulative probabilities (column (3), Table 6-1). Note that the sum of all incremental probabilities is 1.0.

- Estimate damages associated with floods of each return period (columns (1) and (4), Table 6-1).

- Compute expected damages associated with each incremental probability, $D(P) \times \Delta P$ (column (5), Table 6-1).

- Estimate total (expected annual) damage as indicated by Eq. (6-4) (column (5), Table 6-1).

Example 6-1: A flood control project involves channelization, levee construction, and erosion protection. The cost of the project (in present-day dollars) is $4.5 million. The discount (interest) rate for the project life of 50 yr is 7%. Estimated flood damages likely to be prevented by the project are shown in column (4) of Table 6-1. The estimated operation and maintenance cost in present-day dollars is $25,000 per year. Compute the B/C ratio, ignoring price escalation and also considering price escalation at 3% per year.

Solution: The damages that would have occurred due to floods of different probabilities but would be prevented by the project are shown in columns (2) and (4) of Table 6-1. The frequency or incremental probability of each flood event or prevented damage is shown in column (3). Column (3) is the difference of two successive values of P in column (2). The expected damage is the product of columns (3) and (4) and is shown in column (5).

The estimated total annual prevented damage is $427,500. To simplify the analysis, it is assumed that this average annual benefit is a deterministic (rather than probabilistic) estimate.

Ignoring price escalation, Eq. (6-1) gives $CA = 4.5 \times 0.07 \, (1.07)^{50} / [(1.07)^{50} - 1] \times 10^6 = \$326{,}069$.

Adding annual operation and maintenance cost, total annual cost $= CA + OM = \$326{,}069 + \$25{,}000 = \$351{,}069$.

B/C ratio $= 427{,}500/351{,}069 = 1.22$.

With price escalation, the dollar value of project benefits and operation maintenance costs would increase. The capital cost has already been incurred. So price escalation may not affect its annualized value. Using Eq. (6-4) with $j = 0.03$, adjusted annual benefit $= 427{,}500 \times (1.03) \times 0.07 \times [(1.07)^{50} - (1.03)^{50}] / [\{(1.07)^{50} - 1\} (.07 - .03)] = 427{,}500 \times 1.588 = \$678{,}870$.

Similarly, adjusted annual operation and maintenance cost $= \$25{,}000 \times 1.588 = \$39{,}700$.

Total annual cost $= \$39{,}700 + \$326{,}069 = \$365{,}769$.

Adjusted B/C ratio $= 678{,}870/365{,}769 = 1.86$.

Evaluation of Water Resources Engineering Project Alternatives

The B/C ratio is a reasonable index for comparative evaluation of different project alternatives or for evaluation of the economic viability of a project. An additional refinement in economic analyses of project alternatives where additional capital investment may increase project benefits is to evaluate incremental costs and economical benefits of various project alternatives.

Example 6-2: Evaluate the economic feasibility of six alternatives of a flood control project with incremental costs and benefits indicated in Table 6-2.

Solution: Based on B/C ratios shown in Table 6-2, alternative 3 is the preferred option. However, if the economic capacity of the project sponsors permits, alternative 4 may provide an additional annual benefit of $130,000 with additional annual cost of $100,000 and may be worth consideration. This is indicated by an incremental benefit/incremental cost ratio of greater than unity.

A subjective method for comparative evaluation of various project alternatives where both monetary and intangible benefits and costs are to be considered is a combination of the delphi and fuzzy set approaches (Prakash 1991). In the delphi approach, a panel of experts is constituted that comprises a water resources engineer; representatives of the owner/operator, beneficiaries, and impacted communities of the project; a member of the regulatory agency; and an environmental scientist. The composition of the panel may vary depending on the type of project and relative importance of different evaluation factors. The panelists participate in several delphi sessions.

In the first session, all factors for comparative evaluation of the project alternatives are listed, and the impacts associated with each alternative are discussed and identified. Each panelist independently assigns a weight to each evaluation factor such that the sum of all weights is 1.0; that is,

$$\Sigma W(i) = 1.0 \qquad (6\text{-}6)$$

where $W(i)$ = fractional weight assigned to evaluation factor i, and the summation is over all evaluation factors. The weights assigned by each panelist along with his or her rationale are reviewed by the entire panel to arrive at a consensus on weights. Based on these discussions, each panelist is asked to revise his weights. The process is repeated until a set of agreeable weights is determined for all evaluation factors. This results in a column vector $W(i)$.

Table 6-2. Computation of incremental costs and benefits

Alternative	Annualized cost (dollars)	Incremental cost (dollars)	Annualized benefits (dollars)	Incremental benefit (dollars)	B/C ratio	Incremental benefit/ incremental cost
1	220,000	0	250,000	0	1.14	0
2	258,000	38,000	300,000	50,000	1.16	1.31
3	310,000	52,000	420,000	120,000	1.35	2.31
4	410,000	100,000	550,000	130,000	1.34	1.30
5	568,000	158,000	600,000	50,000	1.06	0.32
6	670,000	102,000	680,000	80,000	1.01	0.78

In the second delphi session, each panelist assigns to each alternative a score that corresponds to each evaluation factor on a scale of 0 to 10. A score of 10 represents the most favorable and 0 the most adverse impact. As in the first session, the scores of all panelists are revised based on mutual discussions until a set of agreeable scores is obtained for each alternative for all evaluation factors. For evaluation factors with quantitative values (e.g., capital cost, annual benefits, etc.), an arbitrary score of 5.0 is assigned to the median value for all alternatives. Any increments or decrements with respect to the median value are represented by incremental additions or subtractions to the median score of 5.0. These increments are also determined in this delphi session.

Based on the information about weights and scores, a fuzzy set evaluation matrix is tabulated. The evaluation factors form the columns and the number of alternatives form the rows of this table. Each row of the table (matrix) includes scores assigned to an alternative with respect to each evaluation factor. The weights assigned to each evaluation factor are included in a column (i.e., as a column vector). The product of the tabular matrix with the column vector of weights gives a column vector containing the weighted score of each alternative. Each entry in a row is multiplied by the corresponding weight in the column vector, and all the products are added to get the weighted average score for the alternative in that row. This can be accomplished manually or on a spreadsheet (e.g., Excel). The weighted scores represent the ranking of each alternative.

Example 6-3: Nine alternatives and eight evaluation factors have been identified for a water resources engineering project. The weights and scoring criteria are determined in several delphi sessions and are shown in Tables 6-3, 6-4, and 6-5. Perform a comparative evaluation of the alternatives and identify the preferred options using the two sets of weights indicated in Table 6-3.

The failure consequences include the cost of damages to utilities and facilities resulting from structural failures (e.g., breach in a dam, levee, or water storage tank). Social impacts include disruption of communities, relocation of houses, schools, hospitals, and other social facilities.

Solution: The scoring criteria for tangible evaluation factors determined in the delphi sessions are given in Table 6-4.

Scoring criteria for constructibility and other intangible factors are included in Table 6-5.

The resulting scores for each alternative for the eight evaluation factors are shown in Table 6-6.

The weighted average score for alternative 1 is $9 \times 0.2 + 3.5 \times 0.2 + 9 \times 0.2 + 9 \times 0.1 + 9 \times 0.05 + 7 \times 0.05 + 9 \times 0.1 + 9 \times 0.1 = 7.8$.

Proceeding in the same manner, the weighted average scores of the nine alternatives are shown in Table 6-7.

In this case, the least costly alternative (alternative 1) is indicated to be the preferred option.

If economic benefits of the project are given high weight, as indicated by the alternate weights in Table 6-3, then the weighted scores will be as shown in Table 6-8.

In this case, the most expensive alternative turns out to be the preferred option. This demonstrates that the method takes into account subjective factors that the decision maker may have to consider in selecting the preferred course of action.

Table 6-3. Description of alternatives and weights of evaluation criteria

| | Annualized (million dollars) | | | | Intangibles | | | |
| | | | | | Impacts | | | |
Alt.	Cost	Benefit	Failure consequences	Constructibility	Ecological	Water quality	Social	Other users
1	1.0	1.05	1.1	Easy	Almost none	Minimal	Almost none	Almost none
2	1.2	1.22	1.6	Easy	Moderate	Moderate	Minimal	Minimal
3	1.4	1.35	2.1	Easy	Moderate	Moderate	Minimal	Minimal
4	1.6	1.52	2.6	Moderate	Moderate	Minimal	Moderate	Moderate
5	1.9	1.6	3.1	Moderate	Minimal	Minimal	Moderate	Moderate
6	2.2	1.75	3.5	Difficult	Minimal	Minimal	Moderate	Moderate
7	2.6	2.0	4.0	Difficult	Moderate	Significant	Significant	Significant
8	3.3	2.4	4.6	Very difficult	Significant	Significant	Significant	Significant
9	4.3	3.0	4.7	Very difficult	Significant	Minimal	Significant	Significant
Wt.	0.2	0.2	0.2	0.10	0.05	0.05	0.1	0.1
Alt. Wt.	0.05	0.5	0.05	0.08	0.08	0.08	0.08	0.08

Alt. = alternative; Wt. = weight assigned to the evaluation factor; Alt. Wt. = alternative weight assigned to the evaluation factor.

Table 6-4. Scoring criteria for tangible factors

Cost (million dollars)	Score	Benefits (million dollars)	Score	Failure consequences (million dollars)	Score
0.5– 1	9	0.5–0.99	3	1.0–1.4	9
1.1–1.25	8	1.0–1.19	3.5	1.5–1.9	8
1.3–1.5	7	1.2–1.29	4	2.0–2.4	7
1.1– 1.75	6	1.3–1.49	4.5	2.5–2.9	6
1.8–2.0	5	1.5–1.7	5	3.0–3.2	5
2.1–2.4	4.5	1.71–1.8	6	3.3–3.8	4.5
2.5–3.0	4.0	1.81–2.1	7	3.9–4.4	4
3.1–4.0	3.5	2.15–2.5	8	4.5–5	3.5
>4.0	3	>2.5	9	>5	3

Table 6-5. Scoring criteria for intangible factors

Constructibility		Other intangibles	
Factor value	Score	Factor value	Score
Easy	9	Almost none	9
Moderate	6	Minimal	7
Difficult	5	Moderate	5
Very difficult	4	Significant	4

Table 6-6. Evaluation matrix

| Alternative | Evaluation factors | | | | | | | | Weights |
	1	2	3	4	5	6	7	8	
1	9	3.5	9	9	9	7	9	9	0.2
2	8	4	8	9	5	5	7	7	0.2
3	7	4.5	7	9	5	5	7	7	0.2
4	6	5	6	6	5	7	5	5	0.1
5	5	5	5	6	7	7	5	5	0.05
6	4.5	6	4.5	5	7	7	5	5	0.05
7	4	7	4	5	5	4	4	4	0.1
8	3.5	8	3.5	4	4	4	4	4	0.1
9	3	9	3.5	4	4	7	4	4	

Table 6-7. Weighted average scores, Case 1

Alternative	Weighted average score
1	7.8
2	6.8
3	6.5
4	5.6
5	5.3
6	5.2
7	4.75
8	4.6
9	4.85

Table 6-8. Alternate weighted average scores, Case 2

Alternative	Weighted average score
1	6.09
2	5.44
3	5.59
4	5.34
5	5.4
6	5.77
7	5.66
8	5.95
9	6.665

CHAPTER 7

ENVIRONMENTAL ISSUES AND MONITORING

Introduction

Water is a natural resource essential for the sustenance of life. It is continually depleted and replenished through the hydrologic cycle. However, a large number of competing uses makes equitable appropriation of water for different uses extremely complex. Development of water resources for one or a few selected uses impacts its availability for other potential uses. One type of use deemed as consumptive in one form may be nonconsumptive in the holistic framework of the hydrologic cycle. For instance, water diverted for irrigation may be considered consumptive as far as the surface water resources of a reservoir system are concerned because that water would not be readily available for other uses within a certain local environment. However, the portion of diverted water that is lost as seepage and reappears as return flow in a different water body may constitute nonconsumptive use in the sense of an overall water balance in nature. Nonavailability of water of required quantity and quality for uses other than those for which it is developed under a water resources engineering project results in impacts on other environmental features.

To ensure optimum and equitable use of the available quantity and quality of water in a region, it is necessary to evaluate all conceivable beneficial and adverse environmental impacts of a water resources engineering project. The term "environment" includes both natural and physical environment and the relationship of people with that environment (e.g., land; ground- and surface water regime; air; ecosystem; living organisms; geological and archaeological features; historic artifacts; cultural, social, economic, and political units; and environmental values at the site). Potential redistribution of water resources among different types of past, present, and future uses following the development of a water resources engineering project may result in conflicts among competing interests. A viable method to resolve or minimize such conflicts is to prepare an environmental assessment (EA) or environmental impact statement (EIS) before decision making. Impact is a change or consequence, which may be positive or negative, that results from the project.

In the United States, evaluation of environmental impacts of dams and other water resources engineering projects is governed by the following:

- National Environmental Policy Act (NEPA) of 1969, as amended (Public Law 91-190, 42 U.S.C. 4321 et seq.)

- Environmental Quality Improvement Act of 1970, as amended (42 U.S.C. 4371 et seq.)

- Executive Order 11514, Protection and Enhancement of Environmental Quality (March 5, 1970, as amended by Executive Order 11991, May 24, 1977)

NEPA establishes policy, sets goals, and provides means for implementing the policy and contains action-forcing provisions to ensure that federal agencies act according to the letter and spirit of the act. NEPA procedures ensure that environmental information about the project is available to public officials and citizens before decisions are made and before actions are taken. Accurate scientific analysis, expert agency comments, and public scrutiny are essential to implementing NEPA. The federal agencies are responsible for the following:

- Interpreting and administering the policies, regulations, and public laws related to NEPA
- Implementing procedures to make the NEPA process more useful to decision makers and the public
- Emphasizing real environmental issues and alternatives
- Integrating the requirements of NEPA with other planning and environmental review procedures
- Encouraging and facilitating public involvement in decisions that affect the quality of the human environment
- Using the NEPA process to identify and assess reasonable alternatives to proposed actions
- Using all practicable means to restore and enhance the quality of the human environment and to avoid or minimize any possible adverse effects upon the quality of the human environment (CFR 40, Part 1500)

The term "environmental impacts" includes all beneficial or adverse impacts or alterations in the hydrologic regime attributable to the construction, modification, and operation of the dam, diversion of flow, or implementation of other water resources engineering activities. Impacts can be structural; nonstructural; ecosystem-related; aesthetic; archaeological; water-, sediment-, and soil-related; economic (commerce-related); streamflow- and water-quality–related; geomorphic; socioeconomic; cultural; and recreational. EISs have to be prepared using an interdisciplinary approach that ensures integrated use of the natural and social sciences and the environmental design arts. In addition, different states have developed guidelines or regulations for the preparation of EISs for dams and environmental permit requirements for other water resources engineering projects in their jurisdictions. These guidelines or regulations have to be followed concurrent with federal guidelines.

Similar guidelines or regulations have been established by the World Bank and other organizations that provide assistance for the construction and operation of water resources engineering projects. In addition, many countries have developed their own guidelines for the preparation of an environmental impact report (EIR) or EIS for water resources engineering projects.

In some countries, including the United States, a system of water rights has been established to facilitate the distribution and use of ground- and surface water. There are two types of water rights: riparian and appropriative. Withdrawal of water from streams is governed by appropriated or riparian water rights that generally are administered by state agencies. Under riparian rights, the owner of land adjacent to a stream is entitled to reasonable and beneficial use of the natural flow of the stream without change in quality or quantity. The adjacent land is called riparian land. Owners of upstream riparian water rights may not significantly increase or decrease the natural flow to the disadvantage of the downstream water rights owner. Appropriated water rights establish quantities and priorities for water withdrawal for beneficial use by each owner of the water right, provided sufficient quantities are available after satisfying the demands of senior water rights. They are based on the concept of first in time, first in right. The water rights may be transferred from one entity or owner to another. Also, the points of withdrawal may be transferred from one location to another using appropriate legal and administrative procedures. When available water is short, the holder of senior water rights may use his or her entire allocation, whereas holders of junior water rights have to live with the shortage. Water resources engineering projects, which may jeopardize senior water rights, are not permitted without proper compensation. Appropriative water rights have been established in most western states of the United States. In many cases, groundwater is considered tributary to surface water with which it is in hydraulic connection.

Environmental Impact Statement

Definition and Format of an EIS

An EIS is a document that describes the impacts on the environment that will result from implementing a proposed project. It also describes impacts of conceivable projects that may be alternatives to the proposed action, as well as plans to mitigate the impacts. Mitigation means minimization or elimination of negative impacts. "No action" also is treated as an alternative to the proposed project. Once a project plan is submitted to a regulatory agency, the agency prepares an EA, which evaluates if the proposed project is likely to have significant environmental impacts and whether or not an EIS is needed. If the EA determines that an EIS is not needed, the agency issues a finding of no significant impact (FONSI) that briefly explains why the action will not have significant impacts on the environment. Depending on the size, setting, sensitivity, and details of the proposed project, it may require an EIS, may require an EA but not an EIS, or may require neither an EA nor an EIS.

A typical EIS document includes the following components:

1. Cover sheet, which includes the title of the action; its location; EIS designation (i.e., final or draft); responsible federal agency and cooperating agencies; agency point of contact, with name, address, and phone number; date that comments on the EIS are due; and a one-paragraph abstract.

2. Summary (no more than about 15 pages), which includes a narrative summary of the proposed action and alternatives, conclusion, areas of controversy, and significant issues to be resolved.

3. Table of contents, which lists headings and subheadings, figures, tables, abbreviations, and scientific or other symbols used.

4. Sections that describe the following:

    - Purpose of and need for the action.
    - Alternatives, including the proposed action and no action, alternatives not rigorously explored, environmental consequences of alternatives in a comparative form, preferred alternative, and mitigation measures.

- Description of the affected environment
- Environmental consequences, including direct effects and their significance; indirect effects and their significance; conflicts with other federal, state, local, or native American tribe plans; energy requirements and conservation potential; natural or depletable resource requirements and conservation potential; impacts on historic and cultural resources; and mitigation measures.

5. List of authors, including names and qualifications.
6. Distribution list, which includes identification of agencies from which official comment is requested, identification of officials and organizations from whom comment is solicited, those who requested copies of the EIS (although identification of citizens is not necessary), and location of public access copies.
7. Index of environmental topics.
8. Appendices that contain material prepared in support of the EIS, analysis to support effects, and analytic computations relevant to the decision.

Content of an EIS

An EIS for a typical water resources engineering project provides the following information:

- Description of the proposed action, in sufficient detail to permit accurate assessment of the environmental impacts.
- Discussion of probable impacts on the riverine and human environment and means to mitigate adverse environmental impacts.
- Identification of adverse environmental impacts that cannot be avoided.
- Alternatives that might avoid some or all of the adverse environmental impacts, including cost analyses and environmental impacts of these alternatives.
- Assessment of the cumulative long-term impacts, including their relationship to short-term use of the riverine environment versus the environment's long-term productivity.
- Any irreversible or irretrievable commitment of the resources that might result or that might curtail beneficial use of the riverine environment and water resources.
- Discussion, objections, or comments by federal, state, and local agencies, private organizations, and individuals and how they are addressed.

The EIS also documents the following:

- Applicable water quality standards and stream segment classification
- Proposed/committed minimum flow releases
- Measures to protect and improve water quality
- Impacts on water quality
- Impacts on fish and wildlife and botanical resources, including threatened and endangered species, and measures to mitigate impacts on these resources

- Identification of impacted historical and archeological sites eligible for inclusion in the National Register of Historic Places
- Measures to enhance or create recreational opportunities
- Impacts on wetlands and land use and measures to mitigate such impacts
- Implementation or construction schedule for any proposed measures, including sources of financing for implementation

The EIR or EIS is site specific, and the level of detail it provides must be commensurate with the complexity of the possible environmental impact. The EIR or EIS includes statements/commitments related to monitoring programs during construction and operation, environmental tradeoffs, research and development, and restoration measures that will be taken routinely or as the need arises. It documents provisions for pre- and post-monitoring of significant environmental impacts of the project, including programs for monitoring changes in operational phases and measures for detecting and modifying noise levels, monitoring air and water quality, inventorying key species in food chains, and detecting induced changes in weather. In addition, the EIR or EIS identifies federal, regional, state, and local regulations and codes that must be complied with in the construction, maintenance, and operation of the project.

In the case of certain relatively less sensitive projects or relatively small projects, an EA may be adequate. In general, an EA is a less detailed document than an EIR or EIS.

Environmental Impacts of Water Resources Engineering Projects

Environmental Impacts of Dams

Impacts of dams on the environment include impacts on channel reaches both downstream and upstream of the dam. In particular, these may include impacts on water quality, river morphology, flooding, fisheries, living marine resources, wildlife, threatened and endangered species, vegetation, cultural resources, land use, recreation, aesthetics, historical and archaeological features, and socioeconomics.

General environmental impacts on the riverine environments downstream of a dam or other hydraulic structures that must be analyzed and evaluated include the following (ASCE 1989; CFR 18, Chapter 1 and Part 380; Prakash 2002):

- Accumulation of stabilized and armored riverbed material downstream of the dam caused by successive clear water sluicing and lack of substrate redeposition.
- Occurrence of a relatively low (depressed) and constant thermal regime of water released from a deep reservoir intake.
- Profuse accumulation of algae caused by increases in transparency, nutrients, bed instability, and absence of fine sediments.
- Reduced species diversity in the macroinvertebrate community, which may be attributed to lack of water temperature variations necessary for completion of important life history events, such as egg hatching and maturation.
- Increased macroinvertebrate biomass and low diversity associated with flow constancy and/or organic loading from the reservoir.
- Impacts on threatened or endangered species listed by federal and state agencies.

- Impacts on federally designated wild and scenic rivers.

- Impacts on groundwater levels downstream of the dam due to long-term seepage from the impoundment.

- Impacts on migratory fish due to blockage of passage and relatively low adaptability of fish to fish ladders and fishways.

- Impacts associated with dissolved oxygen (DO) deficit due to biological and chemical processes in the reservoir that consume oxygen.

- Impacts associated with higher dissolved concentration of metals, such as iron and manganese, because of the increase in their solubility in a relatively low DO environment.

- Impacts associated with supersaturation of water released from the reservoir, resulting in gas-bubble disease in fish population. When water plunges to a great depth downstream of a dam at the end of a spillway, it traps air. The increase in depth during the plunge causes more gases to dissolve than would occur at the surface. When this supersaturated water rises from the bottom of the plunge to near the water surface in the downstream channel, gases begin to come out of solution, resulting in gas-bubble disease. Water in deeper reservoirs is sometimes saturated with dissolved gases. When this water is released through a deep reservoir intake, it is exposed to much lower pressure and higher temperature in the downstream channel where gases may begin to come out of solution, resulting in gas-bubble disease.

- Modification of the river regime due to alteration in the natural flow hydrograph, including alterations in floodplains, wetlands, islands, and sandbars, and impacts on aquatic population associated with these features.

- Lower assimilative capacity of the downstream channel for wastewater discharges because of flow diversions and regulation at the dam.

- Increased risk of flooding with attendant loss of life and property due to potential dam failure.

- Beneficial impacts associated with primary, secondary, and tertiary tangible and intangible benefits of the dam related to the downstream environment.

Environmental impacts on the hydrologic environment upstream of a dam or other hydraulic structures that must be evaluated include the following:

- Areas and features inundated or submerged by the impoundment (such as utilities, archaeological and historical monuments, religious sites, old cemeteries, farmland, and any sites either listed or determined to be eligible for inclusion in the National Register of Historic Places), and impacts on land use, wildlife habitat, forests, minerals, and unique ecosystems.

- Population displaced and resulting impacts on human health and safety and on aesthetic and cultural values and living standards.

- Increased flood levels and frequency of flooding due to hydraulic obstruction in the river channel caused by the dam.

- Potential for landslides and enhanced soil erosion along the shoreline of the reservoir.

- Impacts on flora and fauna in the reservoir area and in the upstream channel.
- Potential for degradation of water quality due to long-term interaction between sediment deposited and water stored in the reservoir.
- Beneficial impacts associated with primary, secondary, and tertiary tangible and intangible benefits of the dam related to the upstream environment.

Environmental Impacts of Other Water Resources Engineering Projects

Other water resources engineering activities that result in adverse impacts on the environment include discharges of polluted water from point and non-point sources. Point source discharges that impact riverine environments include sanitary wastewater, industrial wastewater, and storm water runoff from construction or mining areas, industrial facilities, roads, and parking lots through individual outfalls. Non-point sources of water pollution include runoff from agricultural areas and forests.

It is reported that non-point sources, which mainly include agricultural runoff, contribute two-thirds of the total water-quality–related impacts on riverine environments and one-half of the water-quality–related impacts on lakes and reservoirs (NRCS 1996; Ojima et al. 1999). Non-point source discharges from agricultural areas contain contaminants from fertilizers, herbicides, pesticides, livestock wastes, salts, and sediments. This results in increased salinity, nutrient loading, turbidity, and siltation of streams.

Another significant non-point source of environmental impacts on surface water quality is surface runoff by way of overland flow from abandoned mine waste piles and old tailings disposal areas. Rainwater infiltrating through such facilities or entering abandoned underground mine workings enters the groundwater environment as contaminated water (e.g., acid mine drainage or water containing high levels of dissolved solids), which eventually reach nearby streams. This type of impact may pertain to old coal mining areas, copper or other mineral mines, and uranium mines. In certain instances, uranium mill tailings disposed in river floodplains have become exposed to river flows and required remediation. If overland flow or seepage from such facilities is discharged through one or more outfalls, it may constitute a point source of stream pollution.

Similar types of impacts result from surface runoff and seepage from old abandoned industrial facilities.

Remediation or mitigation of impacts resulting from past mining and industrial activities has been undertaken through remediation programs implemented by governmental agencies, actions taken by private entities, and principal responsible parties identified by the regulatory agencies. In the United States, hydraulic impacts associated with existing and new mines are controlled and regulated by the federal Office of Surface Mining for coal mines; U.S. Nuclear Regulatory Commission for uranium mines; and federal and state agencies for environmental protection for other minerals through a system of permits, regulations, and inspections. Similar regulations are enforced by government agencies in many other countries.

Mining and other developmental (including urban and industrial) activities require relocation or diversion of streams to maximize ore recovery, minimize flooding, or enable construction of contiguous facilities. Relocation of streams has negative impacts on stream biota. Existing depressions and pits are sometimes incorporated into the relocated channel. This may result in decrease in lotic (flowing water) habitat and increase in lentic (still water) habitat in the stream.

Industrialization also results in water-quality–related impacts on the hydrologic environment. Widespread use of smokestacks and chimneys emanate flue gases with particulate matter containing different chemical constituents. These particles get deposited on tree leaves, twigs, and hill slopes or are kept in suspension in the atmosphere. When moist air passes by them, they form tiny nuclei for the condensation of water vapor. The chemicals are dissolved by water droplets and reach the ground as precipitation and the hydrologic environment with surface runoff. Because the droplets are predominantly acidic in character, this phenomenon is known as acid rain.

In the United States, under Section 303(d) of the 1972 Clean Water Act, each state must identify waters that are not expected to meet applicable water quality standards with existing point and non-point source discharges. The state must then prioritize these impaired waters, taking into account the severity of the pollution and the uses to be made of such waters. After prioritizing, it must set total maximum daily loads (TMDLs) that will result in each water body meeting ambient standards, accounting for seasonal variations and including a margin of safety (USEPA 1991b; Eheart et al. 1999). In order to set a TMDL, both the Criteria Continuous Concentration (CCS) and Criteria Maximum Concentration (CMC) must be considered. The U.S. Environmental Protection Agency (EPA) defines the CCC and CMC, respectively, as the four-day average and one-hour average concentrations of a pollutant in ambient water that should not be exceeded more than once every 3 yr on average. Because there are differences in toxicological effects on aquatic organisms, these two concentrations vary depending on the pollutant of concern. In order to translate these criteria into water-quality–based permit limits, suitable waste load allocation models are used to determine the fate and transport of any pollutants that enter a water body. These models perform multiple discharge waste load allocations for point source discharges under design streamflow conditions, which is usually taken to be the 7-day, 10-yr average low flow (7Q10) of the stream. Waste loads contributed by non-point sources may be assumed constant at the existing/ambient levels.

Withdrawal of water from a river for irrigation, municipal, or industrial uses may result in reduced assimilative capacity of the downstream channel and reduction in the instream flows necessary to sustain the aquatic population and ecosystem in the downstream reaches.

Return flows from agricultural areas enter stream channels by way of groundwater flow and impact riverine water quality. Generally, these return flows contribute high concentration of salts, pesticides, herbicides, and fertilizers in the riverine environment.

Evaluation and Analysis of Impacts

For the evaluation of environmental impacts of water resources engineering projects, federal and state agencies may require various types of analyses, physical and computer modeling, field monitoring of various parameters, and field surveys of biota under pre- and post-project conditions both upstream and downstream of a facility (e.g., TCEQ 2002). Additional analyses and monitoring may be required to address comments provided by different entities during public hearings associated with the approval of the EIS or EIR.

Evaluation of pre- and post-project flooding conditions in the river upstream and downstream of a dam or levee requires floodwater surface profile computations for pre- and post-project design flood hydrographs for the upstream and downstream channel reaches. Evaluation of flood risks associated with failure of dams or levees requires dam-break analysis for plausible postulated failure scenarios and development of emergency evacuation and warning plans for dams located upstream of populated areas and levees proposed to protect populated areas.

Evaluation of impacts on groundwater levels downstream of a dam may require groundwater modeling with and without the reservoir. These analyses may be used to determine if seepage from the dam would result in waterlogging and increased dampness, salinity, mosquito population, and incidence of diseases downstream of the dam.

Water quality monitoring and analyses are generally required to evaluate impacts on dissolved oxygen (DO), temperature, and other parameters of concern for specific river reaches and to evaluate impacts on the assimilative capacity of the river after construction of a dam, intake structure, water division project, or stormwater discharge facility.

Degradation of water quality may result due to water sediment interaction in the impoundment or other modifications in the hydrologic regime. Construction of a dam or other water resources engineering project may result in loss of artifacts of historic, religious, or ethnic significance. Often, such impacts are brought out in public hearings related to the EIS of the proposed projects and must be addressed.

Analytical studies are required to determine instream flow requirements for navigation, recreation, assimilative capacity, and fish and aquatic life in the channel downstream of a dam, intake structure, or water diversion project. A number of analytical methods are used to determine instream flow requirements for different aquatic species (ASCE 1989).

Field surveys are required to determine pre- and post-project aquatic population and biodiversity both upstream and downstream of a dam, intake structure, or water diversion project and to evaluate potential impacts on the ecosystem. Field surveys also are required to identify the locations and sizes of existing islands, sandbars, and meanders in the river channel and existing vegetation and animal populations sustained by these features. Geomorphologic studies are required to determine potential impacts of the construction and operation of the project on these features and the ecosystem supported by them. Field monitoring is required to determine pre- and post-project water quality of the stream.

Topographic surveys are required to identify areas, households, and utilities and monetary values of property that may be inundated or dislocated after the construction of a water resources engineering project. The cost of compensation for unavoidable impacts, insurance against damages or adverse impacts, and replacement costs for lost features attributable to a project are included in the computation of the benefit-cost ratio for the project. When a benefit-cost analysis is prepared, it must include analyses of unquantified environmental impacts, values, and amenities. Impacts for which monetary compensation cannot be calculated should be expressed in adequately defined qualitative and descriptive terms.

According to federal and state regulatory guidelines, dams are divided into different size categories, based on their storage capacities and heights, and into different hazard categories, based on potential damage to life and property due to failure. Different design and/or rehabilitation criteria are specified for different combinations of hazard and size classifications (e.g., low, significant, or high hazard potential combined with small, intermediate, or large size of a dam) (NRC 1985). For dams that do not or cannot meet specified design criteria, or where consequences of dam failure are more severe than those without failure, methods have been developed to perform risk analyses and determine costs for indemnifying the affected entities against damages attributable to potential dam failure (ASCE 1988). These indemnification costs, along with non–dollar-denominated environmental consequences of dam failure, may be included in the benefit-cost analysis for a dam.

In the United States, consensus is growing for the demolition of a number of dams where the benefits of removal are deemed to exceed those of operation and maintenance of the facilities. Analysis of the environmental impacts (including benefit-cost analysis) for such dams revealed that, even with fish passages, these dams have been a primary cause of fish mortality in the river system. The reservoirs have inundated riparian communities and natural wetlands, which were important in the cycling of water, nutrients, sediment, organic matter, and aquatic and terrestrial organisms in the riverine ecosystem. Cultural resources important to the local people were inundated, made inaccessible, or buried.

The natural transport of coarse sediment downstream is halted by dams, which may render the downstream channel unusable or less friendly to fish. Construction of dams results in a lowered or degraded river channel on the downstream side. The loss of river bed material may degrade anadromous fish habitat, allow vegetation to be established on gravel islands and floodplains, reduce natural river meandering, lower flood stages, and curtail the formation of slower moving side channels, periodic wetlands, and riparian areas. Many species require slower moving water, riparian vegetation, or a functional estuary to spawn or rear, all of which are impacted by the elimination of natural sediment transport. Reservoirs affect water quality by acting as large settling basins during floods, landslides, or other events, which will normally produce surges of turbidity downstream. Because of the reservoir, turbidity during such events becomes less intense but longer lasting. These impacts should be evaluated in both quantitative and qualitative terms.

In the United States, impacts of point source discharges on the hydrologic environment are evaluated and controlled through National Pollutant Discharge Elimination System (NPDES) permits issued under the Clean Water Act (33 U.S.C. 1251). NPDES permits are issued by the USEPA or state agencies responsible for environmental protection. Generally, these permits and regulations include discharge of sanitary wastewater, industrial wastewater, and storm water runoff from industrial and construction areas. In addition, impacts of the discharges of dredged or fill material into waters of the United States are controlled by the U.S. Army Corps of Engineers (USACE) through permits issued under Section 404 of the Clean Water Act (33 U.S.C. 1344), and those of the discharges from mining areas are controlled by the Office of Surface Mining, U.S. Nuclear Regulatory Commission, and other federal and state agencies.

The impacts of water withdrawals and entry of agricultural return flows on river water quality are estimated by water quality modeling for surface water bodies, groundwater flow and transport modeling, and water quality monitoring both upstream and downstream of the points of withdrawal and in the stream reaches receiving return flows.

Hydrologic impacts associated with point and non-point sources of pollution from abandoned mines and industrial facilities are evaluated by water quality monitoring of ground- and surface water, surface water runoff and erosion modeling, surface water quality modeling, and groundwater flow and transport modeling.

To evaluate water quality or thermal impacts due to point source discharges to a stream, usually mixing zones are permitted that comprise a specified length, width, and flow section (approximately 25 to 33% of the cross-sectional area or volume of flow) of the river. Normal water quality standards are not applicable within the mixing zones. Dispersion analyses are conducted to demonstrate that no more than a specified section of the river is impacted by the mixing zone under the 7Q10 condition in the river. Diffusers are incorporated into the outfall design to minimize the size of the mixing zone. Usually, mixing zones are not allowed in waters containing sensitive species (e.g., endangered species), fish spawning areas, bathing beaches, bank fishing areas, boat ramps, and other public access areas and must allow for a zone of passage for aquatic life in which water quality standards are met.

Environmental Monitoring

Introduction

The commitments included in the EIR or EIS for a particular project constitute the basis for post-construction and post-mitigation monitoring requirements. These documents identify environmental impacts of the project that cannot be mitigated and mitigation measures for impacts that can be mitigated by structural or nonstructural means, along with monitoring plans to evaluate the progression of unmitigable impacts and effectiveness of mitigation measures for mitigable impacts. Approval of the EIR or EIS by appropriate agencies is required before permits for construction, operations, or relicensing of the facilities are issued. Usually, these documents form a part of the permit application for the facility.

The EIR or EIS is a public document approved by appropriate regulatory agencies after a series of public hearings and technical reviews. The commitments for mitigation and monitoring of specific environmental impacts stipulated in these documents become legally and politically binding on the owners and operators of dams and other facilities. The performance of the mitigation measures and progression of unmitigable impacts of the project during the operating period are routinely scrutinized by the regulatory agencies, impacted public, and various nongovernmental organizations (NGOs) or environmental groups. Additional suspected post-mitigation impacts of the project (not identified in the EIR or EIS) can be and often are reported by the above-mentioned groups in the form of petitions, notices, or complaints. After adequate analysis and evaluation, the resolution of these complaints may constitute the basis of additional mandated environmental monitoring for the project.

Thus, post-mitigation environmental monitoring requirements are generally site-specific, depending on the sensitivity and environmental significance of the impacted environmental parameter (e.g., aquatic habitat, vegetation, morphology, hydrology, water quality, and socio-economics).

Components of Environmental Monitoring

The main components of an environmental monitoring plan include monitoring of fish and wildlife, habitat, and water quality.

Monitoring of fish and wildlife includes species composition, abundance, distribution, and habitat use of fish and benthic invertebrates; species composition, abundance, distribution, and habitat use of shorebirds, herons, waterfowl, and eagles; and species composition, abundance, and distribution of herpetofauna, and habitat use of amphibians and reptiles.

Monitoring of habitat includes aquatic and terrestrial monitoring. Aquatic monitoring includes water depth, water velocity, substrate size and composition, and quantity of large woody debris. Terrestrial monitoring includes plant species composition of riparian and floodplain vegetation and wetlands; wetland surface area, volume, duration, and depth; number, area, and elevation of unvegetated sandbars; groundwater elevations; floodplain land cover; and floodplain geomorphology and hydroperiod.

Water quality monitoring includes water quality parameters of concern in the region, including DO, temperature, turbidity, and nutrients for river reaches, reservoirs, and wetland habitats.

Environmental water quality parameters for various streams in the United States are monitored and published by the U.S. Geological Survey (USGS). Another source of water quality data of streams is the Storage and Retrieval (STORET) system maintained by the USEPA. Commonly monitored water quality parameters include specific conductance, pH, temperature, turbidity, DO, fecal coliform, fecal streptococci, hardness, calcium, magnesium, sodium, potassium, arsenic, barium, cadmium, chromium, cobalt, copper, iron, lead, manganese, mercury, selenium, silver, zinc, bicarbonate, carbonate, sulfate, chloride, fluoride, silica, total dissolved and suspended solids, nitrogen, phosphorus, organic carbon, PCBs, aldrin, chlordane, DDD, DDT, DDE, di-eldrin, di-azinon, endrin, ethion, heptachlor, lindane, malathion, methoxychlor, methyl parathion, methyl tri-thion, parathion, toxaphene, trithion, phytoplankton, and periphyton.

For projects under the jurisdiction of the U.S. Army Corps of Engineers (USACE), a policy has been established for conducting environmental compliance assessments with respect to the provisions for environmental monitoring stipulated in the EIS (USACE 1999). Internal assessments for compliance with environmental monitoring commitments are conducted annually by project/facility personnel. External assessments are conducted on a 5-yr cycle by outside contractors or personnel not employed at the facility. Records of the provisions of the EIS, amendments to the EIS, or other decisions pertaining to environmental mitigation are kept by the appropriate authority. Upon request from interested agencies or the public, reports on the progress of required mitigation, committed environmental monitoring, and other provisions have to be made available to them. Monitoring is required to ensure that the decisions and committed environmental monitoring, including mitigation measures, are implemented.

Some agencies have implemented ambient lake monitoring programs (ALMPs) to characterize trends in the environmental condition of lakes, determine causes of environmental degradation, monitor progress of restoration programs, and update lake classification/prioritization systems with respect to environmental quality and restoration. This involves collection and analysis of water quality and sediment samples as well as field observations for water color, weather, sediment deposition, algae, macrophytes, and other significant aspects of the lake. Ambient trend monitoring is conducted to determine long-term trends in water quality and to evaluate restoration programs. For this purpose, lakes are monitored once during the spring runoff and turnover period (April or May), three times during the summer (June, July, and August), and once during the fall turnover (September or October). The parameters analyzed include suspended solids, nutrients, chlorophyll, DO and temperature profiles, and other field parameters for water samples collected from 1 ft below the surface and 2 ft above the bottom at the deepest location. Intensive lake-specific monitoring is conducted at some lakes. Lakes are selected for Phase I (diagnostic), Phase II (evaluation), or Phase III (post-restoration evaluation) of intensive monitoring. For Phases I and II, monitoring is generally conducted twice per month from May to September and monthly from October to April for a 1-yr period prior to implementation (Phase I) and for 1 or more yr during and following implementation (Phase II). Phase III monitoring usually consists of 3 yr of sampling from 1 ft below the surface, mid-depth (at deeper lakes), and 2 ft above the bottom at the deepest location. Parameters monitored include suspended solids, nutrients, chlorophyll, DO and temperature profiles, and major biological resources (e.g., phytoplankton, fish, aquatic vegetation, and sometimes zooplankton and benthos). Sediment samples are analyzed for solids content, nutrients, persistent organics, and heavy metals in the initial diagnostic period. In addition, fish contaminant sampling and analyses also are conducted during this initial phase.

For dams and other international waters projects, the World Bank requires an EA or EIS, which are public documents that spell out unmitigable impacts and associated monitoring plans that must be scrutinized by the bank staff and by others in public hearings. The commitments for environmental monitoring stipulated in these documents are binding on the owners or operators of dams and related facilities. The description of the environmental monitoring programs must include reasons for and costs of monitoring and institutional arrangements for monitoring, evaluating monitoring results, and initiating necessary action to limit adverse impacts disclosed by monitoring. It is imperative that there is an institution to monitor compliance and institute enforcement action when and where needed.

The World Bank has developed guidelines for the design and implementation of monitoring and evaluation of projects funded by the Global Environment Facility (WB 1996). All World Bank-funded projects must include plans for monitoring and evaluation

(M&E), and projects that address environmental degradation must include environmental performance indicators (EPIs) and socioeconomic performance indicators (SEIs) in their M&E plans. The selection of EPIs is project specific. In general, an EPI should identify the following:

- Pressure indicators, which measure the underlying forces driving environmental degradation
- State indicators, which measure the quality or state of the environment of the targeted ecosystem
- Response indicators, which measure efforts taken to improve a specific environment or mitigate its degradation
- Socioeconomic indicators, which address how humans affect the environment and how the environment will affect human health, safety and welfare, subsistence base of local communities, economic well-being, educational opportunities, social values, and self esteem

The following are some examples of EPIs:

- Eutrophication involving monitoring nitrogen, phosphorus, and biochemical oxygen demand concentrations in water
- Toxic contamination, including monitoring of heavy metals concentration in water
- Monitoring of demand, supply, and quality of freshwater resources
- Monitoring of areas of degraded forests
- Monitoring of fish populations

Additional factors to be monitored for a dam and reservoir project include the following (WB 1998):

- Annual volume of sediment transported into the reservoir
- Water quality at the discharge point of a dam and at various points along the river (including salinity, pH, temperature, electrical conductivity, turbidity, DO, suspended solids, phosphates, and nitrates)
- Hydrogen sulfide and methane generation behind a dam
- Limnological sampling of microflora, microfauna, aquatic weeds, and benthic organisms
- Fisheries assessment surveys (including species and populations) in the reservoir and river
- Species, distribution, and numbers of wildlife
- Vegetation changes (including cover, species composition, growth rates, and biomass) in the upper watershed, reservoir drawdown zone, and downstream areas
- Erosion in the watershed
- Public health and disease vectors

- In- and out-migrations of people
- Changes in economic and social status of resettlement population and people remaining in the river basin

Similar monitoring factors are identified for projects involving flood protection, watershed development, irrigation and drainage, fisheries, agroindustry, natural forest management and plantation development, rangeland and livestock management, and rural roads.

There is no standard monitoring program for hydroelectric projects. The EA or EIS for these projects should include a monitoring plan for those variables included in the list for dam and reservoir projects that are pertinent for each particular site (WB 1998).

Project monitoring is the responsibility of the project management team. Institutional responsibilities for evaluation of performance differ depending on the evaluation designation:

1. Interim evaluation (during project implementation)
2. Terminal evaluation (at the end of the project)
3. Impact evaluation (several years after completion of the project) to measure direct and indirect impacts

The development, implementation, and evaluation of the M&E plan involve the following:

- Initial assessment of existing environmental and socioeconomic conditions and institutional entities for environmental management, monitoring, and evaluation
- Identification of major problems causing environmental degradation
- Development of indicators for performance monitoring
- Development of procedures for data collection, analytical methods, maintenance of monitoring equipment, and regional coordination
- Identification of institutional responsibilities and funding sources for implementation
- Periodic evaluation of project performance and adequacy of the M&E plan

Evaluation of Environmental Significance of Projects

Significance for Fish Habitat Sustainability and Enhancement

Some factors used to evaluate the significance of water resources engineering projects for fish habitat sustainability and enhancement include the following (USDA 1977):

- Acreage of water body that is available for fish movement, spawning, and hatching.
- Whether flow is constant or intermittent throughout the year and whether water levels remain nearly constant or fluctuate greatly from time to time.
- Water quality, including temperature, pH, DO, dissolved carbon dioxide, and concentrations of chemicals toxic to fish and humans.
- Percentage of total stream length occupied by the following:
  - Pools: Stream sections that are deeper and usually wider than normal with appreciably slower current than immediately upstream and downstream stream reaches and with streambed consisting of a mixture of silt and coarse sand.

- Riffles: Stream sections that contain gravel and/or rubble in which surface water is slightly turbulent and current is swift enough so that the surface of gravel and rubble is kept fairly free from sand and silt.

- Flats: Stream sections with current too slow to be classified as a riffle and too shallow to be classified as a pool. In a flat, streambed is usually composed of sand or finer materials, with occasional coarse rubble, boulders, or bedrock.

- Cascades: Stream sections without pools, consisting primarily of bedrock with little rubble, gravel, or other such material; the current is usually more swift than in riffles.

From the standpoint of fish habitat, a stream section with about 35% of area occupied by pools and 35% by riffles may be classified as excellent.

- Area or length of water body that is affected by shade due to vegetation (e.g., trees and shrubs). Generally, larger area or length of the water body under shade is better for fish habitat.

- Turbidity of water body. If the bottom of the water body can be distinctly seen through about 1.2 m of water or more, the water body may be classified as excellent from the standpoint of aquatic habitat.

Significance for Recreational Activities

Some factors used to evaluate the viability of water resources engineering projects for recreational activities include the following (USDA 1977):

- Proximity to population centers. Projects within a radius of about 80 km may provide good potential for recreational activities.

- Access. Projects or stream sections within about 1.5 km of an all-weather road may be good for recreational activities.

- Size of water body. The size of the water body should be sufficient to support public recreational activities.

Some factors to evaluate the recreational potential of a water resources engineering project depend on the type of recreational activity. Recreational activities associated with a water resources engineering project and factors to determine the viability of each recreational activity are as follows:

- Fishing: Factors affecting the viability of recreational fishing include abundance of fish population, accessibility, scenic setting, existence of riffles and pools, and overall diversity of landscape.

- Swimming: Desirable environmental factors for swimming include good water quality, pH between 6.5 and 8.3, coliform count below 800, and clear water with minimum inflow of 2,450 l/day per bather, water depth of about 1.5 m or more, and shoreline slope of less than 10%, preferably 2 to 4%.

- Boating and canoeing: Water depth in boating streams should be at least 0.6 m for rowboats and about 1 m for boats with outboard motors. The width should be at least 2.5 times the length of permitted boats. Water depths in canoeing streams may be as low as 15 cm for short reaches and 45 cm for a major portion. The stream width may be at least 1.8 m. The average flow may be about 3 m^3/s, although this may vary with width, depth, and slope of the stream.

- Hiking and walking: Over-water walkways and pedestrian bridges enhance hiking and walking utility of a water body. Trails along water bodies should be about 1.25 m in width with desirable longitudinal slopes of less than 10%.

Remedial Investigation and Feasibility Studies (RI/FS)
Definition and Scope

Industrial activities during the early and middle part of the twentieth century have resulted in soil, groundwater, and surface water contamination at or in the vicinity of a number of sites. In many cases, this has posed threats to human health in particular and the environment in general. To address the issues related to the remediation of these contaminated sites, the United States Congress enacted the Comprehensive Environmental Response, Compensation, and Liability Act (CERCLA, 42 U.S.C. 9601, 1980) and Resource Conservation and Recovery Act (RCRA, 42 U.S.C. 6921, 1976). RIs imply all field investigations and analytical studies required to identify the nature and extent of contamination that should be remediated. FSs include all laboratory, field, and office studies and experiments required to evaluate the feasibility of potential alternatives to eliminate or reduce the identified contamination to acceptable levels. The goal of remedial investigations may be to identify measures to limit soil, surface water, and groundwater contamination to acceptable maximum contaminant levels (MCL); to limit contamination to acceptable risk-based levels; or to achieve resource protection to pristine conditions. Acceptable contaminant levels, cleanup objectives, or goals for soil, surface water, and groundwater remediation under specific environments have been developed by the USEPA and state organizations responsible for environmental quality. These agencies also have developed MCLs deemed acceptable for different classes of ground- and surface waters.

In general, the requirements and main components of an RI/FS include the following (USEPA 1989b):

- Definition of remediation problem and objectives
- Design of soil, groundwater, and surface water monitoring programs
- Soil boring and installation of monitoring wells and surface water monitoring stations
- Surface water, soil, and sediment sampling and analysis
- Groundwater monitoring and analysis
- Contaminant movement (plume) analysis in groundwater and surface water
- Hydrogeologic investigations to determine aquifer characteristics
- Identification of potential exposure pathways and potential receptors of contaminated ground- and surface water traversing or in the vicinity of the site
- Site characterization
- Evaluation of remediation strategies and technologies

- Identification and design of preferred remediation method or combination of methods
- Agency approval of proposed remediation method

A significant part of investigations related to the RI/FS of contaminated sites focuses on soil and groundwater contamination. Detailed remedial investigations may be preceded by a series of preliminary studies. These studies involve preliminary assessment of potential for contamination at the site using available maps, site inspections, history of operations at the site, and available documents from the operator and government agencies. Detailed remedial investigations related to contaminated soils include the following (USEPA 1996c):

1. Development of a Conceptual Site Model (CSM): The CSM is a three-dimensional picture or description of site conditions that illustrates contaminant distributions in soils, contaminant release mechanisms, exposure pathways and migration routes, and potential receptors of contaminants released from soils. Development of a CSM involves data collection (historical records, maps, background information, and state soil surveys); data analysis to identify sources of contamination, affected media, and potential migration routes, and receptors; and site reconnaissance to identify data gaps.

2. Data collection: This involves field inspection to make an assessment of the distribution of contaminants, development of a soil sampling plan, identification of appropriate methods for sampling and chemical analyses, and collection of soil samples.

3. Analysis of soils data: This includes analysis of soil samples, identification of contaminants, delineation of spatial and depth-wise extent of contamination, and determination of soil characteristics (e.g., organic carbon content, texture, dry density, and hydraulic conductivity).

Remedial investigations related to groundwater include the following (USEPA 1989b):

- Characterization of chemicals of concern, which includes identification of specific compounds and their physical state, extent of contamination, solubility, adsorption, degradation, toxicity, density, viscosity, and concentrations.
- Site characterization, which includes:
  - Identification of the subsurface medium
  - Distances of on-site and off-site water supply wells, streams, and property boundaries
  - Unsaturated and saturated thickness of porous medium at the site and along potential pathways
  - Infiltration, net recharge, and water content of unsaturated soil zones
  - Porosity, bulk density, hydraulic conductivity, chemical characteristics, dispersion characteristics, and organic carbon content of porous media
  - Depth to groundwater in unsaturated soil zone
  - Saturated thickness, hydraulic gradient, and geologic features of saturated soil zones

The sources of information for a groundwater remediation investigation may include handbooks, records maintained by owners and operators of facilities and state and federal organizations, previous studies, and site-specific field data. Based on an evaluation of existing information, a groundwater sampling plan must be developed to collect additional data to fill in data gaps and delineate the horizontal and vertical extent of the existing contaminant plume and potential pathways of groundwater contamination. Monitoring wells required to collect field data should include sufficient number of upgradient wells to assess background concentrations of the chemicals of concern; downgradient wells to evaluate the potential for on-site and off-site contaminant migration and to serve as compliance wells; and on-site wells to delineate the extent of groundwater contamination.

Statistical methods may be used to determine the adequacy of soil and groundwater sampling and to analyze soil and groundwater monitoring data (USEPA 1989c). For instance, if the number of possible sample locations is large, a binomial distribution (Haan 1977) may be used to determine if at least one of the samples is from the most contaminated wells or hot spots. If a total of N soil samples is available from a site, n ($< N$) samples may be selected at random and analyzed. Then the probability can be calculated that at least one of the n soil samples is from the worst k samples (or hot spots) in the entire population of N samples using the hypergeometric distribution.

Example 7-1: Groundwater samples from 30 monitoring wells are tested from a site. Estimate the probability, P, that at least one of these 30 wells represents the worst 10% of contaminated locations at the site.

Solution: In this case, the number of possible sample locations (population) is large, so a binomial distribution is appropriate with n (selected number of samples) = 30 and p = probability of worst contamination in the population = 0.10. Thus,

P (at least one sample is from the worst 10% of wells) = $1 - (1 - p)^n = 1 - (1 - 0.10)^{30} = 0.96$ (Haan 1977).

Example 7-2: A total of 100 soil samples is available from square grids on a site. Only 30 randomly selected samples from the total number of samples are analyzed. Estimate the probability, P, that at least one of the selected samples represents the 10 worst hot spots at the site.

Solution: In this case, the population (i.e., 100 sample locations) is known. So, a hypergeometric distribution may be appropriate (Haan 1977). For hypergeometric distribution, N = 100, n = 30, and k = 10. Then,

P (probability that at least one of the 30 randomly selected samples is from the 10 worst spots) = 1 − probability that none of the 30 samples is from the 10 worst hot spots.

According to hypergeometric distribution, the probability that x of n samples randomly selected from a population of N belong to the group of worst k samples = $p(x, N, n, k)$ = $[k!/\{x! (k - x)!\}] [(N - k)!/\{(n - x)! (N - k - n + x)!\}]/[N!/\{n! (N - n)!\}]$ = $\{1 \times 90!\ 30!\ 70!\}/\{30!\ 60!\ 100!\}$ = $\{90!\ 70!\}/\{60!\ 100!\}$ = $[(1.48571 \times 10^{138}) \times (1.19785 \times 10^{100})]/[(8.32098 \times 10^{81}) \times (9.33262 \times 10^{157})]$ = $1.77966/77.65654$ = 0.023 (note that $x = 0$). The values of factorials may be obtained from standard tables (e.g., Abramowitz and Stegun 1972). Thus,

$P = 1 - 0.023 = 0.98$.

The uncertainty involved in the estimated treatment cost and volume of contaminated soil or groundwater based on a certain number of samples from a site also can be estimated using statistical methods (Spurr and Bonini 1973). The computational steps are illustrated in Example 7-3.

Example 7-3: A total of 100 soil samples is available from square grids on a site. Each sample represents approximately 2,000 tons of soil weight, and the cost of remediation is $1,000 per ton. Chemical analysis indicated that 30 of the samples are above permissible contaminant levels. Estimate the uncertainty in the cost of remediation at the 95% confidence level and also the maximum uncertainty cost.

Solution: The total number of samples = $N = 100$. Total weight of investigated soil = $2,000 \times 100 = 200,000$ tons.

The proportion of contaminated samples = $p = 30/100 = 0.30$.

Standard error of proportion $p = S_p = \sqrt{\{(p(1-p)/N\}} = \sqrt{\{(0.30(1-0.30)/100\}} = 0.0458$.

The highest standard error of proportion = $S_p(\max) = 0.5/\sqrt{N} = 0.5/\sqrt{100} = 0.05$.

For a confidence level α (i.e., 95%), the upper confidence limit of $p = p_u = p + t_{\alpha,N} \cdot S_p$ where $t_{\alpha,N} = t$-statistic for confidence level α for sample size N taken from statistical tables (Spurr and Bonini 1973) = 1.65 (for $\alpha = 95\%$ and $N = 100$). Thus, $p_u = 0.30 + 1.65 \times 0.0458 = 0.3756$.

Measure of uncertainty = $M_u = p_u - p = t_{\alpha,N} \cdot S_p = 0.0756$, and $M_u(\max) = 1.65 \times 0.05 = 0.0825$.

Cost of uncertainty at 95% confidence level = $0.0756 \times 200,000 \times 1,000 = \$15,120,000$.

Cost of maximum uncertainty = $0.0825 \times 200,000 \times 1,000 = \$16,500,000$.

Notice that the cost of uncertainty reduces with the increase in number of samples because both S_p and $t_{\alpha,N}$ reduce with the increase in number of samples. Separate plots may be made between cost of uncertainty at different confidence levels and number of samples, and a decision may be made about the required number of samples to have reasonable cost of uncertainty at an acceptable confidence level.

Feasibility studies related to remediation of contaminated soils and groundwater include comparative evaluation of different technologies including removal or extraction, in-situ treatment, natural attenuation, and other appropriate technologies (see the section in this chapter entitled "Relevant Hydrologic Processes and Simulations").

Instances of surface water or sediment contamination related to or resulting from soil and groundwater contamination occur in the following situations:

- Where plumes of contaminated groundwater discharge to surface water bodies, the concentrations of diluted contaminants may be of concern with respect to their impacts on aquatic biota, and the total mass flux of contaminants entering the receptor ecosystem may be significant with respect to the assimilative capacity of the receiving surface water body.

- Where contaminated soils from industrial sites are eroded during rainstorms, transported to nearby surface water bodies, and deposited on the bed and floodplains of streams, in marshes and wetlands, or behind obstructions, such as dams.

- Where discharge of contaminated storm water from industrial sites results in deposition of contaminated sediments on the bed and banks of nearby surface water bodies.
- Where sediments contaminated with biological or other types of contaminants get deposited behind dams, in marshes, or in wetlands, and analyses are required to estimate the volumes of deposited sediments and modeling required for the transport and distribution of sediments removed by dredging or flushing into the downstream channels and floodplains.

Data required for hydrologic, hydraulic, and risk analyses related to contamination of surface waters and sediments differ from site to site. In general, these data include the following:

- Topographic (contoured) maps indicating the locations of surface water bodies with respect to the site
- Surface soil characteristics, including erodibility, slopes, vegetation cover, soil conservation practices, permeability, and grain size distribution
- Chemicals of concern and their adsorption/desorption characteristics and solubility in water
- Chemical character (i.e., concentrations of chemicals of concern) of surface soils
- Flow, stages, cross sections, and water quality of water bodies in the site vicinity and water bodies likely to be the receptors for site-related contamination
- Particle size distribution of streambed sediments and suspended sediments
- Water quality, bathymetry, and particle size distribution and depths of sediments at the bed of impoundments in the site vicinity
- Precipitation data for the site and watersheds of nearby streams
- Results of groundwater flow and contaminant transport modeling and/or monitoring indicating the flux of contaminated groundwater entering the impacted surface water body
- TMDL or other discharge limitations on the stream (e.g., Section 303(d) listing or specific use designation) (USEPA 1991b)

Regulatory Perspective

The CERCLA of 1980, as amended by the Superfund Amendments and Reauthorization Act of 1986 (SARA), authorizes the federal government to respond to releases or threatened releases of hazardous substances, pollutants, or contaminants into the environment. In pursuance of the provisions of this act, all contaminated sites designated for in-depth evaluation through a RI/FS are included on the National Priorities List (NPL). A part of the FSs for these sites is the development of a risk assessment that projects health-related impacts attributable to the site. The risk assessment is based on the results of a site exposure assessment. This exposure assessment evaluates the type and extent of contaminant release from the site to the environmental media, transport and transformation of the contaminants following the release, and implications of the resulting contact with exposed aquatic and human populations.

In addition, regulatory agencies have the authority to issue no further remediation (NFR) letters for sites (not included in the NPL) enrolled in voluntary cleanup programs after the site owner/operator demonstrates, through proper investigation and remediation, that the remediated site is not a threat to human health or the environment. Procedures have been developed to evaluate risks to human health posed by contaminant releases from these sites and develop remediation objectives to protect surface waters, groundwater, sediments, and the ecosystem. The objective of these procedures is to provide adequate protection of human health and the environment based on exposure risks. Man-made or natural human exposure pathways include a physical condition that may pose risk to human health due to the presence of a contaminant of concern, an exposure route, and receptor activity at the point of exposure.

Some regulatory agencies have developed a tiered approach to corrective action objectives for contaminated sites. Tier 1 evaluation for the development of corrective action objectives compares contaminant concentrations detected at the site to the corresponding prescribed remediation objectives for residential and industrial/commercial properties. Tier 2 evaluation uses a risk-based analysis, and Tier 3 evaluation allows alternative parameters and factors to be used in the analysis for developing remediation objectives. Tier 2 and Tier 3 evaluations require more detailed analyses than Tier 1 evaluations (e.g., IPCB 2001).

For both NPL and voluntary cleanup sites, regulatory agencies have to be convinced, by technical analyses and field data, about the adequacy of the remedial investigations, completeness of the feasibility study, and appropriateness of the selected remedial actions.

As far as contamination of surface water and sediments and exposure risk to human health and the environment is concerned, the hydrologist has to conduct hydrologic/hydraulic and transport analyses and probabilistic risk analyses to assess the potential risk of contaminants reaching a point of potential exposure. The toxicologist or health scientist has to evaluate the resulting risk to human health, and the biologist has to evaluate the risk to aquatic life.

With a view to protect the quality of surface waters from pollution due to point and nonpoint sources, the U.S. Congress, in 1972, enacted the Clean Water Act. Section 303(d) of the Clean Water Act requires states to determine whether surface water bodies within their borders meet water quality standards. For surface water bodies that do not or cannot meet a particular standard, even with existing effluent limits, states are to identify the water body as impaired and determine the maximum load (or TMDL) of that pollutant that the water body can accept and still meet water quality standards. Then, the state is to allocate that TMDL among those sources discharging to the water body, with the goals of reducing pollutants being discharged to the impaired water body and bringing it into compliance with the standards (USEPA 1991b). Through hydrologic, hydraulic, and water quality modeling, the hydrologist has to determine the extent of remediation that will limit the loading in a stream or lake to that prescribed by the state agencies.

Relevant Hydrologic Processes and Simulations

Hydrologic Processes

Hydrologic processes relevant to RI/FS of contaminated sites include the following:

1. Rainfall-runoff: Quantifies precipitation, surface runoff, erosion potential of rainfall-runoff events, and portion of precipitation that is transmitted to groundwater as infiltration or recharge.

2. Infiltration and groundwater transport: Defines that portion of precipitation that is available to desorb, dissolve, and transport soluble contaminants through vadose and saturated soil zones.

3. Surface water transport of miscible contaminants: Includes mixing, advection, dispersion, and adsorption or desorption of contaminants entering a surface water body by groundwater seepage, wastewater discharge, or storm water discharge.

4. Sediment transport: Includes erosion and deposition of contaminated sediments from surface soils, their transport through streams or by overland flows, and deposition in deltas, reservoirs, wetlands, marshes, and estuaries.

The following steps of analyses are required to simulate the rainfall-runoff process and to evaluate the extent of contaminated sediment deposition in a surface water body resulting from the aforementioned processes:

- Obtain or synthetically generate hourly (or 15-min) rainfall sequences for the period of analysis. Alternatively, for a simplistic analysis, estimate duration and rainfall depths for discrete significant storm events for each year. Significant storm events may be defined as those generating sufficient rainfall intensity to cause discernible soil erosion. The USEPA defines 0.1 in. to be the minimum rainfall depth capable of producing the rainfall/runoff characteristics necessary to generate a sufficient volume of runoff for meaningful sample analysis for storm water (40 CFR Parts 122, 123, and 124; Federal Register Vol. 55, No. 222, Friday, November 16, 1990, Page 48018; USEPA 1991d). A storm event capable of causing discernible soil erosion should be greater and more intense.

- Estimate volumes and rates of surface runoff and soil erosion from the site during discrete storm events.

- Estimate erosion during nominal rain events (rainfall events that may not produce discernible soil erosion) from water quality data for suspended sediments.

- Compute the quantity of sediment transported through the stream with the surface runoff and soil erosion input from the site area, including erosion and deposition in the streambed, floodplains, and downstream reservoirs.

- Estimate contaminated/uncontaminated portions of sediment deposited in the stream and reservoirs by prorating based on the contaminated and uncontaminated portions of the watershed.

- Verify the reasonableness of estimated extent of sediment contamination using soil sampling data from selected locations and extrapolation by kriging or other interpolation/extrapolation procedures.

With respect to the discharge of groundwater from contaminated sites and other discharges from point and non-point sources contaminated with miscible constituents, regulatory agencies responsible for water quality specify acute and chronic regulatory mixing zones (ARMZ and CRMZ) for receiving surface water bodies within which normal water quality standards are not applied (Hutcheson 1998). A mixing zone is that region of the surface water body downstream of the point of discharge of contaminants where physical mixing occurs in all directions until the constituents in the discharge achieve uniform concentrations in the receiving water body. The expected uniform concentration of chemicals of concern must be within prescribed MCLs, and the total mass flux of these chemicals must be within the permissible mass loading for that water body. The size of the mixing zone should be kept to a minimum and should allow safe passage, protection, and propagation of aquatic organisms. A commonly used criterion to define the limit of complete mixing or mixing zone is that the concentration at any point in the cross section at the boundary of the mixing zone should not deviate by more than 5% from the mean value for that cross section (Fischer et al. 1979). In cases where discharge of groundwater from contaminated sites occurs in lakes and reservoirs, the total mass flux of contaminants and concentrations in the surface water body during critical periods with respect to the spawning and growth of aquatic habitat may be limiting considerations.

Most water quality standards specify the use of 7Q10 to evaluate the impacts of point and non-point discharges on the water quality of streams. The steps of analyses generally used to evaluate the extent of surface water contamination resulting from groundwater discharges from contaminated sites include the following:

- Define the source of contaminated groundwater entering the stream or lake in terms of concentrations of chemicals of concern, mass flux, and dimensions of the source area. Depending on specific site conditions and degree of sophistication of the required analysis, approximate the source configuration as a point, line (vertical or horizontal), or plane (vertical or horizontal) source, or as a source of irregular shape defined by rectangular grids or triangular, rectangular or isoparametric elements.

- Define hydraulic and water quality characteristics of the receiving surface water body (e.g., cross sections, flows, flow velocities, ambient water quality, and dispersion/diffusion parameters for streams, and inflows, outflows, storage capacity, stratification potential, and ambient water quality for lakes and reservoirs). The extent of required data depends on the sophistication of the desired analyses.

- Obtain climatic data required for water quality modeling for lakes and reservoirs.

- Perform advective-dispersion analysis with desorption/adsorption for the transport of contaminants entering the stream with the groundwater discharge through the impacted stream and/or lake.

- Estimate the mixing zone dimensions and total mass flux and compare with the assimilative capacity of the surface water body established by the regulatory agencies.

Usually, adequate site-specific data are not available to calibrate and verify the accuracy of the results of modeling and analyses related to the contamination of surface water and sediments. The sophistication of the selected modeling or analytical techniques has to be commensurate with the amount of available site-specific data. So, in most cases, the results of the analyses and modeling have limited accuracy. However, the objective of an RI/FS is to collect and develop sufficient information to identify principal sources, nature, and extent of contamination; identify principal responsible parties (PRPs) for contamination; apportion responsibilities for remediation; and develop feasibility-level details for potential methods of remediation. The detailed design and costs for remediation have to be determined later based on additional data on the extent of contamination and cleanup levels agreed upon with the regulatory agencies and results of pilot tests for the effectiveness of the selected remediation technique.

Hydrologic models simulate surface and subsurface watershed hydrologic processes and resulting land surface runoff, groundwater interaction, and non-point source contaminant loading. These hydrologic models serve as a basis for various environmental analyses, especially when dealing with watershed management issues and evaluation of impacts of upstream sources on receiving water bodies. These models rely on the representation of land-based processes and human activities occurring within the drainage watershed to estimate flow, sediment, and other pollutant loading. In general, many of these models are designed to provide an integrated prediction of water balance and pollutant movement from point and non-point sources. These predictions are often done over a long period of time (sometimes on the order of decades) to capture the effects of changes that are implemented over longer time horizons.

The volume and hydrograph of surface runoff is estimated using event-based models such as HEC-1 (USACE 1991a) and HEC-HMS (USACE 2002) or continuous flow simulation models such as HSPF (USEPA 1991a), NWSRFS (NWS 1998; Prakash and Dearth 1990), and SSAR (USACE 1986). Potential sediment erosion during discrete storm events can be estimated using the Modified Universal Soil Loss Equation (MUSLE). The average annual sediment erosion can be estimated using the Universal Soil Loss Equation (USLE). Depending on the length and steepness of the pathway of surface runoff, the entire quantity of sediment eroded from the site area may not reach the receiving stream. To account for sediment deposition along the pathway of surface runoff, a sediment delivery ratio (SDR) should be estimated and used (USEPA 1988a). For event-based simulations, models that simulate both surface runoff and sediment erosion may be used. As an example, the SEDIMOT II (Wilson 1984) model is an event-based model that can compute surface runoff volume and hydrograph, sediment erosion using USLE or MUSLE and SDR, sediment volume and concentrations reaching a receptor (e.g., sedimentation basin), and sediment deposition in the receptor (see Chapter 3).

Surface water hydrologists have to estimate the portion of precipitation that is likely to infiltrate through the vadose and saturated soil zones. The groundwater hydrologist simulates the transport of surface contaminants with infiltrating rainwater and groundwater flow to nearby surface water bodies. Groundwater inflow to a surface water body may occur through seeps along the banks under, above, or below the water surface elevation in the surface water body. These groundwater inflows can be simulated as point, line, or plane sources located at the bank of the surface water body (see Chapter 4).

Vertical infiltration through the vadose and saturated soil zones is a relatively slow process. Usually, average annual infiltration is used to evaluate the potential for downward movement of contaminants with rainwater infiltration. The rate of vertical transport depends on the vertical hydraulic conductivity of the vadose and saturated soils at the site. Hydraulic conductivity is a function of soil moisture and so is difficult to determine because of temporally and spatially variable soil moisture conditions. Estimation of average annual infiltration is generally based on empirical relationships and available regional information. The Curve Number Method of the Soil Conservation Service (SCS), with curve number selected to reflect average annual surface runoff conditions for the site area, may be used to estimate overall precipitation loss due to interception, evaporation, and infiltration (ASCE 1996). Empirical methods (e.g., Maxey-Eakin infiltration coefficients) may be used for specific site conditions (Avon and Durbin 1994). For river basins where information on average annual surface runoff and precipitation is available, the difference may provide an estimate of overall precipitation loss. In cases where continuous flow simulation is conducted using the models described previously, the soil moisture accounting component of the models provides an estimate of infiltration. The average annual infiltration rate is used to simulate the movement of dissolved contaminants from surface sources to the groundwater table through the vadose zone (Prakash 1996, 2000a). Contaminant transport through the saturated zone is simulated using one-, two-, or three-dimensional groundwater flow and transport models (see Chapter 4).

Components that are relevant to surface water quality models include hydrologic, hydrodynamic, and sediment transport processes. Hydrodynamic models simulate the transport of storm water and contaminated sediments through surface water systems. The hydrodynamic model provides water surface elevations, horizontal and vertical velocity, shear stress, and turbulent diffusion to predict the transport of suspended sediment and dissolved and adsorbed contaminants. The contaminant transport model simulates the movement of dissolved and sediment-sorbed contaminants by advection, particulate settling, turbulent diffusion, advection, molecular diffusion, hydrodynamic dispersion, and biological mixing.

Sediment transport modeling includes simulation of erosion, transportation, and deposition due to water, wind, and gravity. Erosion may expose contaminated soils, which may be subject to exposure pathways due to additional erosion by wind and water. Transport of contaminated sediments may impair environmental quality by movement of adsorbed contaminants to other potential receptors (e.g., rivers, lakes, and estuaries). Deposition of contaminated sediments may impact river floodplains, marshes, and wetlands and may threaten groundwater environment due to infiltration and leaching of contaminants.

Remediation Methods

Commonly used methods for soil and groundwater remediation include removal, treatment, and natural attenuation.

Removal

Removal is a conventional contaminated sediment remediation method. It involves transport, treatment, and disposal of contaminated sediments and extraction and disposal of contaminated groundwater. The viability of physical removal of sediments depends on the nature of the sediments, types of contaminants, depth of deposition, thickness and volume of sediments, and availability of required equipment. Dredging and transport are appropriate when the environmental effects of no action are unacceptable; environmental conditions such as wave action, flooding, or erosion prohibit leaving the sediment in place; or sediment lies in navigation waterways and must be removed. Common methods of dredging include mechanical, hydraulic, and pneumatic dredging. Mechanical dredging involves sediment dredging or excavation using conventional earth-moving equipment. Hydraulic dredging uses portable, handheld, plain suction, cutter-head, dustpan, and hopper dredges. Pneumatic dredging includes airlift dredges and pneuma dredges.

Treatment

In-situ or ex-situ treatment includes containment (by capping, solidification and stabilization, and ground freezing), soil washing, chemical treatment including chemical extraction and destruction/conversion, biological treatment, and incineration. Capping involves placement of clean material over the top of contaminated sediments. Solidification and stabilization and ground freezing immobilize contaminated sediments by treating them with reagents to solidify or fix them. Soil washing is a water-based volume reduction process in which contaminants are extracted and consolidated into a small residual portion of the original volume using physical or chemical means. Chemical treatment includes dechlorination and oxidation. It adds chemical reagents to the sediments in order to destroy, detoxify, or remove the contaminants. Biological treatment includes aerobic and anaerobic biological treatments. Incineration involves burning of contaminants at high temperatures and disposal of the resulting ash and slag.

Remediation by Natural Attenuation (RNA)

RNA involves progressive and consistent reduction in contaminant concentrations in soils and groundwater in the site vicinity due to molecular diffusion, mechanical dispersion, dilution by recharge due to rainwater infiltration, volatilization, sorption, chemical (abiotic) reactions (e.g., hydrolysis/substitution, elimination, and oxidation/reduction), and aerobic and/or anaerobic biodegradation, including dehalogenation. Some of the major advantages of RNA include minimal site disturbance, continued use of the site during remediation, low capital cost, and applicability to soils and groundwater below buildings and other areas that may not be accessible. In addition, RNA can be used as a supplement to other remediation techniques.

Evaluation of the viability of RNA requires modeling and determination of contaminant degradation rates and pathways to demonstrate that natural processes of contaminant degradation will reduce contaminant concentrations to acceptable levels before potential exposure pathways are completed. Groundwater monitoring and sediment sampling must be continued for sufficient time to confirm that degradation is occurring at rates consistent with established cleanup objectives. Feasibility of RNA as an acceptable remediation measure can be demonstrated through relatively inexpensive field data collection, laboratory analyses, and modeling during the site characterization phase. If it is found that RNA is not feasible, the data collected (with or without additional field work) can be used to design supplemental or other remediation measures. However, RNA requires soil and groundwater monitoring for a relatively long period of time both before and after remediation and may require source removal and/or establishment, approval, and long-term maintenance of Groundwater Management Zones (GMZs). The GMZ is a three-dimensional region containing groundwater being managed to mitigate impairment caused by the release of contaminants of concern at a remedial site.

Investigative steps to demonstrate the viability of natural attenuation at a site include the following:

- Site characterization, including source identification, plume delineation, and determination of aquifer characteristics
- Assessment of the feasibility of source or free product removal
- Collection of data for soil and groundwater contamination for several periods of time and comparison of contaminant concentrations for various periods
- Laboratory analyses to demonstrate biodegradation of the contaminants of concern
- Fate and transport modeling to demonstrate that contaminant concentration will be attenuated by natural processes before impacting any potential receptors with or without source removal
- Risk-based analysis to demonstrate that there is no potential threat of contaminant exposure in the short or long term
- Practicability of continuous groundwater monitoring for sufficiently long periods of time in the future
- Evaluation of the viability of establishing and maintaining GMZs for specified time periods in the future

Remediation Technologies

Commonly used remediation technologies for contaminated soils involve the following treatments:

1. In situ Biological Treatment: This includes removal of volatile and semi-volatile contaminants (VOCs and SVOCs) from soils using biodegradation or bioventing methods.

2. In situ Physical/Chemical Treatment: This includes soil flushing (to remove VOCs and metals), soil vapor extraction (to remove VOCs), thermally enhanced vapor extraction (to remove VOCs and SVOCs), and solidification or stabilization techniques (to remove SVOCs and metals).

3. Ex situ Physical/Chemical Treatment: This includes dehalogenation (to remove VOCs and SVOCs), soil washing (to remove VOCs, SVOCs, and metals), soil vapor extraction (to remove VOCs), low-temperature thermal desorption (to remove VOCs), and incineration (to remove VOCs and SVOCs).

Remediation technologies for groundwater, surface water, and leachate treatment include the following:

1. In situ Biological Treatment: This includes using bioremediation with oxygen enhancement to remove VOCs and SVOCs.

2. In situ Physical/Chemical Treatment: This includes air sparging to remove VOCs; dual-phase extraction to remove VOCs; passive treatment walls to remove VOCs, SVOCs, and metals; slurry walls to contain migration of VOCs, SVOCs, and metals; and vacuum vapor extraction to remove VOCs, SVOCs, and metals.

3. Ex situ Physical/Chemical Treatment: This includes air stripping to remove VOCs and SVOCs; carbon adsorption to remove VOCs, SVOCs, and metals; and ultraviolet oxidation to remove VOCs and SVOCs.

Remediation technologies for dissolved phase plumes include the following:

1. Exposure risk assessment for humans and environment and development of risk-based corrective action (RBCA) plans (IPCB 2001)

2. RNA including monitoring of the progress of natural attenuation

3. Accelerated biodegradation of dissolved phase contaminants using various types of catalysts

4. Permeable barriers (e.g., funnel and gate system using iron filings) to funnel impacted groundwater to a defined area for treatment

5. Pump-and-treat system to extract contaminated groundwater for treatment and disposal

The following factors are considered in the selection of treatment methods:

- Source and nature of contaminants
- Physical characteristics and location of the site
- Use of nearby water bodies for such purposes as water supply, navigation, recreation, industrial, municipal discharge, or a combination of these
- Quality and quantity of contaminated sediments
- Physical properties of sediments or soils (e.g., fine-grained or coarse-grained)
- Organic matter content of sediments or soils

- In situ water content of sediments or soils
- Mobility and biological availability of contaminants in sediments or soils (e.g., low or high pH conditions)
- Salinity of sediments (elutriates) and water. Elutriate is water obtained from drainage of saturated/submerged sediments.
- Sulfide content of sediments or soils
- Amount and type of cations and anions
- Amount of potentially reactive iron and manganese in sediments or soils
- Contaminant characteristics (e.g., tendency to adsorb onto a sediment particle or solubility in water)
- Chemical characteristics and three-dimensional extent of the plume of contaminated ground water
- Hydrogeologic characteristics of units where the plume of contaminated water is located.

REFERENCES

Abramowitz, M., and I.A. Stegun (1972). *Handbook of mathematical functions*, Dover, New York, 1,046 pp.
Abt, S.R., J.F. Ruff, and R.J. Wittler (1991). "Estimating flow through riprap." *J. Hydraul. Eng.*, 117(5), 670–675.
Adams, E.E. (1982). "Dilution analysis for unidirectional diffusers." *J. Hydr. Div.*, 108(HY3), 327–342.
ASCE (1959). "Time of concentration for overland flow." *Civil Engineering*, March 1959.
ASCE (1976). "Design and construction of sanitary and storm sewers." *ASCE Manuals and Reports on Engineering Practice No. 37*, New York, 332 pp.
ASCE (1988). *Evaluation procedures for hydrologic safety of dams*, New York, 95 pp.
ASCE (1989). *Civil engineering guidelines for planning and designing hydroelectric developments*, New York.
ASCE (1996). "Hydrology handbook." *ASCE Manuals and Reports on Engineering Practice No. 28*, 2d Ed., New York, 784 pp.
Anderson, M.P., and W.W. Woessner (1992). *Applied groundwater modeling*, Academic, San Diego, 381 pp.
ASTM (1995). "Standard guide for risk-based corrective action applied at petroleum release sites." *EI 1739-95*, West Conshohocken, Pa., 51 pp.
Avon, L., and T.J. Durbin (1994). "Evaluation of the Maxey-Eakin method for estimating recharge to ground-water basins in Nevada." *Water Resources Bulletin*, 30(1), Feb. 1994, 99–111.
Baehr, A.T. (1987). "Selective transport of hydrocarbons in the unsaturated zone due to aqueous and vapor phase partitioning." *Water Resources Research*, 23(10), 1923–1938.
Bair, E.S., A.E. Springer, and G.S. Roadcap (1992). "An analytical flow model for simulating confined, leaky confined, or unconfined flow to wells with superposition of regional water levels." CAPZONE, IGWMC, Colorado School of Mines, Golden, Colo.

Bansal, M.K. (1971). "Dispersion in natural streams." *J. Hydr. Div.*, 97(HY11), 1867–1886.
Barfield, B.J., R.C. Warner, and C.T. Haan (1981). *Applied hydrology and sedimentology for disturbed areas*, Oklahoma Technical, Stillwater, Okla., 603 pp.
Batu, V. (1998). *Aquifer hydraulics: A comprehensive guide to hydrogeologic data analysis*, Wiley, New York, 727 pp.
Bear, J. (1979). *Hydraulics of groundwater*, McGraw-Hill, New York, 569 pp.
Bouwer, H., and R.C. Rice (1976). "A slug test for determining hydraulic conductivity of unconfined aquifers with completely or partially penetrating wells." *Water Resources Research*, 12(3), 423–428.
Bouwer, H. (1989). "The Bouwer and Rice slug test—an update." *Ground Water*, 27(3), May–June 1989, 304–309.
Bouwer, H., J.T. Back, and J.M. Oliver (1999). "Predicting infiltration and groundwater mounds for artificial recharge." *J. Hydrol. Eng.*, 4(4), 350–357.
Bradley, C., and D.J. Gilvear (2000). "Saturated and unsaturated flow dynamics in a floodplain wetland." *Hydrological Processes*, 14(16–17), Nov.–Dec. 2000, 2945–2958.
Bras, R.L. (1990). *Hydrology, an introduction to hydrologic science*, Addison-Wesley, Reading, Mass., 643 pp.
Brater, E.F., H.W. King, J.E. Lindell, and C.Y. Wei (1996). *Handbook of hydraulics*, McGraw-Hill, New York.
Brooks, R.H. and A.T. Corey (1964). "Hydraulic properties of porous media." *Hydrology Papers*, No. 3, Colorado State University, Fort Collins, Colo.
Campbell, D.B., and P.C. Johnson (1984). "RCC dam incorporates innovative hydraulic features." *Water resources development: Proc. of the conf. of the Hydraulics Div.*, August 14–17, 1984,138–142.
CAP (2001). "Culvert Analysis Program." http://www.waterengr.com/freeprog.htm.
Carslaw, H.S., and J.C. Jaeger (1984). *Conduction of heat in solids*, Oxford Univ. Press, New York, 510 pp.
Central Board of Irrigation and Power (CBIP) (1971). "Manual on river behaviour control and training." *Publication No. 40*, New Delhi, India, 432 pp.
Chamani, M.R., and N. Rajaratnam (1994). "Jet flow on stepped spillways." *J. Hydraul. Eng.*, 120(2), 254–259.
Chanson, H. (1994). "Comparison of energy dissipation between nappe and skimming flow regimes on stepped chutes." *J. Hydraulic Research*, 32(2), 213–218.
Chapra, S.C. (1997). *Surface water-quality modeling*, McGraw-Hill, New York, 844 pp.
Charbeneau, R.J. (2000). *Groundwater hydraulics and pollutant transport*, Prentice Hall, Upper Saddle River, N.J., 593 pp.
Chin, D.A. (1985). "Outfall dilution: The role of a far-field model." *J. Environ. Eng.*, 111(4), 473–486.
Chow, V.T. (1959). *Open-channel hydraulics*, McGraw-Hill, New York, 680 pp.
Chow, V.T., ed. (1964). *Handbook of applied hydrology*, McGraw-Hill, New York.
Coastal Engineering Research Center (CERC) (1984). *Shore protection manual.* Dept. of the Army, Corps of Engineers, Waterways Experiment Station, Vicksburg, Miss.
Code of Federal Regulations 18 (CFR 18) (1999). Chapter 1, Conservation of Power and Water Resources, and Part 380, Regulations Implementing the National

Environmental Policy Act, Office of the Federal Register, National Archives and Records Admin., U.S. Government Printing Office, Washington, D.C.

Code of Federal Regulations 18 (CFR 18) (1999). Chapter 1, Federal Energy Regulatory Comm., U.S. Dept. of Energy, Washington, D.C.

Code of Federal Regulations 40 (CFR 40), Part 1500 (1999). Office of the Federal Register, National Archives and Records Admin., U.S. Government Printing Office, Washington, D.C.

Cooper, H.H., J.D. Bredehoeft, and I.S. Papadopulos (1967). "Response of a finite-diameter well to an instantaneous charge of water." *Water Resources Research*, 3(1), 1st qtr. 1967, 263–269.

Creager, W.D., and J.D. Justin (1950). *Hydroelectric handbook*, Wiley, New York.

Crippen, J.R. (1982). "Envelope curves for extreme flood events." *J. Hydr. Div.*, 108(HY10), 1208–1212.

Dalton, F.E., and R.S. La Russo (1979). "Chicago's TARP solves problems in big way." *Water & Wastes Engineering*, Technical Publishing Co.

Davis, C.V., and K.E. Sorensen, eds. (1970). *Handbook of applied hydraulics*, McGraw-Hill, New York.

Delleur, J.W. (1999). *The handbook of groundwater engineering*, CRC, Boca Raton, Fla.

Domenico, P.A., and G.A. Robbins (1985). "A new method of contaminant plume analysis." *Ground Water*, 23(4), 476–485.

Domenico, P.A. (1987). "An analytical model for multidimensional transport of a decaying contaminant species." *J. Hydrology*, 91, 49–58.

Domenico, P.A., and F.W. Schwartz (1998). *Physical and chemical hydrogeology*, 2d Ed., Wiley, New York, 506 pp.

Donovan, D.J., and T. Katzer (2000). "Hydrologic implications of greater groundwater recharge to Las Vegas Valley, Nevada." *JAWRA*, 36(5), Oct. 2000, 1133–1148.

Dragun, J. (1988). "The soil chemistry of hazardous materials." *Hazardous Materials Control Research Institute*, Greenbelt, Md.

Driscoll, F.G. (1989). *Groundwater and wells*, 2d Ed., Johnston Filtration Systems, Inc., St. Paul, Minn., 1,089 pp.

Duffield, G.M., and J.O. Rumbaugh (1989). "Aquifer test solver, AQTESOLV." Geraghty & Miller, Inc., Reston, Va., 134 pp.

Eheart, J.W., A.J. Wildermuth, and E.E. Herricks (1999). "The effects of climate change and irrigation on criterion low streamflows used for determining total maximum daily loads." *JAWRA*, 35(6), Dec. 1999, 1365–1372.

Federal Emergency Management Agency (FEMA) (1993). "Flood insurance study guidelines and specifications for study contractors." *FEMA 37*, Washington, D.C.

Fetter, C.W. (1999). *Contaminant hydrogeology*, 2d Ed., Prentice Hall, Upper Saddle River, N.J., 500 pp.

Fetter, C.W. (2001). *Applied hydrogeology*, 4th Ed., Prentice Hall, Upper Saddle River, N.J., 598 pp.

Fiering, M.B., and B.B. Jackson (1971). "Synthetic streamflows." *Water Resources Monograph 1*, American Geophysical Union, Washington, D.C., 98 pp.

Fischer, H.B., J.E. List, R.C.Y. Koh, J. Imberger, and N.H. Brooks (1979). *Mixing in inland and coastal waters*, Academic, New York, 483 pp.

Foster, G.R., K.G. Renard, D.C. Yoder, D.K. McCool, and G.A. Weesies (1996). *RUSLE user's guide.* Soil and Water Conservation Soc., 173 pp.

Fread, D.L. (1988). "The NWS DAMBRK model." Hydrologic Research Lab., Office of Hydrology, National Weather Service, NOAA, Silver Spring, Md.

Freeze, R.A., and J.A. Cherry (1979). *Groundwater,* Prentice Hall, Upper Saddle River, N.J., 604 pp.

Gale Research Co. (1985). *Climates of the states.* Volumes 1 and 2, Book Tower, Detroit, Mich., 1,572 pp.

Gerbert, W.A., D.J. Graczyk, and W.R. Drug (1989). *Average annual runoff in the United States, 1951–1980,* Hydrologic Investigations Atlas, U.S. Geological Survey, Reston, Va.

Glover, R.E. (1985). *Transient ground water hydraulics.* Water Resources Publications, Highlands Ranch, Colo.

Golze, A.R., ed. (1977). *Handbook of dam engineering,* Van Nostrand Reinhold, New York, 793 pp.

Graf, J.B. (1995). "Measured and predicted velocity and longitudinal dispersion at steady and unsteady flow, Colorado River, Glen Canyon Dam to Lake Mead." *Water Resources Bulletin,* 31(2), 265–281.

Haan, C.T. (1977). *Statistical methods in hydrology,* Iowa State Univ. Press, Ames, Iowa, 378 pp.

Hampton, D.R. (1990). "Monitoring of free products in wells: Purposes and pitfalls." *4th national outdoor action conf. on aquifer restoration, ground water monitoring and geophysical methods,* Assoc. of Ground Water Scientists and Engineers & U.S. Environmental Protection Agency, Las Vegas, Nev., May 14–17.

Hawkins, R.H., A.T. Hjelmfelt, and A.W. Zevenbergen (1985). "Runoff probability, storm depth, and curve numbers." *J. Irrig. Drain. Div.,* 111(4), Dec. 1985, 330–340.

Hershfield, D.M. (1961). "Rainfall frequency atlas of the United States." *Technical Paper No. 40,* U.S. Dept. of Commerce, Weather Bureau, Washington, D.C., 115 pp.

Huff, F.A., and J.R. Angel (1989). "Frequency distributions and hydroclimatic characteristics of heavy rainstorms in Illinois." *ISWS/BUL-70/89,* Illinois State Water Survey, Champaign, Ill., 177 pp.

Hutcheson, M.R. (1998). "Implementation of acute criteria for conservative substances." *JAWRA,* 34(5), 1025–1033.

Illinois Pollution Control Board (IPCB) (2001). "Tiered Approach to Corrective Action Objectives (TACO)." *35 Ill. Adm. Code Part 742-R97-12(A),* Bureau of Land, Springfield, Ill.

Javandel, I., and C.F. Tsang (1986). "Capture-zone type curves: A tool for aquifer cleanup." *Ground Water,* 24(5), Sep.–Oct. 1986, 616–625.

Johnson, P.C., C.C. Stanley, M.W. Kemblowski, D.L. Buyers, and J.D. Colthart (1990). "A practical approach to the design, operation, and monitoring of in-situ soil-venting systems." *Ground Water Monitoring Review,* 10(2), 159–178.

Johnson, T.L. (1999). "Design of erosion protection for long-term stabilization." *NUREG-1623, draft report,* U.S. Nuclear Regulatory Comm., Washington, D.C.

KYPIPE2 and KYPIPE3 (1992). *Hydraulic network analysis program,* Civil Engineering Software Center, Univ. of Kentucky, Lexington, Ky.

Leps, T.M. (1973). *Flow through rockfill, in embankment-dam engineering*, R.C. Hirschfeld and S.J. Poulos, eds., Wiley, New York, 87–107.

Levy, B., and R. McCuen (1999). "Assessment of storm duration for hydrologic design." *J. Hydrol. Eng.*, 4(3), 209–213.

Linsley, R.K., J.B. Franzini, D.L. Freyberg, and G. Tchobanoglous (1992). *Water-resources engineering*, 4th Ed., McGraw-Hill, New York, 841 pp.

Long, J.C.S., and P.A. Witherspoon (1985). "The relationship of the degree of interconnection to permeability in fracture networks." *J. Geophysical Research*, 90(B4), March 10, 1985, 3087–3098.

Long, J.C.S., J.S. Remer, C.R. Wilson, and P.A. Witherspoon (1982). "Porous media equivalents for networks of discontinuous fractures." *Water Resources Research*, 18(3), June 1982, 645–658.

Low, H.S. (1989). "Effect of sediment density on bed-load transport." *J. Hydraul. Eng.*, 115(1), 124–138.

Lyman, W.J., W.F. Reehl, and D.H. Rosenblatt (1984). *Handbook of chemical property estimation methods*, McGraw-Hill, New York.

Maidment, D.R., ed. (1993). *Handbook of hydrology*, McGraw-Hill, New York, 1,404 pp.

Martin, J.L., and S.C. McCutcheon (1999). *Hydrodynamics and transport for water quality modeling*, CRC, Boca Raton, Fla., 794 pp.

Maynord, S.T., J.F. Ruff, and S.R. Abt (1989). "Riprap design." *J. Hydraul. Eng.*, 115(7), 937–949.

Mays, L.W., ed. (1999). *Hydraulic design handbook*, McGraw-Hill, New York.

McCuen, R.H.M., S.L. Wong, and W.J. Rawls (1984). "Estimating urban time of concentration." *J. Hydraul. Eng.*, 110(7), 887–904.

McCuen, R.H.M. (1998). *Hydrologic analysis and design*, 2d Ed., Prentice Hall, Upper Saddle River, N.J., 814 pp.

Miller, J.F. (1963). *Probable maximum precipitation and rainfall frequency data for Alaska*, U.S. Dept. of Commerce, Weather Bureau, Washington, D.C.

Monsonyi, E. (1963). *Water power development*, Hungarian Academy of Sciences, Budapest, Hungary.

Muellenhoff, W.P., A.M. Soldate, D.J. Baumgartner, M.D. Schuldt, L.R. Davis, and W.E. Frick (1985). "Initial mixing characteristics of municipal ocean discharges." *EPA/600/3-85/073*, U.S. Environmental Protection Agency, Washington, D.C.

National Bureau of Standards (NBS) (1972). "American national standard, building code requirements for minimum design loads in buildings and other structures." *ANSI, A58.1-1972*.

National Oceanic and Atmospheric Administration (NOAA) (1973). *Precipitation-frequency atlas of the western United States*, NOAA Atlas 2, Volumes I to XI, Silver Spring, Md.

National Research Council (NRC) (1985). *Safety of dams, flood and earthquake criteria*, National Academic Press, Washington, D.C., 321 pp.

National Resources Conservation Service (NRCS) (1996). *State of the land for the Northern Plains region*, Northern Plains Regional Office, Lincoln, Neb.

National Weather Service (NWS) (1998). *National Weather Service River Forecast System (NWSRFS) model, user's manual*, Office of Hydrology, National Weather Service, Silver Spring, Md. (www.nws.noaa.gov/oh/hrl/nwsrfs/users_manual).

Nelson, J.D., S.R. Abt, R.L. Volpe, D. Van Zyl, N.E. Hinkle, and W.P. Straub (1986). "Methodologies for evaluating long-term stabilization of uranium mill tailings impoundments." *NUREG/CR-4620, ORNL/TM-10067*, for U.S. Nuclear Regulatory Comm., Washington, D.C., 145 pp.

Ojima, D., L. Garcia, E. Elgaali, K. Miller, T.G.F. Kittel, and J. Lackett (1999). "Potential climate change impacts on water resources in the great plains." *JAWRA*, 35(6), Dec. 1999, 1443–1454.

Pankow, J.F., and J.A. Cherry (1996). *Dense chlorinated solvents and other DNAPLs in groundwater: History, behavior, and remediation*, Waterloo, Portland, Ore., 522 pp.

Peterka, A.J. (1958, 1978). "Hydraulic design of stilling basins and energy dissipators." *Engineering Monograph No. 25*, U.S. Bureau of Reclamation, Denver, Colo., 222 pp.

Pinder, G.F., J.D. Bredehoeft, and H.H. Cooper (1969). "Determination of aquifer diffusivity from aquifer response to fluctuations in river stages." *Water Resources Research*, 5(4), Aug. 1969, 850–855.

Ponce, V.M. (1989). *Engineering hydrology, principles and practices*, Prentice Hall, Englewood Cliffs, N.J., 640 pp.

Potter, M.C., and D.C. Wiggert (1991). *Mechanics of fluids*, Prentice-Hall, Englewood Cliffs, N.J., 692 pp.

Prakash, A., and G. Dearth (1990). "Streamflow simulation using deterministic model." *J. Irrigation and Drainage Engineering*, 116(4), July/Aug. 1990, 566–580.

Prakash, A. (1977). "Convective-dispersion in perennial streams." *J. Environ. Eng.*, 103(EE2), 321–340.

Prakash, A. (1978). "Optimal sequence of incremental precipitation." *J. Hydr. Div.*, 104(HY12), Dec. 1978, 1668–1671.

Prakash, A. (1982). "Groundwater contamination due to vanishing and finite size continuous sources." *J. Hydr. Div.*, 108(4), 572–590.

Prakash, A. (1983). "Deterministic and probabilistic perspectives of the PMF." *Proc. of the conf. on frontiers in hydraulic engineering*, ASCE/MIT, Cambridge, Mass., August 9–12, 1983, 535–540.

Prakash, A. (1984). "Groundwater contamination due to transient sources of pollution." *J. Hydraul. Eng.*, 110(11), 1642–1658.

Prakash, A. (1987). "Current state of hydrologic modeling and its limitations." *Flood hydrology*, V.P. Singh, ed., Reidel, Dordrecht, The Netherlands, 1–16.

Prakash, A. (1991). "Evaluation of rehabilitation alternatives for small hydropower plants." *Water Power 91*, 1884–1893.

Prakash, A. (1992a). "Implications of design uncertainty in benefit-cost analysis." *Water Forum 92*, New York.

Prakash, A. (1992b). "Design basis flood for rehabilitation of existing dams." *J. Hydraul. Eng.*, 118(2), 291–305.

Prakash, A. (1995). "Analysis of hydraulic barriers for ground water in stream-aquifer systems." *Proc. of the int. symp. on groundwater management*, San Antonio, Tex., August 14–16, 1995, 337–342.

Prakash, A. (1996). "Desorption of soil contaminants due to rainwater infiltration." *J. Hydraul. Eng.*, 122(9), 523–525.

Prakash, A. (1997). "Estimating diffusivity of aquifers with sloping water tables."

Proc. of the 27th congress of the Int. Assoc. for Hydraulic Research, San Francisco, Calif., August 10–15, 1997, 15–20.

Prakash, A. (1999). "Risk-based analysis of remediation requirements." *Proc. of the int. water resources engineering conf.,* Seattle, Wash., August 8–12, 1999.

Prakash, A. (2000a). "Analytical modeling of contaminant transport through vadose and saturated soil zones." *J. Hydraul. Eng.,* 126(10), 773–777.

Prakash, A. (2000b). "Evaluation of bank protection methods." *Proc. of ASCE's joint conf. on water resources engineering and water resources planning and management,* Minneapolis, Minn., July 30–August 2, 2000.

Prakash, A. (2002). "Environmental issues of construction and demolition of dams." *Proc. of EWRI/ASCE conf. on managing water resources extremes, Water Resources Planning and Management Council,* Roanoke, Va., May 19–22, 2002.

Quimpo, R.G. (1968). "Stochastic analysis of daily river flows." *J. Hydr. Div.,* 94(HY1), Jan. 1968, 43–57.

Rai, D., and J.M. Zachara (1984). "Chemical attenuation rates, coefficients, and constants in leachate migration." Battelle, Pacific Northwest Lab., Richland, Wash.

Renard, K.G., G.R. Foster, G.A. Weesies, and J.P. Porter (1991). "RUSLE: Revised Universal Soil Loss Equation." *J. Soil Water Conservation,* 46(1), 30–33.

Rouse, H. (1950). *Engineering hydraulics,* Wiley, New York, 1,039 pp.

Seo, I.W., and T.S. Cheong (1995). "Predicting longitudinal dispersion coefficient in natural streams." *J. Hydraul. Eng.,* 124(1), 25–31.

Simons, D.B., and F. Senturk (1976, 1992). *Sediment transport technology,* Water Resources Publications, Fort Collins, Colo., 919 pp.

Simons, D.B., R.M. Li, and W.T. Fullerton (1981). "Theoretically derived sediment transport equations for Pima County, Arizona." Prepared for Pima County DOT and Flood Control District, Ariz.

Singh, B. (1967). *Fundamentals of irrigation engineering,* Nem Chand & Bros., Roorkee, India, 532 pp.

Soil Conservation Service (SCS) (1954). "Handbook of channel design for soil and water conservation." *SCS-TP-61,* Stillwater Outdoor Hydraulic Lab., Stillwater, Okla.

Soil Conservation Service (SCS) (1978). *Water management and sediment control for urbanizing areas,* Columbus, Ohio.

Sorensen, R.M. (1985). "Stepped spillway hydraulic model investigation." *J. Hydraul. Eng.,* 111(12), 1461–1472.

Spurr, W.A., and C.P. Bonini (1973). *Statistical analysis for business decisions,* Richard D. Irwin, Inc., Homewood, Ill., 724 pp.

Streeter, V.L. (1971). *Fluid mechanics,* McGraw-Hill, New York, 751 pp.

Sudicky, E.A., and R. Therien (1999). *Variably-saturated groundwater flow and transport in discretely fractured porous media, FRAC3DVS,* Waterloo Hydrogeologic, Inc., Waterloo, Ont., Canada.

Sudicky, E.A. (1988). *Parallel crack model, CRAFLUSH,* Waterloo Center of Groundwater Research, Univ. of Waterloo, Waterloo, Ont., Canada.

SURGE5 (1996). Civil Engineering Software Center, Univ. of Kentucky, Lexington, Ky.

Swamee, P.K., and A.K. Jain (1976). "Explicit equations for pipe-flow problems." *J. Hydr. Div.,* 102(HY5), May 1976.

Tchobanoglous, G., and F.L. Burton (1991). *Wastewater engineering, treatment, disposal, and reuse*, McGraw-Hill, New York, 1,334 pp.

Texas Commission on Environmental Quality (TCEQ) (2002). *Guidelines for preparation of environmental, social, and economic impacts statements*, TCEQ Rules, Chapter 261, Austin, Tex. (www.tnrcc.state.us/oprd/rules).

Thompson, J.R. (1964). "Quantitative effect of watershed variables on the rate of gully head advancement." *Transactions, American Society of Agricultural Engineers*, 7(1), St. Joseph, Mich., 54–55.

Todd, D.K. (1980). *Ground water hydrology*, Wiley, New York, 535 pp.

Tschantz, B.A., and R.M. Mojib (1981). "Application of and guidelines for using available dam break models." *Water Resources Research Center*, Univ. of Tennessee, Knoxville, Tenn., 84 pp.

Tullis, J.P., N. Amanian, and D. Waldron (1995). "Design of labyrinth spillways." *J. Hydraul. Eng.*, 121(3), 247–255.

U.S. Army Corps of Engineers (USACE) (1960). "Routing of floods through river channel." *EM-1110-2-1408*, Washington, D.C.

U.S. Army Corps of Engineers (USACE) (1970, 1994). "Hydraulic design of flood control channels." *EM-1110-2-1601*, Washington, D.C.

U.S. Army Corps of Engineers (USACE) (1971a). *Monthly streamflow simulation, HEC-4, user's manual*, Hydrologic Engineering Center, Davis, Calif.

U.S. Army Corps of Engineers (USACE) (1971b). "Dewatering and groundwater control for deep excavations." *Technical Manual No. 5-818-5*, Washington, D.C.

U.S. Army Corps of Engineers (USACE) (1974). *Dimensionless graphs of floods from ruptured dams*. Hydrologic Engineering Center, Davis, Calif., 60 pp.

U.S. Army Corps of Engineers (USACE) (1977). "Guidelines for calculating and routing a dam-break flood." *Research Note No. 5*, Hydrologic Engineering Center, Davis, Calif.

U.S. Army Corps of Engineers (USACE) (1978). *Water Quality for River-Reservoir Systems (WQRRS)*, Hydrologic Engineering Center, Davis, Calif., 288 pp.

U.S. Army Corps of Engineers (USACE) (1979). "Feasibility studies for small scale hydropower additions." *DOE/RA-0048*, Hydrologic Engineering Center, Davis, Calif.

U.S. Army Corps of Engineers (USACE) (1981). *Reservoir system analysis for conservation, HEC-3*, Hydrologic Engineering Center, Davis, Calif.

U.S. Army Corps of Engineers (USACE) (1982). *Simulation of flood control and conservation systems, HEC-5*, Hydrologic Engineering Center, Davis, Calif.

U.S. Army Corps of Engineers (USACE) (1984). *Shore protection manual, volumes I and II*, Coastal Engineering Research Center, Waterways Experiment Station, Vicksburg, Miss.

U.S. Army Corps of Engineers (USACE) (1985). "Hydropower engineering manual." *EM-1110-2-1701*, Engineering Design, Washington, D.C.

U.S. Army Corps of Engineers (USACE) (1986). *Streamflow Synthesis and Reservoir Regulation (SSAR)*. U.S. Army Engineer Div., North Pacific, Portland, Ore.

U.S. Army Corps of Engineers (USACE) (1989). "Sedimentation investigation of rivers and reservoirs." *EM-1100-2-4000*, Engineering Design, Washington, D.C.

U.S. Army Corps of Engineers (USACE) (1991a). *Flood hydrograph package, HEC-1, user's manual*, Hydrologic Engineering Center, Davis, Calif.

U.S. Army Corps of Engineers (USACE) (1991b). *Simulation of flood control and conservation systems, HEC-5, user's manual*, Hydrologic Engineering Center, Davis, Calif.

U.S. Army Corps of Engineers (USACE) (1991c). *Surface water profiles, HEC-2*, Hydrologic Engineering Center, Davis, Calif.

U.S. Army Corps of Engineers (USACE) (1991d). *Scour and deposition in rivers and reservoirs, HEC-6*, Hydrologic Engineering Center, Davis, Calif.

U.S. Army Corps of Engineers (USACE) (1992). *Interior Flood Hydrology package, HEC-IFH*, Hydrologic Engineering Center, Davis, Calif.

U.S. Army Corps of Engineers (USACE) (1994). "Hydraulic design of flood control channels." *EM-1110-2-1601*, Washington, D.C.

U.S. Army Corps of Engineers (USACE) (1995). *HEC Floodflow Frequency Analysis, HEC-FFA*, Hydrologic Engineering Center, Davis, Calif.

U.S. Army Corps of Engineers (USACE) (1998). *River Analysis System, HEC-RAS*, Hydrologic Engineering Center, Davis, Calif.

U.S. Army Corps of Engineers (USACE) (1999). *Engineer regulations, civil works, environmental compliance assessments and environmental management program planning* (www.usace.army.mil/inet/usace-docs/eng-regs/), Washington, D.C.

U.S. Army Corps of Engineers (USACE) (2002). *Hydrologic Modeling System, HEC-HMS*, Hydrologic Engineering Center, Davis, Calif.

U.S. Army Engineer Research and Development Center (USAERDC) (2003). *Effect of riprap on riverine and riparian ecosystems*, Vicksburg, Miss.

U.S. Army Engineer Waterways Experiment Station (USAEWES) (1977). *Hydraulic design criteria*, Vicksburg, Miss.

U.S. Army Engineer Waterways Experiment Station (USAEWES) (1983). *Techniques for reaeration of hydropower releases*, Technical Report E-83-5, Vicksburg, Miss.

U.S. Bureau of Reclamation (USBR) (1966). "Effect of snow compaction on runoff from rain on snow." *Engineering Monograph No. 35*, Denver, Colo.

U.S. Bureau of Reclamation (USBR) (1971). *A procedure to determine sediment deposition in a settling basin*, Denver, Colo., 8 pp.

U.S. Bureau of Reclamation (USBR) (1977). *Design of small dams*, Denver, Colo., Second edition (revised reprint), 816 pp.

U.S. Bureau of Reclamation (USBR) (1978). *Design of small canal structures*, Denver, Colo., 435 pp.

U.S. Bureau of Reclamation (USBR) (1984). *Computing degradation and local scour*, Denver, Colo., 48 pp.

U.S. Bureau of Reclamation (USBR) (1987). *Design of small dams*, Denver, Colo., Third edition, 860 pp.

U.S. Dept. of Agriculture (USDA) (1959). "The SAF stilling basin." *Agriculture Handbook No. 156*, Agriculture Research Service, St. Anthony Falls Hydraulics Lab., Minneapolis, Minn., 16 pp.

U.S. Dept. of Agriculture (USDA) (1966). "Procedures for determining rates of land damage, land depreciation, and volume of sediment produced by gully erosion." *Technical Release No. 32, Geology*, Soil Conservation Service.

U.S. Dept. of Agriculture (USDA) (1969). "Summary of reservoir sediment deposition." *Misc. Pub. No. 1143*, Agricultural Research Service, 64 pp.

U.S. Dept. of Agriculture (USDA) (1972, 1985). *National engineering handbook, section 4, hydrology*, Soil Conservation Service, Washington, D.C.

U.S. Dept. of Agriculture (USDA) (1976). "A Water Surface Profile computer program for determining flood elevation and flood areas for certain flow rates, WSP2." *Technical Release No. 61*, Soil Conservation Service, Washington, D.C.

U.S. Dept. of Agriculture (USDA) (1977). "Design of open channels." *Technical Release No. 25*, Oct. 1977, Soil Conservation Service, Washington, D.C.

U.S. Dept. of Agriculture (USDA) (1981). "Simplified dam-breach routing procedure." *Technical Release No. 66*, Design Unit, Soil Conservation Service, Washington, D.C.

U.S. Dept. of Agriculture (USDA) (1982). *Wind erosion equation, technical notes.* Resource Conservation Planning-WY-2, Soil Conservation Service, Casper, Wyo.

U.S. Dept. of Agriculture (USDA) (1983a). "Computer program for project formulation hydrology." *Technical Release 20*, Soil Conservation Service, Washington, D.C.

U.S. Dept. of Agriculture (USDA) (1983b). "Colorado wind erosion guide." *Agronomy Technical Note 53*, Soil Conservation Service, Denver, Colo.

U.S. Dept. of Agriculture (USDA) (1986). "Urban hydrology for small watersheds." *Technical Release 55*, Soil Conservation Service, Washington, D.C.

U.S. Dept. of Commerce (USDOC) (1961). "Generalized estimates of probable maximum precipitation and rainfall frequency data for Puerto Rico and Virgin Islands." *Technical Paper No. 42*, Washington, D.C.

U.S. Dept. of Commerce (USDOC) (1962). "Rainfall-frequency atlas of the Hawaiian Islands." *Weather Bureau Technical Paper No. 43*, Washington, D.C.

U.S. Dept. of Commerce (USDOC) (1965). "Probable maximum and TVA precipitation over the Tennessee River Basin above Chattanooga." *Hydrometeorological Report No. 41*, Washington, D.C.

U.S. Dept. of Commerce (USDOC) (1969). "Interim report, probable maximum precipitation in California." *Hydrometeorological Report No. 36*, Washington, D.C.

U.S. Dept. of Commerce (USDOC) (1977). "Probable maximum precipitation estimates, Colorado River and Great Basin drainages." *Hydrometeorological Report No. 49*, Silver Spring, Md.

U.S. Dept. of Commerce (USDOC) (1978). "Probable maximum precipitation estimates, United States east of the 105th Meridian." *Hydrometeorological Report No. 51*, Washington, D.C.

U.S. Dept. of Commerce (USDOC) (1982). "Application of probable maximum precipitation estimates, United States east of the 105th Meridian." *Hydrometeorological Report No. 52*, Silver Spring, Md.

U.S. Dept. of Commerce (USDOC) (1983). "Probable maximum precipitation and snowmelt criteria for southeast Alaska." *Hydrometeorological Report No. 54*, Silver Spring, Md.

U.S. Dept. of Commerce (USDOC) (1988). "Probable maximum precipitation estimates, United States between the Continental Divide and the 103rd Meridian." *Hydrometeorological Report No. 55A*, Silver Spring, Md.

U.S. Dept. of Commerce (USDOC) (1994). "Probable maximum precipitation, Pacific Northwest states." *Hydrometeorological Report No. 57*, Silver Spring, Md.

U.S. Environmental Protection Agency (USEPA) Region VIII and U.S. Dept. of Agriculture (USDA) (1977). "Preliminary guidance for estimating erosion on areas disturbed by surface mining activities in the interior Western United States, interim final report," *EPA-908/4-77-005*.

U.S. Environmental Protection Agency (USEPA) (1980). "An approach to water resources evaluation of non-point silvicultural sources." *EPA-600/8-80-012*, Environmental Research Lab., Athens, Ga.

U.S. Environmental Protection Agency (USEPA) (1985). "Water quality assessment, a screening procedure for toxic and conventional pollutants in surface and ground water." Part I, *EPA/600/6-85/002a*, and Part II, *EPA/600/6-85/002b*, Environmental Research Lab., Athens, Ga.

U.S. Environmental Protection Agency (USEPA) (1987a). "The Enhanced Stream Water Quality Models, QUAL2E and QUAL2E UNCAS." *EPA/600/3-87/007*, Environmental Research Lab., Athens, Ga.

U.S. Environmental Protection Agency (USEPA) (1987b). "Diffusion in near-shore and riverine environments." *EPA 910/9-87-168*, Region 10.

U.S. Environmental Protection Agency (USEPA) (1988a). "Superfund exposure assessment manual." *EPA/540/1-88/001*.

U.S. Environmental Protection Agency (USEPA) (1988b). "A hydrodynamic and water quality model, WASP4." *EPA/600/3-86/034*, Environmental Research Lab., Athens, Ga.

U.S. Environmental Protection Agency (USEPA) (1989a). *Storm Water Management Model, SWMM*. Environmental Protection Technology Series, Washington, D.C.

U.S. Environmental Protection Agency (USEPA) (1989b). "Guidelines for conducting remedial investigations and feasibility studies under CERCLA." *EPA/540/G-89/004*, Office of Emergency and Remedial Response, Washington, D.C.

U.S. Environmental Protection Agency (USEPA) (1989c). *Statistical analysis of groundwater monitoring data at RCRA facilities, interim final guidance*, Office of Solid Waste, Waste Management Div., Washington, D.C.

U.S. Environmental Protection Agency (USEPA) (1991a). *Hydrologic Simulation Program-Fortran (HSPF)*, Environmental Research Lab., Office of Research and Development, Athens, Ga.

U.S. Environmental Protection Agency (USEPA) (1991b). *Guidance for water quality-based decisions: The TMDL process*, Office of Water, Washington, D.C.

U.S. Environmental Protection Agency (USEPA) (1991c). "Technical support document for water quality-based toxics control." *EPA/505/2-90-001, PB 91-127415*, Office of Water, Washington, D.C.

U.S. Environmental Protection Agency (USEPA) (1991d). *Guidance manual for the preparation of NPDES permit applications for storm water discharges associated with industrial activity*, Office of Water Enforcement and Permits, Washington, D.C.

U.S. Environmental Protection Agency (USEPA) (1992). *A Modular Three-Dimensional Transport Model (MT3D) for simulation of advection, dispersion, and chemical reactions of contaminants in groundwater systems*, National Risk Management Research Lab., Ada, Okla.

U.S. Environmental Protection Agency (USEPA) (1993). *Well Head Protection Area (WHPA) delineation code*, National Risk Management Research Lab., Ada, Okla.

U.S. Environmental Protection Agency (USEPA) (1994). *Dilution models for effluent dis-*

charges, PLUMES, Center for Exposure Assessment Modeling (CEAM), National Exposure Research Lab., Athens, Ga.

U.S. Environmental Protection Agency (USEPA) (1995). *The Hydrologic Evaluation of Landfill Performance (HELP) model,* Risk Reduction Engineering Lab., Office of Research and Development, Cincinnati, Ohio.

U.S. Environmental Protection Agency (USEPA) (1996a). *A hydrodynamic mixing zone model and decision support system for pollutant discharges into surface waters, CORMIX,* Center for Exposure Assessment Modeling (CEAM), National Exposure Research Lab., Athens, Ga.

U.S. Environmental Protection Agency (USEPA) (1996b). *Multimedia Exposure Assessment Model (MULTIMED) for evaluating the land disposal of wastes,* Environmental Research Lab., Athens, Ga.

U.S. Environmental Protection Agency (USEPA) (1996c). "Soil screening guidance, user's guide." *Publication 9355.4-23,* Office of Solid Waste and Emergency Response, Washington, D.C.

U.S. Environmental Protection Agency (USEPA) (1997). *BIOSCREEN,* National Risk Management Research Lab., Ada, Okla.

U.S. Environmental Protection Agency (USEPA) (2000). "Drinking water standards and health advisories." *EPA822-B-00-001,* Office of Water, Washington, D.C.

U.S. Geological Survey (USGS) (1983). "Precipitation-Runoff Modeling System (PRMS): User's manual," *Water Resources Investigations Report 83-4238,* Denver, Colo.

U.S. Geological Survey (USGS) (1994). "Nationwide summary of U.S. Geological Survey regional regression equations for estimating magnitude and frequency of floods for ungaged sites." *1993 Water Resources Investigations Report 94-4002,* Reston, Va., 196 pp.

U.S. Geological Survey (USGS) (2000a). "Estimation of peak stream flows for unregulated rural streams in Kansas." *Water Resources Investigations Report 00-4079,* Lawrence, Kan., 33 pp.

U.S. Geological Survey (USGS) (2000b). *A modular three-dimensional finite-difference ground-water flow model: MODFLOW,* Reston, Va.

U.S. Nuclear Regulatory Commission (USNRC) (1976). "Estimating aquatic dispersion of effluents from accidental and routine reactor releases for the purpose of implementing Appendix I." *Regulatory Guide 1.113,* Washington, D.C.

U.S. Nuclear Regulatory Commission (USNRC) (1977). "Design basis floods for nuclear power plants." *Regulatory Guide 1.59,* Rev. 2, August 1977.

U.S. Nuclear Regulatory Commission (USNRC) (1982). "Literature review of models for estimating soil erosion and deposition from wind stresses on uranium mill tailings covers." *NUREG/CR-2768, PNL-4302.*

U.S. Water Resources Council (USWRC) (1981). "Guidelines for determining flood flow frequency." *Bulletin #17B,* Washington, D.C.

Valliappan, S., and N.K. Naghadeh (1991). "Flow through fractured media." *Computer methods and advances in geomechanics,* Beer, Becker and Carter, eds., Balkema, Rotterdam, The Netherlands.

Vanoni, V.A. (1977). "Sedimentation engineering." *ASCE Manuals and Reports on Engineering Practice No. 54,* New York.

Vetter, C.P. (1940). "Technical aspects of the silt problem on the Colorado River." *Civil Engineering*, 10, Nov. 1940, pp. 698–701.

Watters, G.Z. (1984). *Analysis and control of unsteady flow in pipelines*, 2d Ed., Butterworth, Stoneham, Mass., 349 pp.

Weast, R.C., ed. (1987). *CRC handbook of chemistry and physics*. 68th Ed., CRC, Boca Raton, Fla.

Wenzel, L.K. (1942). "Methods of determining permeability of water bearing materials, with special reference to discharging well methods." *U.S. Geological Survey Water Supply Paper 887*, Washington, D.C.

West Consultants, Inc. (1996). *Riprap design system*, Carlsbad, Calif., 72 pp.

Williams, J.R. (1975). "Sediment yield prediction with universal equation using runoff energy factor." *USDA-ARS, S-40*, U.S Dept. of Agriculture, Washington, D.C., 244–252.

Wilson, B.N., B.J. Barfield, and I.D. Moore (1984). *A hydrology and sedimentology watershed model, SEDIMOT II*, Dept. of Agricultural Engineering, Univ. of Kentucky, Lexington, Ky.

Wood, I.R., and T. Liang (1989). "Dispersion in an open channel with a step in the cross section." *J. Hydraulic Research*, 27(5).

World Bank (WB) (1996). *Monitoring and evaluation guidelines for World Bank-GEF int. waters projects*, Washington, D.C.

World Bank (WB) (1998). *Environmental assessment source book*, Vols. I, II, III, Washington, D.C.

World Meteorological Organization (WMO) (1986). "Manual for estimation of probable maximum precipitation." *Operational Hydrology Report Number 1, WMO Number 332*, Geneva, Switzerland, 269 pp.

Yevjevich, V. (1972a, 1997). *Probability and statistics in hydrology*, Water Resources Publications, Fort Collins, Colo., 302 pp.

Yevjevich, V. (1972b, 1982). *Stochastic processes in hydrology*, Water Resources Publications, Fort Collins, Colo., 276 pp.

Young, M.F. (1982). "Feasibility study of a stepped spillway." *Proc. of the Hydr. Div. specialty conf.*, Jackson, Miss., August 1982.

Zheng, C. (1990). *A modular three-dimensional transport model for simulation of advection, dispersion, and chemical reactions of contaminants in groundwater systems*, U.S. Environmental Protection Agency, Ada, Okla.

Zheng, C., and G.D. Bennett (2002). *Applied contaminant transport modeling*, 2d Ed., Wiley-InterScience, New York, 621 pp.

Zipparro, V.J., and H. Hansen, eds. (1993). *Davis' handbook of applied hydraulics*, McGraw-Hill, New York.

INDEX

7-day average 10-year low flow: *see* flow, low
7Q10 low flow: *see* flow, low

ambient lake monitoring programs 310
antecedent moisture condition 35
apron slopes 230
aprons, launching 226–228
aquifers: classification 147; parameter estimation 184–188; safe yield 169; sample problems 187
atmospheric pressure: *see* barometric pressure

bank storage 171–173; sample problems 173
barometric pressure 166–167; sample problems 167
baseflows 39
basins, detention 216
basins, sedimentation 113
basins, settling 113
basins, stilling 263–270; sample problems 265, 266, 269–270
bends: erodible channels 82–83; flow 82–83
benefit-costs analysis 49–50, 289–294; evaluation of project alternatives 292–297; sample problems 49–50, 292, 293, 294
biochemical oxygen demand 126
BIOSCREEN model 208
Blench curves 93
BOD: *see* biochemical oxygen demand
bridge modification 216

Camp's equation 77
capture zone 157–158
cavitation 134–136
channel transitions 209–212; sample problems 210–212
channelization 216
channels, erodible 72–82; bends 82–83; maximum permissible velocity 72–73; nonscouring velocity 74–78; nonsilting velocity 74–78; sample problems 74, 77–78; sediment stability 75–78
channels, nonerodible 69–72; sample problems 71
channels, vegetated 78–82; sample problems 82
chutes 84–86; HEC-2 model 84–85; HEC-RAS model 84–85; sample problems 85–86
Clark's unit hydrograph 30–31; unit hydrograph ordinates 31

community development projects 5
constant-rate-pumping test 185
contaminant transport 192–205; sample problems 197–199, 200
CORMIX model 114
Creager's equation 44
Crippen's equation 44
culvert analysis program 145
cutoffs 214–215

DAD curves 14
DAMBRK model 144–145
dams 5–6; dam-break flood 99–100; environmental impacts 303–305, 306–308; environmental monitoring 310, 311–312; planning 236–246
degree-day method of snowmelt estimation 37
depth-area-duration curves: see DAD curves
depth-duration-frequency curves 14; defined 14
design precipitation 32–33
detention basins 216
diffusers 136–142; sample problems 139–140, 142
direct method 30
dispersion coefficients 124–126
dispersion, far-field 117–124; QUAL2E model 117; sample problems 121–122, 129–130; WASP4 model 117; WQRRS model 117
dispersion, near-field 114–117; CORMIX model 114
dispersion, open-channel 114–130
dissolved oxygen 126–130, 283; QUAL2E model 126; WASP4 model 126; WQRRS model 126
DO: see dissolved oxygen
draft tubes 280, 282
drains 174–177; sample problems 175
drawdown 170
drop structures 229–236; sample problems 234–236

economic analysis 289–292
efficiency 169–170
Einstein-Strickler-Manning equation 75–76
EIR: see environmental impact
EIS: see environmental impact
energy dissipation 263–270
environmental impact 299–326; of dams 303–305, 306–308; environmental impact reports 300; environmental impact statement 301–303; evaluation and analysis 306–308; evaluation of projects 312–314; of levees 306; recreational activities 313–314
environmental monitoring 308–312; ambient lake monitoring programs 310; components 309–312
environmental performance indicators 310–311
EPIs: see environmental performance indicators
equations: Camp's 77; Creager's equation 44; Crippen's equation 44; Einstein-Strickler-Manning 75–76; Matthai's equation 44; Meyer-Peter-Muller 75; modified universal soil loss 107, 322; revised universal soil loss 106; sediment stability 75–78; Shields shear stress 75; universal soil loss 105, 322
erosion, bank 227–229
erosion control 217–229; reservoirs 242–246; riprap 219–227; structural 218–219
erosion, gully 103; sample problems 104–105
erosion, soil 103–105; sample problems 104–105; on slopes 103–105; see also soil loss
erosion, wind 110–111; sample problems 111
exit gradient 232

failure probabilities 46–48; sample problems 48
fish habitats 312–313

flood bypass 215–216
flood control 212–217; bridge modification 216; channelization 216; cutoffs 214–215; detention basins 216; flood bypass 215–216; flood insurance 216–217; flood proofing 216; groins 214; levees 212–214
flood insurance 216–217
flood proofing 216
floodplain analysis 84
floods 46–48; dam-break 99–100; floodplain analysis 84
floodway 84
flow, along slopes 91
flow classification 67–68
flow, continuous: HSPF model 65; NWSRFS model 64–65; PRMS model 65; simulation models 64–65; SSARR model 65
flow, critical 86–87; defined 68
flow, Darcian 151–153
flow duration analysis 53
flow, free surface 67
flow, groundwater: bank storage 171–173; capture zone 157–158; contaminant transport 192–205; drains 174–177; gas phase transport 203–205; mounding 156–159; one-dimensional steady state 151–155; partially penetrating well 159–160; pumping tests 184–185; radial flow to a single well 155–156; radius of influence 160–161; recharge 162–164; river stages 177–178; sample problems 153, 154–155; seawater intrusion 164–166; slug tests 188–192; steady-state radial 155–170; subsidence 167–169; transient 170–200; unsteady 170–200; unsteady one-dimensional flow 170–171; unsteady radial flow 178–184; well interference 161–162
flow, laminar 100–101; defined 67; in pipes 130–131
flow, low 56–62; sample problems 59–62; water quality 321

flow, nappe 260
flow, non-Darcian 153–155
flow, nonuniform 67
flow, overland 11–12, 100–103; sample problems 101–103
flow, pipe 68, 130–144; diffusers 136–142; laminar 130–131; sample problems 131, 135–136; turbulent 131–136
flow, pressure 67
flow, radial 155–170; sample problems 158
flow, sheet 11–12
flow, skimming 260–261
flow, steady: defined 67; in open channels 68–69
flow, subcritical 86; defined 68
flow, supercritical 86; defined 68; hydraulic jump 91–92
flow, through bends 82–83
flow, turbulent 100; defined 67; in pipes 131–136
flow, uniform 67
flow, unsteady 170–200; dam-break flood waves 99–100; defined 67
flow velocities 11–12
fractured rock 199–200
freeboard 83

gas phase transport 203–205; sample problems 204–205
grade control 229–236
grasses 78–82
groins 214
groundwater: bank storage 171–173; barometric pressure 166–167; capture zone 157–158; contaminant transport 192–205; contamination 317–318; drains 174–177; flow to a single well 155–156; gas phase transport 203–205; mounding 156–159; one-dimensional steady state flow 151–155; partially penetrating well 159–160; porous media 148–151; pumping tests 184–185; radius of influence 160–161; recharge

162–164; river stages 177–178; sample problems 157; seawater intrusion 164–166; slug tests 188–192; steady-state radial flow 155–170; subsidence 167–169; transient flow 170–200; types of aquifers 147; unsteady flow 170–200; unsteady one-dimensional flow 170–171; unsteady radial flow 178–184; well interference 161–162
gully erosion 103; sample problems 104–105

habitat restoration 84
head loss 275–277
HEC-1 model: Clark's unit hydrograph 30–31; combining hydrographs 39–40; dimensionless unit hydrograph 30; kinematic wave method 32; precipitation depths 33; probable maximum flood hydrograph 45; remediation 322; storm runoff hydrographs 62; surface runoff hydrograph development 25
HEC-2 model 144; chutes 84–85; flood control 213, 216; steady flow analysis 83
HEC-3 model 65
HEC-4 model 65
HEC-6 model 145; sediment transport analysis 111
HEC-FFA model 65
HEC-HMS model 64; Clark's unit hydrograph 30–31; kinematic wave method 32; remediation 322; surface runoff hydrograph development 25
HEC-IFH model 213
HEC-RAS model 144; chutes 84–85; flood control 213, 216; steady flow analysis 83
HELP model 207
HSPF model 65
hydraulic conductivity 150–151; sample problems 151
hydraulic jump 91–99, 247

hydroelectric power 270–288
hydrograph, unit 29–32; Clark's 30–31; dimensionless 30; methods 30; Snyder's unit hydrograph 31–32
hydrographs, probable maximum flood 40–46
hydrographs, surface runoff 25–50; baseflows 39; benefit-costs analysis 49–50; Clark's unit hydrograph 30–31; combining 39–40; defined 25; design precipitation 32–33; dimensionless unit hydrograph 30; failure probabilities 46–48; floods 46–48; HEC-1 model 62; HEC-HMS model 64; hydrologic models 62, 64; kinematic wave method 32; models 25, 28, 62, 64; precipitation depths 32–33; probable maximum flood 40–46; probable maximum precipitation 40; rain loads 38; rainfall excess 28; required data 28–29; risk analysis 46–48; roof loads 38–39; runoff curve numbers 35–37; sample problems 39–40; SCS curve number method 33; snow loads 37–39; snowmelt 37–39; Snyder's unit hydrograph 31–32; soil losses 33–37; storm duration 32–33, 41; subwatersheds 29; TR-20 model 64; TR-55 model 64; unit hydrograph 29–32; unit hydrograph methods 30; unit hydrograph ordinates 28
hydropower projects 6
hydrostatic pressure 167

IDF curves 14
indirect method 30
infiltration 284–287, 322
intakes 273–278
intensity-duration-frequency curves: see IDF curves

kinematic wave method 32
Kirpich method 8

levees 212–214; environmental impacts 306

mass curve analysis 54–56; sample problems 54–56
Matthai's equation 44
Meyer-Peter-Muller equation 75
mining projects 5
mitigation measures 308–309
mixing zones 117, 128, 197, 308, 320
models, continuous flow: HSPF model 65; NWSRFS model 64–65, 322; PRMS model 65; SSARR model 65, 322
models, groundwater flow and transport 207–208; BIOSCREEN model 208; HELP model 207; MODFLOW 198, 207; MT3D model 208; MULTIMED model 203, 207–208; WHPA 207
models, hydraulic 25, 144–146; CORMIX model 114; culvert analysis program 145; DAMBRK model 144–145; HEC-2 model 83, 84–85, 144, 213, 216; HEC-6 model 111, 145; HEC-IFH model 213; HEC-RAS model 83, 84–85, 144, 213, 216; NWS Dam-Break model 99; PLUMES 145; QUAL2E model 117, 126, 145; QUAL2E-UNCAS model 145; WASP4 model 117, 126; WQRRS model 117, 126, 146; WSP-2 model 83, 144
models, hydrologic 62–65; for continuous flow simulation 64–65; HEC-1 model 25, 30–33, 39–40, 45, 62, 322; HEC-3 model 65; HEC-4 model 65; HEC-FFA model 65; HEC-HMS model 25, 30–32, 64, 322; for storm runoff hydrographs 62, 64; SWMM model 64; TR-20 model 25, 64; TR-55 model 64
models, sediment transport 25, 108, 113, 322
MODFLOW model 198, 207
modified universal soil loss equation 107, 322

monitoring wells 206–207
MT3D model 208
MULTIMED model 203, 207–208
MUSLE: *see* modified universal soil loss equation

nuclear power projects 6
NWS Dam-Break model 99
NWSRFS model: for continuous flow simulation 64–65; remediation 322

ogee crest 88–90
overland flow 11–12

peak flow estimation 6–25; rainfall intensity 14–19; rational formula 6–7; regression equations 19–21; runoff coefficients 7–8; sample problems 20–21, 24–25; statistical analysis 22–25; time of concentration 8–14
percolation 197
permeability 148–151
phase liquids 192
PLUMES model 145
PMF hydrograph: *see* probable maximum flood hydrograph
PMP hydrograph: *see* probable maximum precipitation hydrograph
porous media 148–151
power plants 270–273; classification 271; dissolved oxygen 283; draft tubes 280, 282; infiltration 284–287; intakes 273–278; sample problems 272–273, 277–278, 279–280, 282, 287; surge tanks 278–280
precipitation depths 32–33; unit hydrograph ordinates 33
PRMS model 65
probable maximum flood hydrograph 40–46; equations 44; sample problems 45
probable maximum precipitation hydrograph 40
project planning 2; environmental

impact evaluation 312–314; evaluation of alternatives 292–297

QUAL2E model 145; dissolved oxygen 126; far-field dispersion 117
QUAL2E-UNCAS model 145

radius of influence 160–161
rain loads 38
rainfall excess 28
rainfall intensity 14–19; sample problems 17–19
rational formula of peak flow estimation 6–7
recharge 162–164; sample problems 163
recreational activities 313–314
regression equations of peak flow estimation 19–21
remediation 314–326; hydrologic processes 319–323; investigations 315–318; methods 323–326; regulations 318–319; sample problems 316–317
report preparation 2–4
reservoirs 5–6; capacity 237; dissolved oxygen 283; environmental monitoring 311–312; erosion protection 242–246; flow duration analysis 53; low flow 56–62; mass curve analysis 54–56; operation studies 50–62; planning 236–246; sample problems 239–242, 245–246; sedimentation 238–242; streamflow sequences 50–53; wind wave heights 242–246
retardance classes 78–81
revised universal soil loss equation 106
Rippl analysis: *see* mass curve analysis
riprap 219–229; bank erosion 227–228; bank protection 228–229; design 219–223; filter design 224; launching apron 226–228; sample problems 222–223; scour prevention 224–226; sizing 223–224; toe protection 228–229

risk analysis 46–48; sample problems 48
river stages 177–178; sample problems 178
roof loads 38–39; sample problems 38–39
runoff coefficients 7–8
runoff curve numbers 35–37
runoff, direct 29
RUSLE: *see* revised universal soil loss

safe yield 169
sample problems: benefit-costs analysis 49–50, 292, 293, 294; channel transitions 210–212; contaminant transport 197–199, 200; diffusers 139–140, 142; drop structures 234–236; erodible channels 74; far-field dispersion 121–122, 129–130; gas phase transport 204–205; groundwater flow 151, 153, 154–155, 157, 158, 163, 165–169, 173, 175, 178, 187, 191; nonerodible channels 71; peak flow estimation 20–21, 24–25; pipe flow 131, 135–136; power plants 272–273, 277–278, 279–280, 282, 287; remediation 316–317; reservoir operation studies 51–52, 54–56, 59–62; reservoirs 239–242, 245–246; riprap 222–223; scour depth 226, 227; sediment transport analysis 113, 114; soil erosion 104–105; soil loss 108–110, 111; spillways 98–99, 249, 255–256, 259–260, 261; steady flow analysis 85–86; stilling basins 265, 266, 269–270; surface runoff hydrographs 38–40, 45, 48; time of concentration 13–14; water hammer pressure 143–144; weirs 98–99; wells 161–162, 182–183; wind erosion 111
saturated zone 200, 203, 322
scour depth 224–226; sample problems 226, 227
scour prevention 224–226

SCS curve number method: precipitation depths 33; time of concentration 8–9
SDR: *see* sediment delivery ratio
seawater intrusion 164–166; sample problems 165–166
sediment delivery ratio 322
sediment stability 75–78
sediment transport analysis 111–114; sample problems 113, 114; SEDIMOT-II model 113
sediment yield analysis 105–110
sedimentation basins 113
sedimentation in reservoirs 238–242
SEDIMOT-II model: remediation 322; sediment transport analysis 113; sediment yield analysis 108; surface runoff hydrograph development 25
SEIs: *see* socioeconomic performance indicators
settling basins 113
sheet flow 11–12
Shields shear stress equation 75
slopes 100–103
slug tests 188–192; sample problems 191
snow loads 37–39; sample problems 38–39
snowmelt 37–39
Snyder's method 9
Snyder's unit hydrograph 31–32
socioeconomic performance indicators 310
Soil Conservation Service Curve Number method: *see* SCS Curve Number method
soil contamination 317–318
soil erosion: sample problems 104–105; on slopes 103–105; *see also* soil loss
soil loss 33–37, 105–110, 322; modified universal soil loss equation 107; revised universal soil loss equation 106; sample problems 108–110, 111; on slopes 103–105; universal soil loss equation 105; *see also* soil erosion

specific capacity 169
specific energy 93
spillways 5–6; chutes 252; conduit 254; earth-cut 246; flow along slopes 91; free overfall 247; fuse-plug 246–247; glory hole 254–256; hydraulic jump 91–99; hydraulics 87–130, 246–263; labyrinth 261–263; morning glory 254–256; ogee 247–250; ogee crest 88–90; open-channel dispersion 114–130; overland flow 100–103; sample problems 98–99, 249, 251–252, 255–256, 259–260, 261; sediment yield analysis 105–110; side channel 250–252; siphon 256, 258–260; soil erosion 103–105; stepped 260–261; trough 252; tunnel 254; unsteady flow 99–100
SSARR model: for continuous flow simulation 65; remediation 322
statistical analysis, peak flow estimation 22–25
steady flow analysis 83–86; sample problems 85–86
step-drawdown test 184–185
storm duration 32–33, 41
stream hydraulics method 9–11
streamflow sequences 50–53; evaluation methods 51–53; sample problems 51–52
studies documentation 2–4
subsidence 167–169, 232; sample problems 168–169
subwatersheds 29
surface runoff hydrographs: *see* hydrographs, surface runoff
surge tanks 278–280
SWMM model 64

tests, pumping 184–185
time of concentration 8–14; defined 8; Kirpich method 8; sample problem 13–14; SCS Curve Number method 8–9; sheet flow equation 11–12; Snyder's method 9; stream hydraulics method 9–11;

U.S. Bureau of Reclamation method 9
toe protection 228–229
TR-20 model: storm runoff hydrographs 64; surface runoff hydrograph development 25
TR-55 model 64

UHO: *see* unit hydrograph ordinates
unit hydrograph 29–32; Clark's 30–31; dimensionless 30; methods 30; Snyder's unit hydrograph 31–32
unit hydrograph ordinates 28, 31, 33
universal soil loss equation 105, 322
U.S. Bureau of Reclamation method 9
USLE: *see* universal soil loss equation

vadose zone 200, 322
vegetation 78–82, 111, 228
velocity: maximum permissible 72–73; nonscouring 74–78; nonsilting 74–78

WASP4 model: dissolved oxygen 126; far-field dispersion 117
wastewater discharges 126
water: extractable 147; hydroscopic 147

water hammer pressure 143–144; sample problems 143–144
water surface profiles 83–86; sample problems 85–86
weirs 87–130; broad-crested 87–88; flow along slopes 91; hydraulic jump 91–99; ogee crest 88–90; open-channel dispersion 114–130; overland flow 100–103; sample problems 98–99; sediment transport analysis 111–114; sediment yield analysis 105–110; sharp-crested 87–88; soil erosion 103–105; unsteady flow 99–100
wells: efficiency 169–170; fully penetrating 178–184; groundwater radial flow 155–156; interference 161–162; monitoring 206–207; partially penetrating 159–160; sample problems 161–162, 182–183; specific capacity 169
WHPA model 207
wind wave 242–246
WQRRS model 146; dissolved oxygen 126; far-field dispersion 117
WSP-2 model 144; steady flow analysis 83

zones: capture 157–158; saturated 200, 203, 322; vadose 200, 322